TUBULAR COMBUSTION

TUBULAR
COMBUSTION

**SATORU ISHIZUKA, DEREK DUNN-RANKIN,
ROBERT W. PITZ, ROBERT J. KEE, YUYIN ZHANG,
HUAYANG ZHU, TADAO TAKENO, MAKIHITO NISHIOKA,
DAISUKE SHIMOKURI**

Ⓜ MOMENTUM PRESS

MOMENTUM PRESS, LLC, NEW YORK

First published by Momentum Press®, LLC
222 East 46th Street, New York, NY 10017
www.momentumpress.net

ISBN-13: 978-1-60650-303-4 (hard back, case bound)
ISBN-10: 1-60650-303-0 (hard back, case bound)
ISBN-13: 978-1-60650-305-8 (e-book)
ISBN-10: 1-60650-305-7 (e-book)

DOI: 10.5643/9781606503058

Cover design by Jonathan Pennell
Interior design by Exeter Premedia Services Private Ltd.,
Chennai, India

10 9 8 7 6 5 4 3 2 1

Printed in the United States of America

CONTENTS

PREFACE

This book introduces fundamental studies and practical applications of tubular combustion which have developed over the past three decades. At the beginning, tubular flames received considerable attention from a fundamental viewpoint, because the flame is dissimilar in shape from other elementary flames such as a flat flame, a cylindrical flame, or a spherical flame. That is, tubular flames are circular in cross section and long in the perpendicular direction. Based on their unusual geometry, tubular flames were investigated experimentally, computationally, and theoretically by many researchers. In their early stage, tubular flame studies were limited to premixed combustion. More recently, however, non-premixed combustion and liquid-fueled heterogeneous combustion have been demonstrated in the tubular geometry, so that the field of tubular flame study was extended to what can now legitimately be called "tubular combustion". Furthermore, in the course of the fundamental studies, it has turned out that tubular flames are robust in stability because of their thermal and aerodynamic advantages. This triggered development of practical tubular combustor, and recently, a variety of tubular burners have been developed for various purposes.

Over the centuries, combustion studies have been made on candle flame, jet flame, coal combustion, and so on, and modern combustion technologies have been built up. Tubular combustion is very new; most works have been made in these three recently past decades. As editors of this book, we thank all of the authors who have contributed, and we pass on the thanks of all the authors to the researchers over the years with whom they have interacted and whose work is reflected in this unique compendium of information on tubular combustion. Tubular combustion has potential for providing new technologies in the field of combustion, and we hope that readers of this book will be interested in this subject and will participate in fundamental study as well as practical applications of tubular combustion.

Satoru Ishizuka and Derek Dunn-Rankin

KEYWORDS

Tubular flame, cylindrical flame, axisymmetric flame, flame sheet, flame stretch, flame curvature, vortex, swirl, Lewis number, similar solution, asymptotic analysis, computational simulation, pressure diffusion, Raman spectroscopy, tubular flame structure, cellular instability, rapidly-mixed combustion, non-premixed combustion, liquid film combustor, microcombustor, tubular flame burners

CHAPTER 1

INTRODUCTION

Satoru Ishizuka

Tubular flame is a flame that is circular in cross section and long in the perpendicular direction. This chapter provides first the background of the research interest in tubular flames. Then, their notable characteristics, which were experimentally found, are briefly introduced. To understand these and other tubular flame characteristics, many fundamental studies have been conducted theoretically, computationally, and experimentally. These fundamental researches are specifically introduced in the next section. In early stages of tubular flame studies, they were limited on premixed combustion. Later, however, they have been extended to non-premixed combustion and to liquid-fueled heterogeneous combustion. Relevant studies on non-premixed tubular flame and miniature liquid-film combustor are briefly introduced in the fourth section together with relevant knowledge on laminar diffusion flames and counterflow diffusion flames. Since tubular flame is robust in stability due to its thermal and aerodynamic advantages, a variety of tubular flame burners have been developed for practical use. In the last section, prototype tubular flame burners and an inherently safe technique of rapidly mixed tubular flame combustion are introduced. After these introductory descriptions, specific subjects on tubular flames and related combustion are presented in more detail in Chap. 2–8 of the book.

1.1 BACKGROUND OF TUBULAR FLAME STUDIES

There are two major motivations for tubular flame study. One is related to flame extinction and the other is related to the modeling of turbulent combustion. In the late 1970s and the 1980s, significant advances were made in the field of flame extinction which improved our understanding. Theoretically, flame stretch, a concept proposed earlier by Karlovitz et al. for describing flame extinction in a flow with velocity gradient,[1] has been generally defined as the time derivative of the logarithm of an infinite element of flame area.[2,3] Furthermore, it has been mathematically derived in an explicit form that flame stretch includes three factors, aerodynamic straining, flame curvature, and flame motion.[4,5] Experimentally, also, basic factors such as heat loss, aerodynamic straining, and flame curvature have been separately examined using a variety of flame-flow geometries. As for turbulent combustion, various models have been proposed, and among them, a stretched vortex flow[6] has attracted keen interest in the modeling. In this section, the background of tubular flame studies is briefly described.

1.1.1 AERODYNAMIC STRAINING

Figure 1.1 shows an example of such experiments, in which a flat premixed flame is stabilized in the forward stagnation region of a flat plate.[7,8] An unburned gas is blowing upward onto the surface. One may expect that increasing a blowing rate pushes the flame toward the surface and, eventually, the flame is extinguished. What causes the flame extinction? While approaching the surface, the flame may suffer from heat loss, which may quench the flame. If this is true, can the flame extinction be prevented by heating the surface with a high-temperature, oxygen–propane torch flame? Although the surface temperature was raised up to 1000°C, the blowing rate for extinction was varied only a little with an increase of the surface temperature.[7] It was suspected that instead of the downstream heat loss, flame stretch, which was caused by the nonuniform straining flow, might extinguish the flame.

To elucidate the flame extinction by stretch alone, a new experimental methodology has been adopted; that is a binary flame system shown in Fig.1.2.[8,9] In this system, two porous plug burners are set opposite to each other and two flat flames are formed between the burners. Due to the symmetry of temperature distribution, the downstream heat loss seems negligible although there may be a radiative heat loss from the hot burned gas.[9]

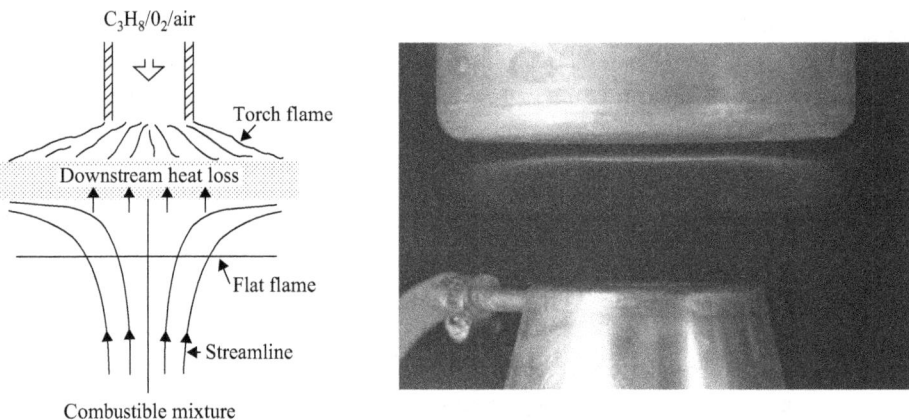

Figure 1.1. Schematic of a flat flame in the forward stagnation region of a flat plate, which can be heated by a propane–oxygen–air torch flame and a picture of flame of a lean propane–air mixture. (*Source*: (left) Ishizuka, S., Characteristics of Tubular Flames, Progress in Energy and Combustion Science 19(1993)187–226,[8] (right) Law, C. K., Ishizuka, S., and Mizomoto, M., Lean-Limit Extinction of Propane–Air Mixtures in the Stagnation-Point Flow, Proceedings of the Eighteenth Symposium (International) on Combustion, The Combustion Institute, Pittsburgh, PA., pp.1791–1798, 1981.[7] Used by permission of Elsevier.)

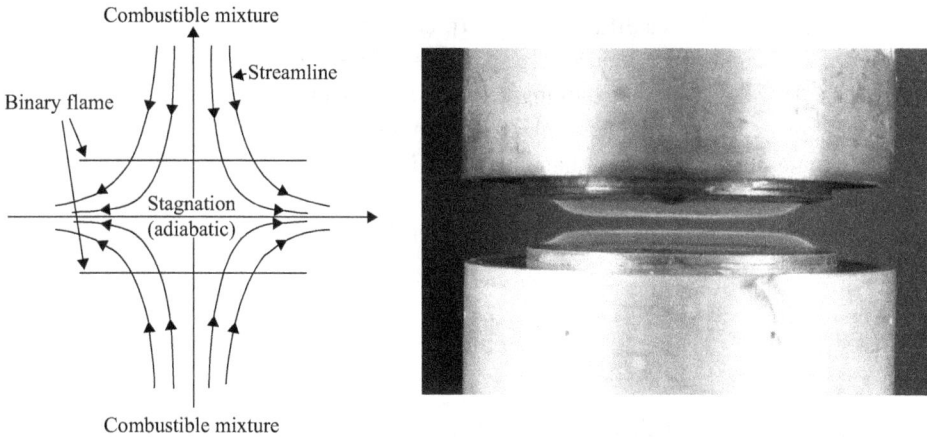

Figure 1.2. Schematic of the binary flame system and a picture of flame of a lean propane–air mixture. (*Source*: (left) Ishizuka, S., Characteristics of Tubular Flames, Progress in Energy and Combustion Science 19(1993)187–226,[8] (right) Ishizuka, S. and Law, C. K., An Experimental Study on Extinction and Stability of Stretched Premixed Flames, Proceedings of the Nineteenth Symposium (International) on Combustion, The Combustion Institute, Pittsburgh, PA., pp.327–335,1982.[9] Used by permission of Elsevier.)

In this experiment, various mixtures were explored in addition to lean propane–air mixtures used in the first experiment of Fig.1.1. Two important findings were obtained. The first was that the flames could be extinguished under the negligible downstream heat loss conditions. The second was that at the time of extinction the flames were separated; however, the separated distance was relatively large in lean propane, which is the case of Fig.1.1, and rich methane–air mixtures, whereas the distance was small or the flames were almost attached at extinction in lean methane and rich propane–air mixtures.[9]

In the experiment of Fig.1.1, the mixtures were lean propane–air mixtures. Thus, it is reasonable to expect that the flame extinction occurs far from the stagnation surface so that the downstream heat loss may not significantly affect the extinction limit. Now, the two distinct behaviors at extinction are understood on the basis of the Lewis number, which is the ratio of thermal to mass diffusivities.[9–11]

Therefore, by adopting a simplified geometry of stretched flame in experiments, aerodynamic straining effects on flame extinction could be clarified and well understood.

1.1.2 FLAME CURVATURE

The next concern is flame curvature. A general invariant expression for flame stretch in a dimensionless form is given by Matalon[4] as

$$\kappa = -\frac{\delta}{S_f^{\,0}} \left\{ \nabla \times \left(\vec{V} \times \vec{n} \right)_{F=0} \cdot \vec{n} + \frac{F_t}{\left| \nabla F \right|} \nabla \cdot \vec{n} \right\}_{F=0} \tag{1.1}$$

in which $\delta = \lambda/\left(\rho_u C_p S_f^0\right)$ is a diffusion length (preheat zone thickness), λ and C_p are respectively the thermal conductivity and the specific heat of the mixture, ρ_u is the density of the unburned gas, S_f^0 is the unstrained burning velocity, $\vec{V}$ is the fluid velocity field, and $\vec{n}$ is a unit vector normal to the flame front. $F\left(\vec{X},t\right) = 0$ defines the flame surface. Similar derivations for flame stretch in a dimensional form have been obtained by Law and Chung[12] and by Candel and Poinsot.[13]

In Eq.1.1, div $\vec{n}(\nabla \cdot \vec{n})$ is the flame curvature, which is given as

$$\nabla \cdot \vec{n} = \frac{1}{R_1} + \frac{1}{R_2} \tag{1.2}$$

where R_1 and R_2 are the principal radii of curvature of the surface and $F_t/|\nabla F|$ is the normal velocity of the flame front v_n, that is,

$$\frac{F_t}{|\nabla F|} = v_n. \tag{1.3}$$

Thus, the second term in Eq.1.1 results from the motion of the moving curved flame front. In the case of a spherically expanding flame, the flow direction is perpendicular to the flame front, and hence, the first term in Eq.1.1 vanishes. The flame experiences stretch only by the second term to yield the flame stretch as

$$\kappa = \frac{\delta}{S_f^0} \frac{dR_f}{dt} \frac{2}{R_f} \tag{1.4}$$

in which $R_f\left(t\right)$ is the flame radius at time t.

In the case of a stationary flame, the second term vanishes. The flame experiences stretch only by the first term. When the flow direction is perpendicular to the flame surface, the flame does not experience stretch; a flat flame stabilized in a one-dimensional flow, and cylindrical and spherical flames stabilized in radially outward flows from line and point sources, respectively, do not experience stretch. When a flame is situated in a nonuniform flow, the flame experiences stretch due to its tangential component of the flow along flame surface. The curl operator ($\nabla \times$) can be expanded as[13]

$$\nabla \times \left(\vec{V} \times \vec{n}\right) = \vec{V} \nabla \cdot \vec{n} - \vec{n} \nabla \cdot \vec{V} - \left(\vec{V} \cdot \nabla\right)\vec{n} + \left(\vec{n} \cdot \nabla\right)\vec{V}. \tag{1.5}$$

Namely, the first term of Eq.1.1 includes a term proportional to the flame curvature $\nabla \cdot \vec{n}$. When a flame is flat such as the flame in Figs 1.1 and 1.2, this curvature term vanishes. If a stationary flame is curved, however, the flame may suffer from a stretch more than the flat flame quantitatively.

The above way of thinking motivated tubular flame studies. In the experiment in Fig.1.2, two flat flames are formed perpendicular to the axis of symmetry. If the direction of the flow is reversed, what will happen? This imagination leads to another flame-flow geometry shown in

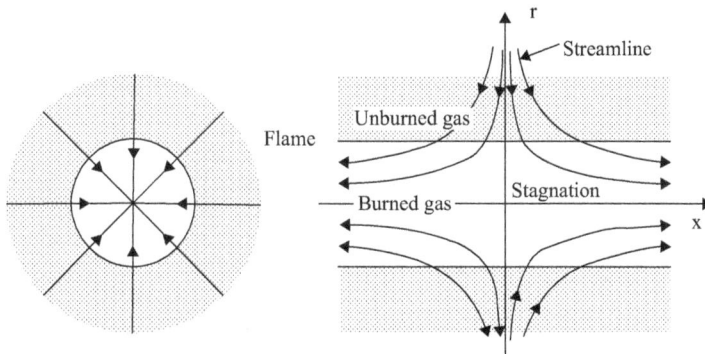

Figure 1.3. Schematic of a tubular flame. (*Source*: Ishizuka, S., Characteristics of Tubular Flames, Progress in Energy and Combustion Science 19(1993)187–226.[8] Used by permission of Elsevier.)

Fig.1.3.[8] The flame surface will be parallel to the axis of symmetry, resulting in the formation of a tubular flame. The flame is stretched along the axis of symmetry and, in addition, the flame has curvature. This flow field can be easily obtained by using a porous cylinder, through which a combustible mixture is injected inward.

1.1.3 ROTATION

Besides the two factors related to flame stretch, there is another motivation for tubular flame studies. That is the modeling of turbulent combustion. Figure 1.4 shows a stretched vortex in the Tennekes' model.[6] This model has been proposed for a small-scale structure of turbulence, and it is considered that a fine vortex of Kolmogorov scale, which is stretched by the Taylor microscale, plays an important role in the fine structure. This vortex element is taken into consideration for turbulent combustion models by Chomiak[14] and by Tabaczynski et al.[15] In the model by Tabaczynski et al., which is shown in Fig.1.5, combustion is assumed instantaneous over the Kolmogorov scale and the burned gas in the vortex tubes is assumed to propagate at a rate of the sum of laminar flame speed S_L plus turbulent intensity u', followed by laminar fashion combustion across the Taylor microscale λ. In the hydrodynamic model by Klimov and Lebedev,[16] the tubular (cylindrical) flame geometry is considered to occur after rapid flame propagation along a vortex axis, although the diameter of the vortex is much larger than that in the Kolmogorov scale. Thus, it is important to investigate experimentally what kind of a flame can be established in such a stretched vortex flow.

This problem is also important from a fluid dynamic viewpoint. As clearly shown, any flow can be locally resolved into a uniform translation, a pure straining motion, and a rigid-body rotation.[17] As for the first, uniform flow, flat, cylindrical and spherical flames are their elementary flames in the rectangular, cylindrical and spherical coordinates, respectively. As for the second, straining flow, the elementary flames are flat flame in rectangular and cylindrical coordinates, as shown in Figs 1.1 and 1.2, and in addition, a tubular flame in the cylindrical

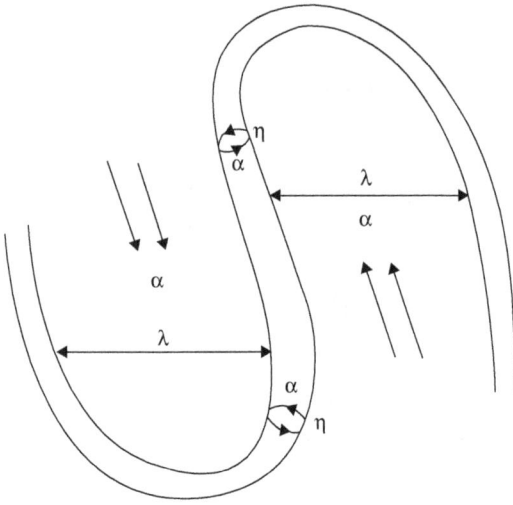

Figure 1.4. Schematic of the proposed small-scale model structure. (*Source*: Tennekes, H. Simple Model for the Small-Scale Structure of Turbulence, Phys. Fluids 11 (1983) 669–670.[6] Used by permission of American Institute of Physics.)

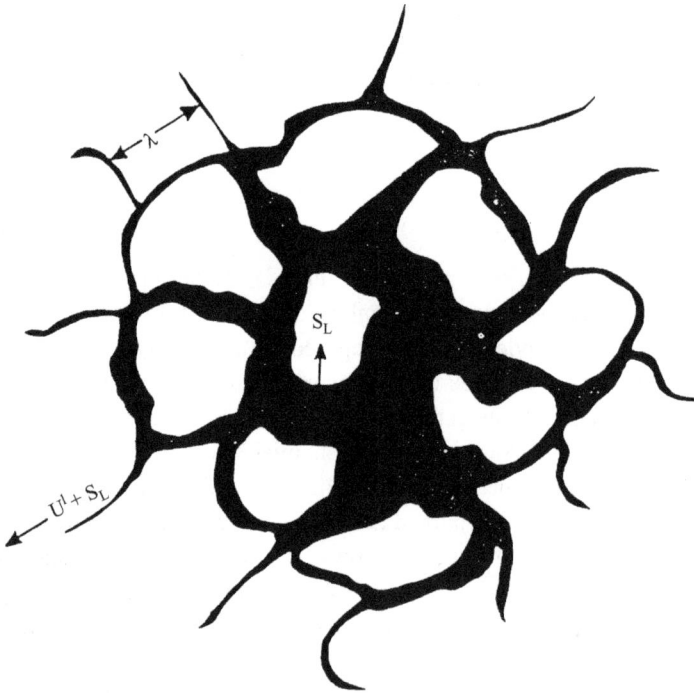

Figure 1.5. Model of a burning, turbulent, small-scale structure. (*Source*: Tabaczynski, R. J., Trinker, F. H., and Shannon, B. A. S., Further refinement and validation of a turbulent flame propagation model for spark-ignition engines, Combust and Flame 39(1980)111–21.[15] Used by permission of Elsevier.)

coordinate as mentioned in Fig.1.3. As for the third flow of rigid-body rotation, however, no flame type has been known experimentally yet.

From the background mentioned above, a tubular flame study has been initiated.

1.1.4 TUBULAR FLAMES

Figure 1.6 shows schematics of the swirl-type burner and the counterflow-type burner to obtain tubular flames.[18] In the swirl-type burner, a long slit is equipped along the axis of the glass tube, through which a combustible premixture is tangentially injected. The diameter and length of the tube are 13.4 mm and 120 mm, respectively, while the slit width is 3 mm. The swirl number roughly obtained is 0.6, and hence, a recirculating flow may not occur; a simple stretched vortex flow is obtained. As shown in Fig.1.7, a flame with a circular cross section and a long length has been established. The burned gas exits from both ends of the glass tube.

Also a tubular shaped flame can be obtained with a counterflow-type burner. Shown is a porous cylinder burner, in which a combustible mixture is normally injected inward through

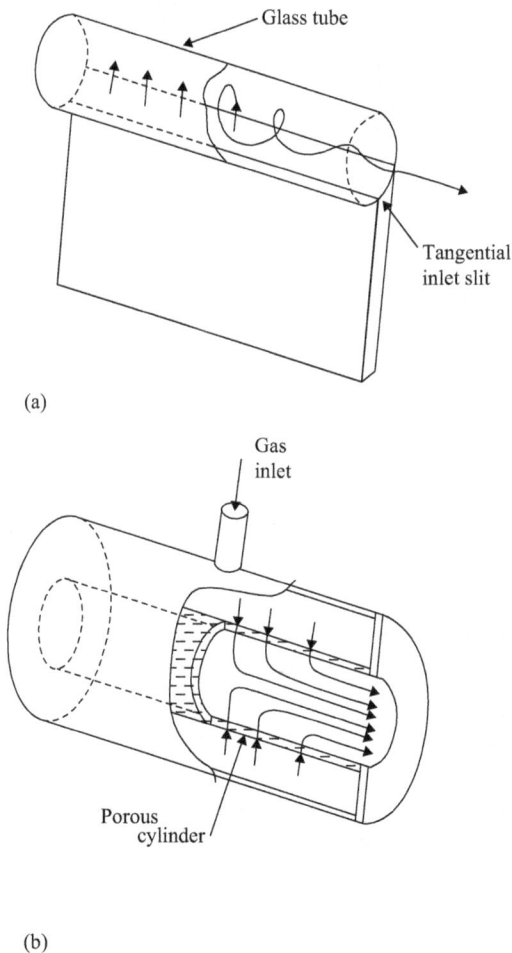

Glass tube

Tangential
inlet slit

(a)

Gas
inlet

Porous
cylinder

(b)

Figure 1.6. Schematics of (a) the swirl-type burner and (b) the counterflow-type burner. (*Source*: Ishizuka, S., On the Behavior of Premixed Flames in a Rotating Flow Field: Establishment of Tubular Flames, Proceedings of the Twentieth Symposium (International) on Combustion, The Combustion Institute, Pittsburgh, PA. pp.287–294, 1985.[18] Used by permission of Elsevier.)

Figure 1.7. Tubular flame in the swirl-type burner (lean methane/air mixture, fuel concentration: 5.4%, tangential velocity: 3 m/s). (*Source*: Ishizuka, S., On the Behavior of Premixed Flames in a Rotating Flow Field: Establishment of Tubular Flames, Proceedings of the Twentieth Symposium (International) on Combustion, The Combustion Institute, Pittsburgh, PA. pp.287–294, 1985.[18] Used by permission of Elsevier.)

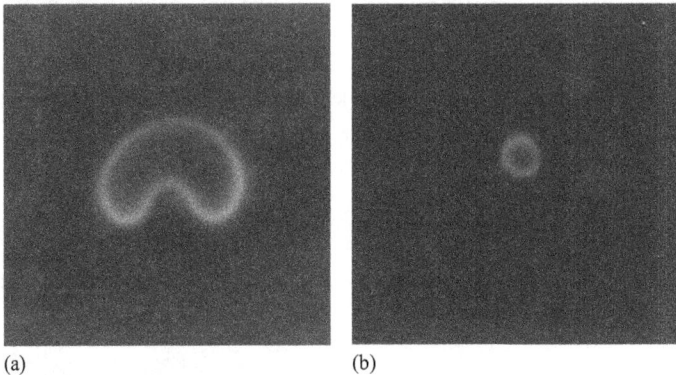

(a)   (b)

Figure 1.8. Appearance of lean methane/air flames when the counterflow burner is mounted (a) horizontally (fuel concentration: 5.3%, injection velocity: 10 cm/s) and (b) vertically (fuel concentration: 4.9%, injection velocity: 10 cm/s). (*Source*: Ishizuka, S., On the Behavior of Premixed Flames in a Rotating Flow Field: Establishment of Tubular Flames, Proceedings of the Twentieth Symposium (International) on Combustion, The Combustion Institute, Pittsburgh, PA. pp.287–294, 1985.[18] Used by permission of Elsevier.)

the porous wall and ejected from both the ends of the burner. The inner diameter of the porous cylinder is 30 mm. Usually, the injection velocity is limited below 30 cm/s because the flow is disturbed and not uniform otherwise. Thus, the flame is prone to be distorted by a buoyancy force working on the hot burned gas. When the burner is mounted horizontally, the flame is distorted as shown in Fig.1.8(a), whereas a tubular shaped flame shown in Fig.1.8(b) can be obtained when the burner is mounted vertically. Since the velocity is limited in a small range, this burner is appropriate to study the tubular flame behaviors at weak flame stretch.

For large rates of blowing, a radial-flow nozzle burner has been designed.[19] Figure 1.9 shows the schematic of the burner. A combustible mixture is introduced through a porous cylinder, accelerated radially inward, and ejected from the outlet of a converging nozzle with a uniform velocity profile. Although the burner is mounted horizontally, a cylindrical flame forms around the axis as shown in Fig.1.10. The extinction limits of the stoichiometric mixtures, which need large stretch rates, can be determined.

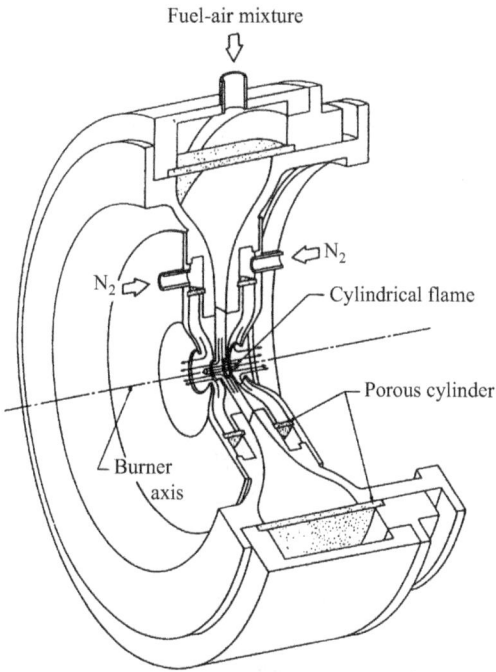

Figure 1.9. Schematic of a radial-flow nozzle burner. (*Source*: Kobayashi, H. and Kitano, M., Extinction Characteristics of a Stretched Cylindrical Premixed Flame, Combustion and Flame 76(1989)285–295.[19] Used by permission of Elsevier.)

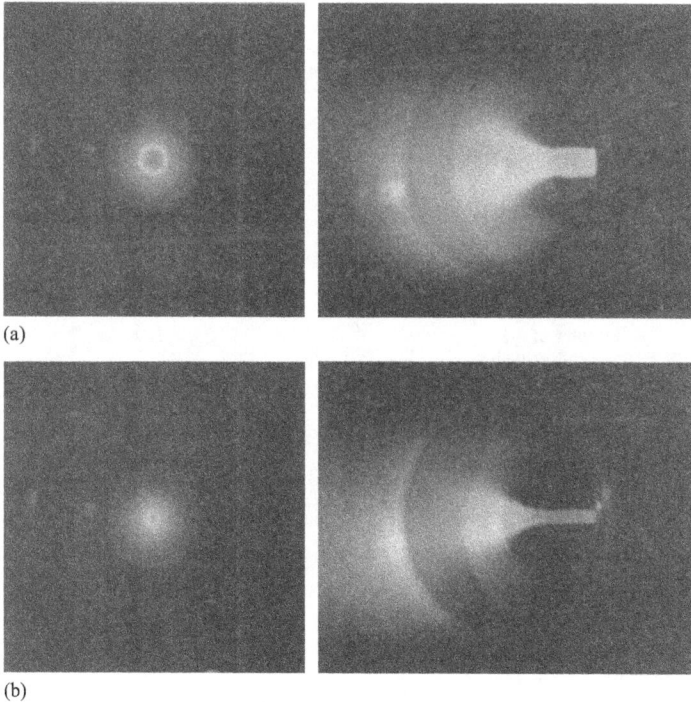

Figure 1.10. Direct photographs of the cylindrical flame: CH_4-air, equivalence ratio = 1.0, D_R = 12 mm, horizontally mounted, (a) U = 2.7 m/s, (b) U = 3.7 m/s. (*Source*: Kobayashi, H. and Kitano, M., Extinction Characteristics of a Stretched Cylindrical Premixed Flame, Combustion and Flame 76(1989)285–295.[19] Used by permission of Elsevier.)

1.2   NOTABLE TUBULAR FLAME CHARACTERISTICS

Compared with other elementary flames, tubular flame has overwhelming stability characteristics. That is, the downstream heat loss is negligible because of symmetry of temperature distribution, and the tubular flame obtained with the swirl-type burner is aerodynamically stable according to the Rayleigh stability criterion.[20] Since tubular flames are formed in nonuniform flows, their characteristics are strongly dependent on the Lewis number. These notable tubular flame characteristics are specifically described in this section.

1.2.1   THERMAL ADVANTAGE

Figure 1.11 shows the radial temperature distributions across the tubular flames of lean methane–air mixtures, in which the fuel concentrations Ω are 5.4 and 5.8% and the tangential velocity V_t is 3 m/s.[18] It is seen that the combustion field is separated into two regions, an outer unburned gas region of low temperature and an inner burned gas region of high temperature. Because of the near symmetrical temperature distribution, there seems minimal or no heat loss behind the flame. The hot gas region inside the flame offers an ideal adiabatic condition for the flame.

Figures 1.12 and 1.13 respectively show the mapping of the various flame configurations for methane–air and propane–air mixtures obtained by the swirl-type burner.[21] The burner inner diameter is 13.6 mm and the slit width and length are 3 mm and 80 mm. The burner has rectangular slits of 3 mm wide and 20 mm long at both sides of the center slit, through which nitrogen

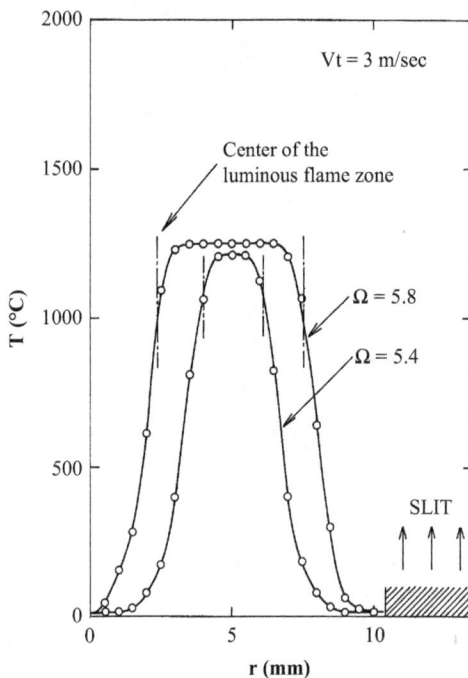

Figure 1.11. Radial temperature distributions across the tubular flames of lean methane-air mixtures (Ω = 5.4 and 5.8%, and V_t = 3.0 m/s). (*Source*: Ishizuka, S., On the Behavior of Premixed Flames in a Rotating Flow Field: Establishment of Tubular Flames, Proceedings of the Twentieth Symposium (International) on Combustion, The Combustion Institute, Pittsburgh, PA. pp.287–294, 1985.[18] Used by permission of Elsevier.)

is injected to avoid the attachment of diffusion flames. For lean methane–air or rich propane–air mixtures, a tubular flame with a uniform flame front can be formed for a wide range of the injection velocities. The fuel concentrations at extinction are close to the flammability limits determined by the standard method.[22,23] This is amazing because the inner diameter of the present burner is 13.6 mm, which is much smaller than the tube diameter of 5 or 10 cm in the standard method.[22,23] For rich methane–air or lean propane–air mixtures, the fuel concentrations at

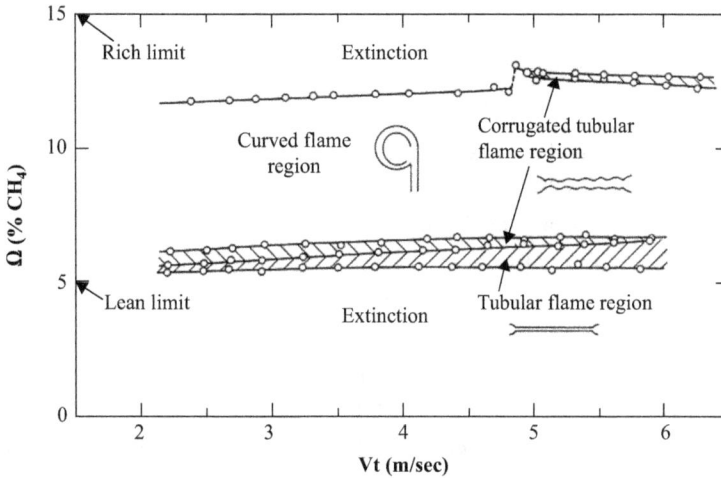

Figure 1.12. Mapping of various flame configurations of methane-air mixtures. (*Source*: Ishizuka, S., An Experimental Study on Extinction and Stability of Tubular Flames, Combustion and Flame 75(1989)367–379.[21] Used by permission of Elsevier.)

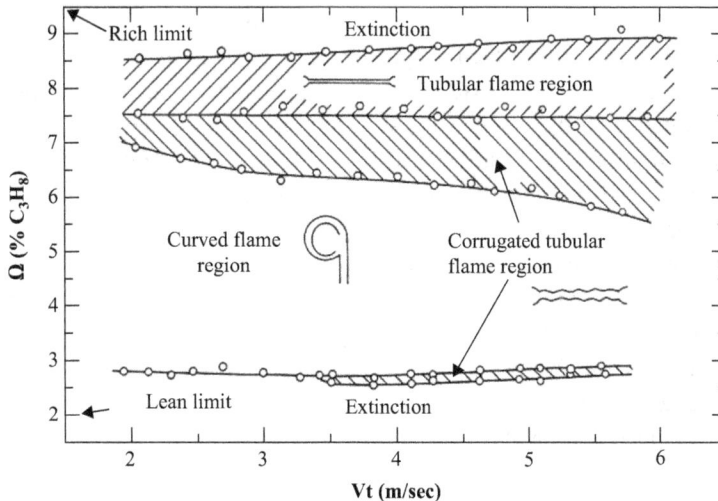

Figure 1.13. Mapping of various flame configurations of propane-air mixtures. (*Source*: Ishizuka, S., An Experimental Study on Extinction and Stability of Tubular Flames, Combustion and Flame 75(1989)367–379.[21] Used by permission of Elsevier.)

Table 1.1. Comparison of the flammability limits determined by the swirl-type burners of different inner diameters with the others. (*Source*: Ishizuka, S., Characteristics of Tubular Flames, Progress in Energy and Combustion Science 19(1993)187–226. Used by permission of Elsevier.)

Reference	Method		CH_4 Lean	Rich	C_3H_8 Lean	Rich	H_2 Lean
This work	Tubular	$D = 13.6$ mm	5.3	13.0	2.6	9.0	4.5
	flame	$D = 18.0$ mm	4.9	13.2	2.3	10.0	3.8
	rotating	$D = 21.0$ mm	4.8	13.7	2.3	10.2	3.7
		$D = 18.0$ mm	4.6	14.0	2.2	-	3.5
Ishizuka	Tubular flame (nonrotating)		4.7	15.1	2.0	9.8	4.2
Zabetakis	Propagating flame		5.0	15.0	2.1	9.5	4.0
Ishizuka and Law	Binary flame		4.8	15.8	2.0	9.7	4.1

extinction are not so close to the flammability limits determined by the standard method. This is due to the Lewis number effect described later.

Table 1.1 summarizes the flammable ranges determined with the swirl and counterflow-type tubular flame burners of different burner diameters.[8] As the burner diameter is increased from 13.6 to 28.0 mm, for example, the minimum fuel concentration in lean methane mixtures is reduced from 5.3 to 4.6%, while the maximum fuel concentration in rich mixtures is increased from 13.0 to 14.0%. It is seen that the flammable ranges of tubular flames are slightly wider in concentration than those determined with the standard, propagating flame method.

Figure 1.14 shows the dilution limits determined with a porous cylinder burner.[24] The dilution limit was determined by varying the methane flow rate while keeping the injection velocity about 5 cm/s and adding the nitrogen ($\chi = [N_2]/([N_2] + [air])$) stepwise. The broken line indicates the stoichiometric mixtures and the dotted line shows the results obtained by the standard method. Both the results with vertically and horizontally mounted 30-mm inner diameter burners are shown. With an increase in χ, the fuel concentration at lean limit is only slightly increased while that at rich limit is markedly decreased, and the flammable range in Ω becomes narrower. The maximum nitrogen concentrations are 42% and 45% for the vertically and horizontally mounted burners, respectively. The flammable ranges determined by means of tubular flames are wider than those with the standard, upward propagating flame, and the peak values (the maximum nitrogen concentrations for flame extinction) are larger than those determined by the standard method. Thus, the counterflow-type tubular flame burner has been adopted for the determination of flammability limits and employed to evaluate the power of the halogenated fire suppressants by the researchers in Fire Research Institute in Japan.[25–27]

1.2.2 AERODYNAMIC ADVANTAGE

Figures 1.15 and 1.16 show the extinction limits of the cylindrical (tubular) flames obtained with the radialflow nozzle burners for methane–air and propane–air mixtures, respectively.[19]

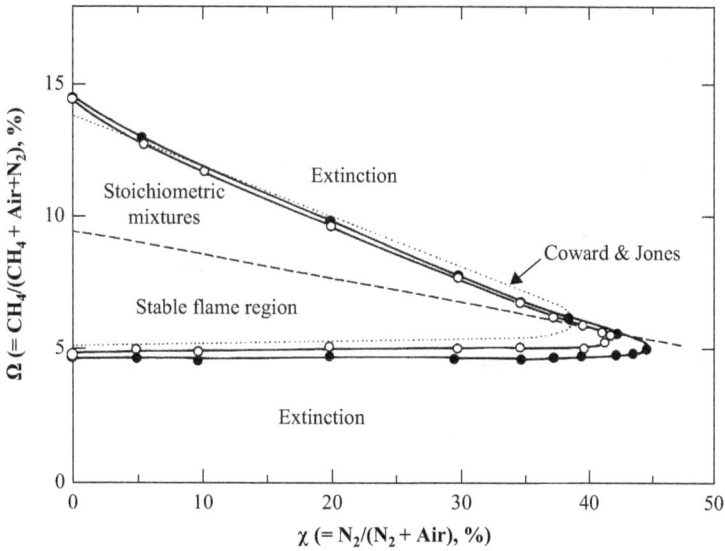

Figure 1.14. Extinction limits of methane flames diluted with nitrogen in the 30-mm burner mounted vertically (open circles) and horizontally (closed circles). (*Source*: Ishizuka, S., Determination of flammability limits using a tubular flame geometry, Journal of Loss Prevention in the Process Industry 4(1991)185–193.[24] Used by permission of Elsevier.)

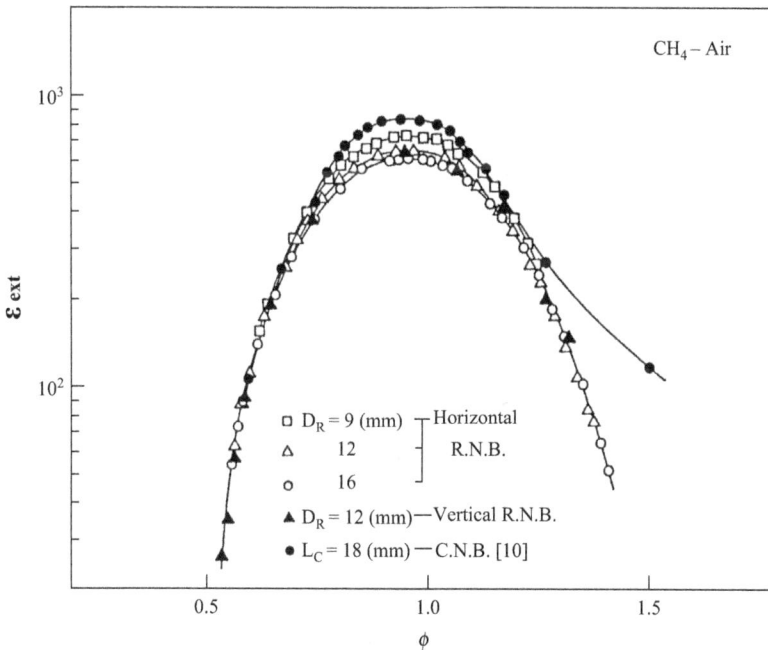

Figure 1.15. Extinction limits of the cylindrical flame and the twin flames for methane-air mixtures. *Abbreviations*: CNB, counterflow nozzle burner; RNB, radial-flow nozzle burner. (*Source*: Kobayashi, H. and Kitano, M., Extinction Characteristics of a Stretched Cylindrical Premixed Flame, Combustion and Flame 76(1989)285–295.[19] Used by permission of Elsevier.)

In these measurements, three kinds of nozzle outlets with different diameters ($D_R = 9, 12, 16$ mm) were prepared and both the horizontal and vertical setup were examined. In these figures, the extinction limits of twin flames established between two opposed axisymmetric premixed gas streams are also plotted, for the measurements of which the outlet diameter D_c was 10 mm and the distance L_c between the two outlets of the counterflow nozzle burner was 18 mm. As a measure of the flame stretch rate, the characteristic velocity gradient ε is taken as $2U/D_R$ for the radial-flow nozzle burner and $2U/L_c$ for the counterflow nozzle burner, where U is the mean flow velocity at the nozzle outlet. The characteristic velocity gradients at the extinction, ε_{ext}, are plotted against the equivalence ratio ϕ.

It is seen that although the extinction limits for the tubular flames determined with different outlet diameters do not coincide each other, their maximum values for methane and for propane are around 6–7×10^2 s^{-1} and around 8–9×10^2 s^{-1}, both of which are smaller than the values of 8.5×10^2 s^{-1} and 1.0×10^3 s^{-1} for the twin flames, respectively.

From Eq.1.1, the stretch rate of tubular flame is given as twice the velocity gradient normal to the flame surface

$$\kappa = \frac{\delta}{S_f^0} \frac{\partial V_z}{\partial z} = \frac{\delta}{S_f^0} \left\{ -\frac{1}{r} \frac{\partial}{\partial r} (rV_r) \right\} = \frac{\delta}{S_f^0} \cdot 2\varepsilon_r \qquad (1.6)$$

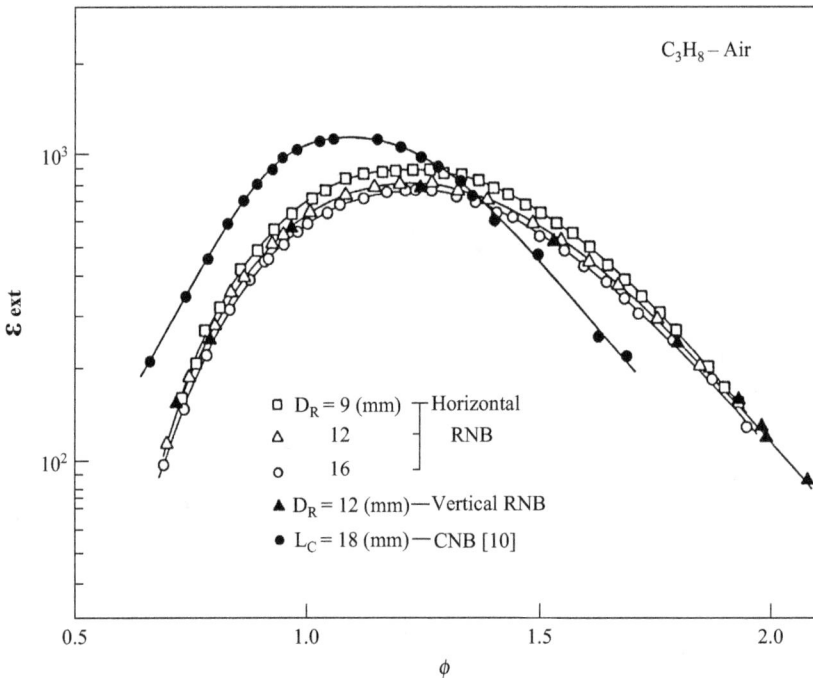

Figure 1.16. Extinction limits of the cylindrical flame and the twin flames for propane-air mixtures. *Abbreviations*: CNB, counterflow nozzle burner; RNB, radial-flow nozzle burner. (*Source*: Kobayashi, H. and Kitano, M., Extinction Characteristics of a Stretched Cylindrical Premixed Flame, Combustion and Flame 76(1989)285–295.[19] Used by permission of Elsevier.)

in which V_z and V_r are respectively the axial and radial velocities in the radial r and axial z coordinates, and ε_r is the velocity gradient in the r-coordinate.[8] In this equation, a mass continuity equation assuming steady, incompressible and cylindrical symmetry is used. The value ε_r corresponds to ε_{ext} in Figs 1.15 and 1.16.[19] On the other hand, the stretch rate of the twin flame is given as

$$\kappa = \frac{\delta}{S_f^0} \frac{\partial V_z}{\partial z} = \frac{\delta}{S_f^0} \cdot \varepsilon_z. \tag{1.7}$$

The velocity gradient in the z-coordinate ε_z corresponds to ε_{ext} in Figs 1.15 and 1.16.[19] Thus, if it is evaluated with the velocity gradient normal to the flame front, that is, ε_r for tubular flame and ε_z for twin flame, tubular flame suffers from a stretch much more than twin flame by a factor of 2.

In Figs 1.15 and 1.16, the maximum velocity gradients of the twin flames are larger than those of the tubular flames, but the difference is at most by a factor of 1.4. In propane–air mixtures, the equivalence ratio at which the velocity gradient at extinction takes its maximum shifts toward the fuel-rich side from the stoichiometric in the case of tubular flame. These extinction characteristics are theoretically studied by taking finite reaction rates and the Lewis numbers for fuel and oxygen as well into consideration.[28] Anyway, the tubular flames are extinguished at a smaller stretch rate than the twin flames if evaluated with the velocity gradient normal to the flame surface.

In the case of the swirling tubular flame, however, the flame extinction limits are significantly extended; the flame is robust in stability. Figure 1.17 shows the extinction limits

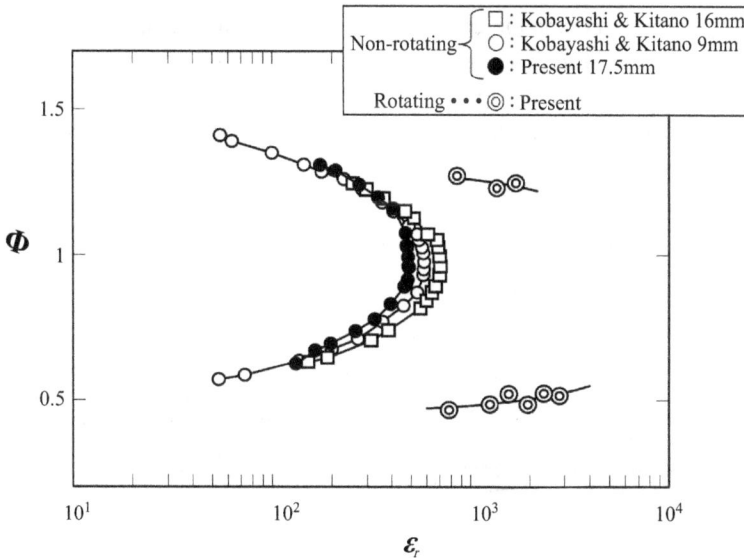

Figure 1.17. Extinction limits of the tubular flames of methane-air mixtures. (*Source*: Ishizuka, S., Mitoma, K., and Shimokuri, D., Extension of blow-off limits of tubular flame by strong rotation, Proceedings of the forty-first symposium (Japanese) on Combustion, pp.17–18, 2003.[29] Used by permission of the Combustion Society of Japan.)

determined with a swirl-type tubular flame burner.[29] The burner's inner diameter is 36 mm. The burner has four tangential slits of 8 mm width and 97 mm length. Fuel is methane. For comparison, the velocity gradients at extinction determined with the radial flow nozzle burners by Kobayashi and Kitano and those determined with a nonrotating tubular flame burner of 17.5 mm in diameter and 50 mm in length at the outlet are also plotted. It is seen that the extinction limit in the swirling tubular flame is significantly widened; especially, the extinction limit for the stoichiometric mixture could not be determined due to the limitation of the fuel- and air-supplying systems used in this experiment.

One reason for the wide extension in stability limit is that the swirling tubular flame is aerodynamically stable according to the Rayleigh stability criterion[20] since the flame consists of an inner hot burned gas region of low density and an outer unburned gas region of high density. Another reason is that when the swirl becomes strong, a recirculation of a hot burned gas occurs, which stabilizes the flame as in the gas turbine combustor.[30]

As an exact solution of the Navier–Stokes equations for a three-dimensional axisymmetric flow, a two-cell vortex solution has been obtained by Sullivan.[31] For reference, the radial velocity gradients for the swirling flame in Fig.1.17 are those of the Sullivan's solution, which has a reverse flow around the axis of rotation, and accordingly has a stagnation point and a stagnation circle on the r-plane as illustrated in Fig.1.18. From the radial velocity profile of the Sullivan's solution, the velocity gradient on the stagnation circle is approximately given as

$$\varepsilon_r \cong 1.64 \frac{V_{r,wall}}{R} \qquad (1.8)$$

in which $V_{r,wall}$ is the mean radial injection velocity and R is the radius of the burner.

Once a recirculating flow is induced, the flame is always sustained by a hot burned gas backward. Figure 1.19 shows the response of a stretched premixed flame, onto which a product stream impinges backward.[2,32] In this figure, the ratio of the rate of heat release of a strained

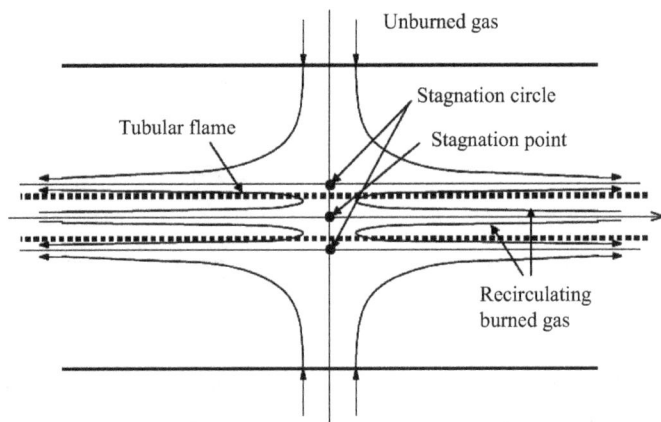

Figure 1.18. Aerodynamic structure of a tubular flame in a strong swirl flow with a recirculating burned gas.

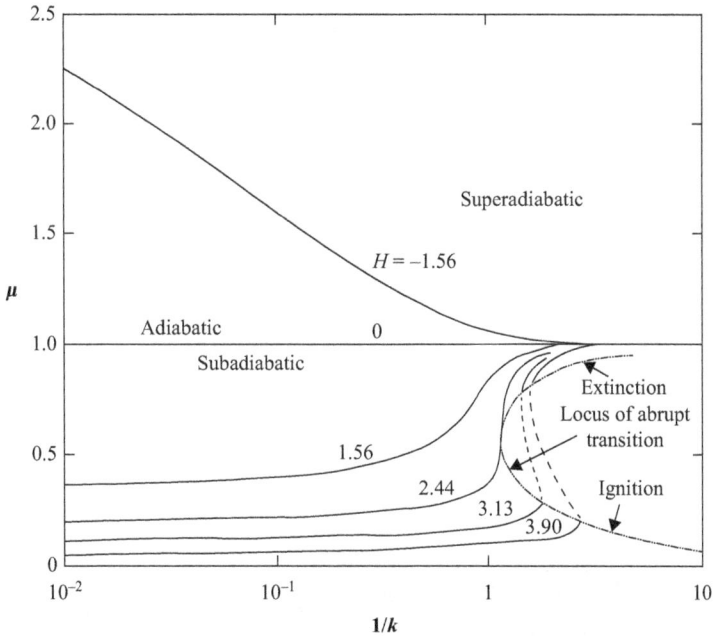

Figure 1.19. The ratio of the rate of heat release of a strained flame to that of an unstrained flames as a function of the reciprocal of the nondimensional strain rate for various values of the nonadiabaticity parameter H. (*Source*: Williams, F. A., Combustion Theory, 2nd ed., p.416, Addison-Wesley, Redwood City, 1985.[2] Used by permission of Perseus Books Group.)

flame to that of an unstrained flame μ is plotted against the reciprocal of the nondimensional strain rate κ for various values of the non-adiabaticity parameter H.

If the product stream temperature is lower than the adiabatic flame temperature (a subadiabatic condition), flame extinction occurs when the strain rate is increased to reach some critical value. If the product stream temperature is higher than the adiabatic flame temperature (a superadiabatic condition), the ratio μ monotonically increases with increasing κ (decreasing κ^{-1}); extinction will never occur.

Thus, it is interesting to note that a swirling tubular flame is robust in stability due to its aerodynamic structure, especially when a recirculation of a hot burned gas occurs.

1.2.3 LEWIS NUMBER EFFECTS

Since the tubular flame is established in a stretched flow field, the flame characteristics are strongly dependent on the Lewis number of a limiting component in the mixture. Figure 1.20 schematically shows the heat and mass fluxes in tubular flame focusing on a stream tube surrounded by two streamlines.[8] It should be noted that mass is conserved within a stream tube. In tubular flame, the axis of rotation is a stagnation streamline, and the flow is diverged in the longitudinal direction while the flow is converged in the radial direction. Since the heat flux

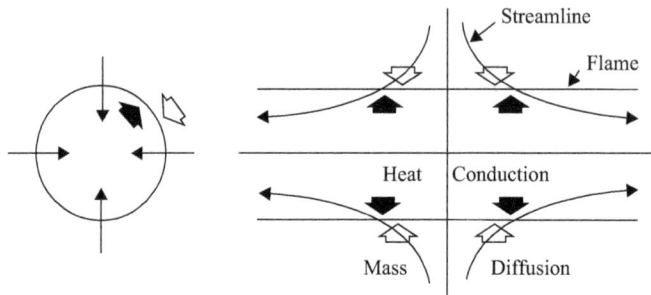

Figure 1.20. Schematics illustrating the directions of the heat conduction and the mass diffusion of a limiting reactant across the flame. (*Source*: Ishizuka, S., Characteristics of Tubular Flames, Progress in Energy and Combustion Science 19(1993)187–226.[8] Used by permission of Elsevier.)

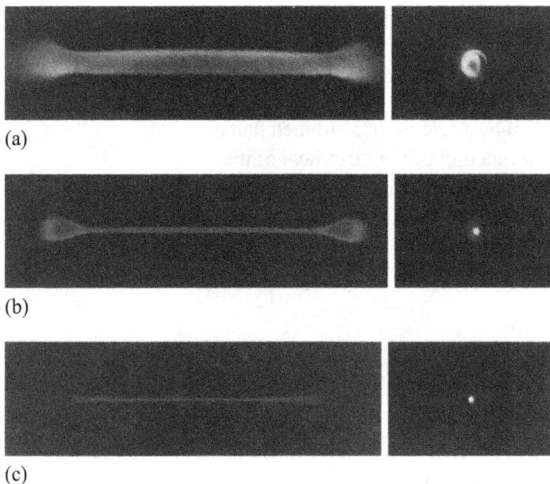

(a)

(b)

(c)

Figure 1.21. Flame configurations of (a) lean propane-air mixture ($\Omega = 3.0\%$), (b) lean methane-air mixture ($\Omega = 5.4\%$), and (c) lean hydrogen-air mixture ($\Omega = 4.9\%$) near the extinction limits ($V_t = 3$m/s) (Ω: the fuel concentration by volume, V_t: the mean tangential velocity). (*Source*: Ishizuka, S., On the Behavior of Premixed Flames in a Rotating Flow Field: Establishment of Tubular Flames, Proceedings of the Twentieth Symposium (International) on Combustion, The Combustion Institute, Pittsburgh, PA. pp.287–294, 1985.[18] Used by permission of Elsevier.)

though conduction and the mass flux through diffusion are perpendicular to the flame front, heat and mass transfers occur through a stream tube. Heat is transported by conduction from the inner burned gas into the outer unburned mixture, whereas reactants are transported by diffusion from the outer unburned gas into the flame zone where they are consumed. Therefore, the enthalpy of the sum of the thermal energy and the chemical energy is not conserved within the stream tube if the thermal diffusivity and the mass diffusivity are not equal.

If the thermal diffusivity κ is larger than the mass diffusivity D, that is $Le \equiv \kappa/D \geq 1$, the flame temperature becomes lower than its original adiabatic value, while it becomes higher for $Le \leq 1$.

Figure 1.21 shows the pictures of a tubular flame near extinction limits.[18] It is seen that the flame diameter is relatively larger for lean propane–air mixtures, whereas it is very small for lean methane–air mixture. In a lean hydrogen–air mixture, the flame is almost converted into a filament. The Lewis number of lean propane–air mixture is larger than unity, whereas those of lean methane–air and lean hydrogen–air mixtures are smaller than unity. Thus, the tubular flame can become smaller in diameter when the mass diffusivity of a deficient, hence limiting component is larger than the thermal diffusivity of the mixture.

1.3 TUBULAR FLAME STUDIES

To understand their characteristics, tubular flames have been studied theoretically and computationally. Elaborate experimental works have been further carried out with the use of advanced laser diagnostics, which enable rigorous comparisons with theories and computations. These studies are introduced in this section.

1.3.1 THEORETICAL STUDIES

In 1984, Mikolatis[33] considered a stretched cylindrical flame, which is formed when a combustible mixture is injected radially outward through a porous cylinder and suction is applied to the outside of the porous insulating disks at both sides. Thus, it should be noted that the inside is the unburned gas and the outside is the burned gas. This flame structure is the reversed one of the tubular flame obtained by the counterflow-type or the swirl-type tubular flame burner, which consists of the inner burned gas region and the outer unburned gas region. Later Tseng, Abhishek and Gore[34] conducted an experiment using an inner porous cylinder from the surface of which a methane–air mixture was injected onto an outer quartz tube and they obtained similar temperature distributions independently of the axial coordinate. Application to a radiant heating tube was specially addressed.

A theoretical study on the tubular flame obtained with the swirl-type burners has been initiated by Prof. Takeno. First, by assuming an incompressible flow and adopting the similar solution developed by Yuan and Finkelstein[35], the flow field and the general response of the flame to the injection Reynolds number and equivalence ratio have been obtained.[36] For example, the longitudinal velocity u and the radial velocity v are analytically obtained as

$$u = \frac{2v_R x}{R} \frac{1}{\xi} \frac{df}{d\xi} \cong \frac{\pi v_R}{R} x$$
$$v = -2v_R \frac{f}{\xi} \cong -\frac{1}{2} \frac{\pi v_R}{R} r$$

(1.9)

in which x and r represent, respectively, longitudinal and radial distances from the center, R and v_R are, respectively, the inner radius of the burner wall and the radial injection velocity at the wall, ξ is the nondimensional radial distance given as $\xi = r/R$, and $f(\xi)$ is the nondimensional stream function, which is approximately given as:

$$f(\xi) \cong \frac{1}{2} \sin\left(\frac{\pi}{2} \xi^2\right) \cong \frac{\pi}{4} \xi^2.$$

(1.10)

Thus, the radial velocity increases almost linearly with an increase in the radial distance r around the center line. It should be noted, however, that the radial velocity in Eq.1.9 gives

$-\pi v_R/2$ at the wall, greater than the value at the boundary; that is, the radial velocity overshoots and approaches the value of $-v_R$ at the wall (see Fig.2.3 in Chap.2). In this overshooting region, the flame is dynamically unstable because the upstream velocity is decreased as the wall is approached; the flame can propagate up to the wall once the balance between the burning velocity and the oncoming upstream velocity is broken down. This can be a trigger source for oscillatory combustion, later described in Chap.8 for large-scale applications.

Next, by means of asymptotic analysis, extinction characteristics have been studied.[37,38] A merit of asymptotic analysis is to find relevant parameters which govern the flame characteristics. For example, it is shown that the quantities at extinction such as extinction injection Reynolds number and extinction flame radius depend on a parameter γ alone, which is given as

$$\gamma = \frac{\beta}{2} \frac{\theta_a - 1}{\theta_a} \frac{Le - 1}{Le}. \qquad (1.11)$$

However, the asymptotic analyses give a larger extinction injection Reynolds number than the numerical solutions and predict finite flame diameter at extinction, although the numerical solution predicts zero extinction flame diameter for very small Lewis numbers less than the critical value. The critical Lewis number in the analysis by Takeno et al. is 0.16.[38] This corresponds with the experimental observation on lean hydrogen–air mixtures; that is, the flame is converted into a luminous rod at extinction, as shown in Fig.1.21(c).

Theoretical studies were further extended to variable density[39,40] and to volumetric heat loss.[41] Besides, the flame response was analyzed under the Hirschfelder boundary[42] and stretch rates were obtained for both the opposed tubular flame burner and the standard tubular flame burner.[43] Very recently, an asymptotic analysis has been made on the diffusive-thermal instability on the tubular flame front,[44] which appears and forms multipetal flames shown in Fig.1.22.[24]

Theoretical studies are described by Tadao Takeno and Mikihito Nishioka in Chap. 2.

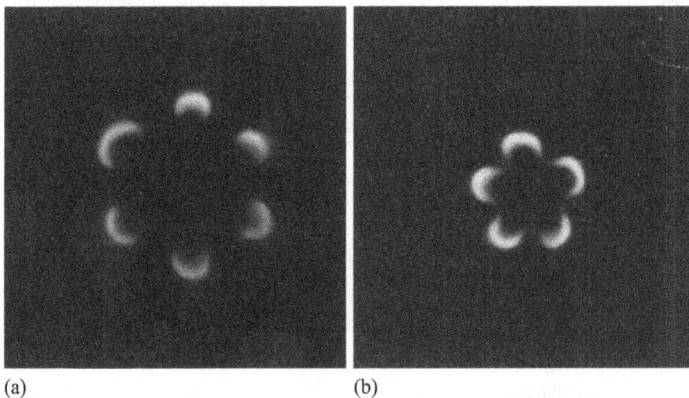

(a)   (b)

Figure 1.22. Appearance of multipetal flames of lean hydrogen–air mixtures obtained with the vertically mounted counterflow-type burner of (a) 30 mm and (b) 15 mm inner diameters. (*Source*: Ishizuka, S., Determination of flammability limits using a tubular flame geometry, Journal of Loss Prevention in the Process Industry 4(1991)185–193.[24] Used by permission of Elsevier.)

1.3.2   COMPUTATIONAL SIMULATIONS

Computational simulations on tubular flames were triggered by a workshop at the University of Leeds in 1990, hosted by Prof. Graham Dixon-Lewis. The first paper with detailed chemistry was presented by Smooke and Giovangigli at the 23rd symposium (International) on Combustion, in which 26 species and 79 reactions were considered for methane–air mixtures and 33 species and 126 reactions for propane–air mixtures.[45] Their results were compared with the experimental results obtained by Kobayashi and Kitano with the radial-flow nozzle burner.[19] Around that time, a numerical study was made by Nishioka et al., in which 18 species and 58 reactions were taken into consideration for methane–air mixtures.[46] Concerning with turbulent combustion under practical conditions, effects of pressures up to 8 atm have been investigated and it has been found that the minimum flame radius is about 0.2 mm at 8 atm, which is one order of magnitude larger than the Kolmogorov microscale encountered in engines. The circumferential velocity, however, was not taken into consideration in this simulation.

In most of computational studies pressure diffusion is ignored. However, under strong rotating flow conditions, pressure diffusion cannot be ignored. In the study by Yamamoto et al.,[47] pressure diffusion was first taken into consideration although a reduced four-step mechanism was used for the swirling methane–air tubular flames. Ten years later, however, it is Zhang et al.[48–50] who discussed fully the pressure diffusion with detailed transport properties and full chemical kinetics. In these calculations, it has been shown that the pressure diffusion effect becomes significant when the flame is situated close to the maximum circumferential velocity position.

For the swirling tubular flame experimentally obtained with a single-slot (see Fig.3.5), a model has been proposed by Kee et al.[51] and their results are compared with the radial distributions of temperature and species obtained experimentally by Sakai and Ishizuka.[52] Pitz and his co-workers[53,54] also have conducted computations to compare their results with experimental results obtained by a laser Raman scattering method.

Computational simulations of tubular flames will be described in detail by Yuyin Zhang, Huayang Zhu, and Robert J. Kee in Chap. 3.

1.3.3   EXPERIMENTAL STUDIES

Although stability limits and distributions of temperature and velocity were determined in the early experiments,[18,19,21] measurements of stable species distributions were quite limited. There were a few literatures, which determined the species distributions with a microsampling probe and gas chromatography.[52] However, using a laser accessible burner, the structure of premixed tubular flame has been investigated in detail.[53,54] Radial distributions of stable species such as methane, hydrogen, propane, oxygen, nitrogen, carbon dioxide, water, and temperature have been successfully determined with a nonintrusive laser Raman scattering technique. Besides these, cellular structure has been experimentally investigated in detail.[55,56]

Raman spectroscopic measurements of tubular flame are described by Robert W. Pitz in Chap. 4.

1.4 RELEVANT STUDIES

In its early stage, tubular flame studies were limited only on premixed combustion. Later, however, relevant studies appeared on non-premixed tubular flame combustion and on miniature liquid-film combustors. These studies are briefly introduced in this section together with relevant knowledge on laminar diffusion flames and counterflow diffusion flames.

1.4.1 TUBULAR NON-PREMIXED, DIFFUSION FLAME STUDIES

For complete understanding of flame characteristics, diffusion flame studies are also important as premixed flame studies. For example, the S-shaped response curve of a counterflow diffusion obtained by Fendell[57] clearly describes the ignition and extinction behaviors of the flame and also gives useful insights on ignition and extinction behaviors of premixed flames in nonuniform flows. As for extinction by sharp flame curvature, similar flame behaviors can be found between premixed flames and diffusion flames.

Figure 1.23 is a picture of flame when a rich propane–air mixture is injected through the nozzle on the flat plate used in the experiment of Fig.1.1.[58] As in the Bunsen burner, diffusion combustion occurred between the excess fuel and the ambient air, which anchors a flame at the rim of the nozzle. Besides, a flat flame is established in the forward stagnation region of the plate. Usually, both the flames are connected. However, as the fuel concentration is increased and reaches some value, local extinction occurs at the connected part due to sharp flame curvature, and the flames are separated as shown in Fig.1.23.

Figure 1.24 shows the mapping of various flame configurations, in which the fuel concentration is plotted against the mean injection velocity from the nozzle.[58] It is seen that local extinction due to sharp flame curvature occurs at an almost constant fuel concentration, here 5.7% propane, independently of the injection velocity. Similar local extinction behaviors can be seen in premixed Bunsen flame tips.[59,60]

Figure 1.25 shows a sequence of photographs of a laminar diffusion flame,[61] in which a hydrogen–nitrogen mixture added with a small amount of propane issues from an inner tube of 14.3 mm in diameter while an air issues from an outer tubes of 27.8 mm in diameter, and the flow rate of hydrogen is successively reduced until the flame is blown off completely. The flow rates of propane, nitrogen, and air are kept constant. As well known by the analysis by Burke and Schumann,[62] the underventilated flame is opened at the tip as shown in Fig.1.25(a),

Figure 1.23. Rim-stabilized smooth flame separated from the plate-stabilized smooth flame. (*Source*: Ishizuka, S., Miyasaka, K., and Law, C. K., Effects of Heat Loss, Preferential Diffusion, and Flame Stretch on Flame-Front Instability and Extinction of Propane/Air Mixtures, Combustion and Flame 45(1982)293–308.[58] Used by permission of Elsevier.)

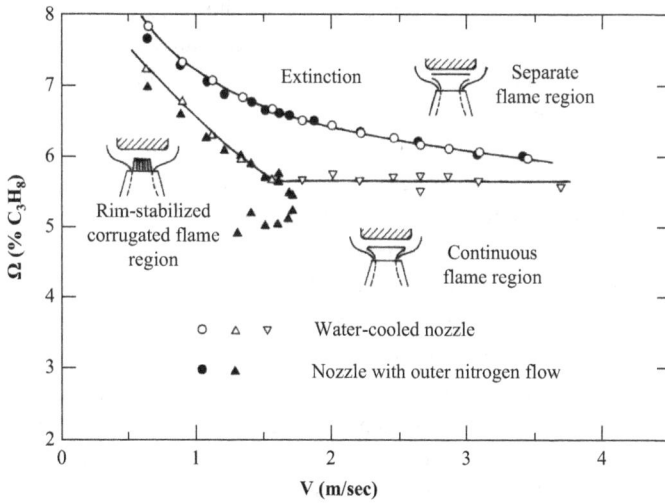

Figure 1.24. Mapping of the flame configurations in the stagnation point flow experiment. (*Source*: Ishizuka, S., Miyasaka, K., and Law, C. K., Effects of Heat Loss, Preferential Diffusion, and Flame Stretch on Flame-Front Instability and Extinction of Propane-Air Mixtures, Combustion and Flame 45(1982)293–308.[58] Used by permission of Elsevier.)

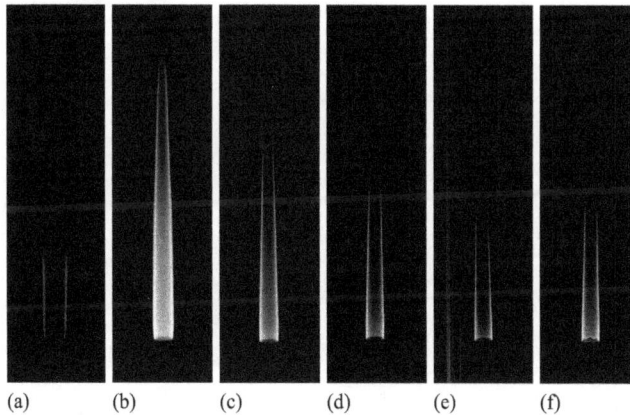

(a)   (b)   (c)   (d)   (e)   (f)

Figure 1.25. Photographs of successive stages in formation of open-tipped hydrogen–propane–nitrogen flame: (C_3H_8: 0.06 l/min, N_2: 5.68 l/min, air: 11.72 l/min, 0°C, 1 atm). (a) Underventilated flame with a pseudo-open tip (H_2: 8.98 l/min, $\Omega = 0.610$). (b) Overventilated flame with a closed tip (H_2: 2.68 l/min, $\Omega = 0.318$) (c) Overventilated flame with a closed tip (H_2: 1.77 l/min, $\Omega = 0.235$) (d) Overventilated flame with a vibrating upper part (H_2: 1.50 l/min, $\Omega = 0.208$). (e) Open-tipped diffusion flame with a uniform base structure (H_2: 1.22 l/min, $\Omega = 0.176$). (f) Open-tipped diffusion flame with a nonuniform base structure (H_2: 1.08 l/min, $\Omega = 0.159$). (*Source*: Ishizuka, S., An Experimental Study on the Opening of Laminar Diffusion Flame Tips, Proceedings of the Nineteenth Symposium (International) on Combustion, The Combustion Institute, Pittsburgh, PA. 319–326, 1982.[61] Used by permission of Elsevier.)

whereas the overventilated flame is closed at the tip as shown in Fig.1.25(b), for which they obtained the flame shape theoretically using the so-called flame sheet approximation and by considering a coupling function. In the present mixture, however, the tip opens when the hydrogen flow rate is gradually decreased.

Figure 1.26 is the mapping of the flame configurations.[61] It is seen that the tip opening occurs at an almost constant hydrogen concentration of the fuel stream independently of the inner stream velocity. This opening occurs not due to the ventilated condition, but due to preferential diffusion of hydrogen in the hydrogen–propane–nitrogen mixture.

This can be confirmed by using helium as a diluent. Figure 1.27 shows the pictures of nitrogen- and helium-diluted diffusion flames obtained in the same double circular pipe.[63] It is seen that the flame is opened at the tip when hydrogen is diluted with nitrogen for an outer oxidizer stream of the standard air (21% O_2 and 79% N_2 by volume), whereas the flame is intensified in burning at the tip and closed when hydrogen is diluted with helium for an outer "helium air" stream of 21% O_2 and 79% He by volume. Thus, preferential diffusion of hydrogen plays an important role in the local flame extinction, which is verified also by the measurements on the species concentration field.[64] Thus, there are very similar behaviors between diffusion flame and premixed flame as for local extinction by sharp flame curvature.

Also, counterflow diffusion flame studies provide useful insights on stretched premixed flame studies. Figure 1.28 shows the variations of flame temperature of counterflow diffusion flames established in the forward stagnation region of a porous cylinder.[65] In a case when hydrogen is injected from a porous cylinder and the oxidizer is diluted with helium, argon or

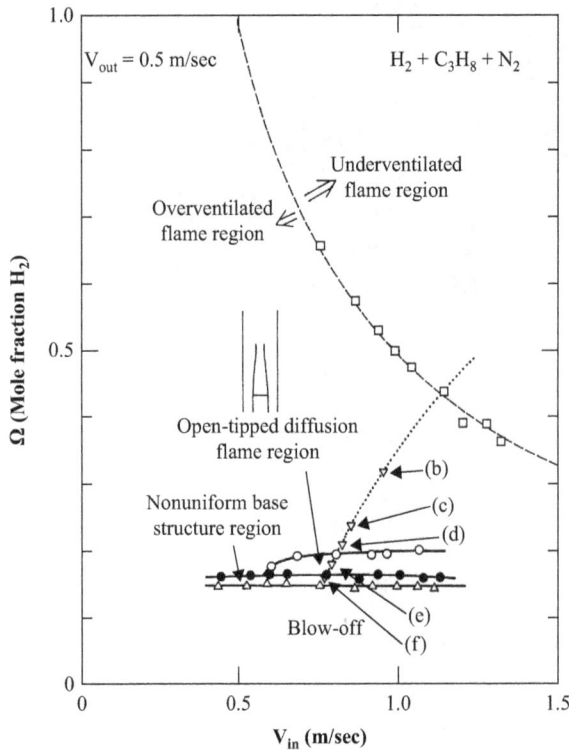

Figure 1.26. Mapping of the flame configurations for hydrogen–propane–nitrogen mixtures. (*Source*: Ishizuka, S., An Experimental Study on the Opening of Laminar Diffusion Flame Tips, Proceedings of the Nineteenth Symposium (International) on Combustion, The Combustion Institute, Pittsburgh, PA. 319–326, 1982.[61] Used by permission of Elsevier.)

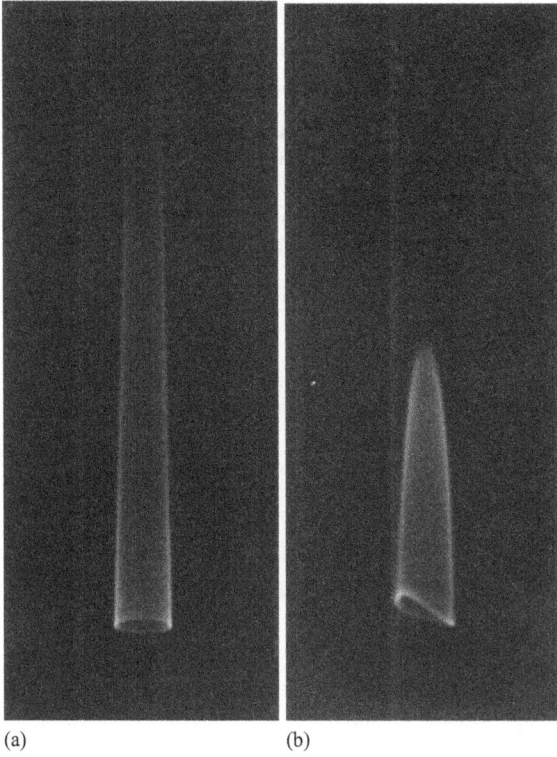

Figure 1.27. Appearance of flames; (a) $H_2/C_3H_8/N_2$ ($\Omega = 0.168$) $\sim O_2/N_2$ ($\beta = 0.21$), (b) $H_2/C_3H_8/He$ ($\Omega = 0.188$) $\sim O_2/He$ ($\beta = 0.21$). (*Source*: Ishizuka, S.,Tip-Opeing of Laminar Diffusion Flame, Proceedings of the Twenty-First Symposium (Japanese) on Combustion, pp.287–289, 1983.[63] Used by permission of the Combustion Society of Japan.)

(a)   (b)

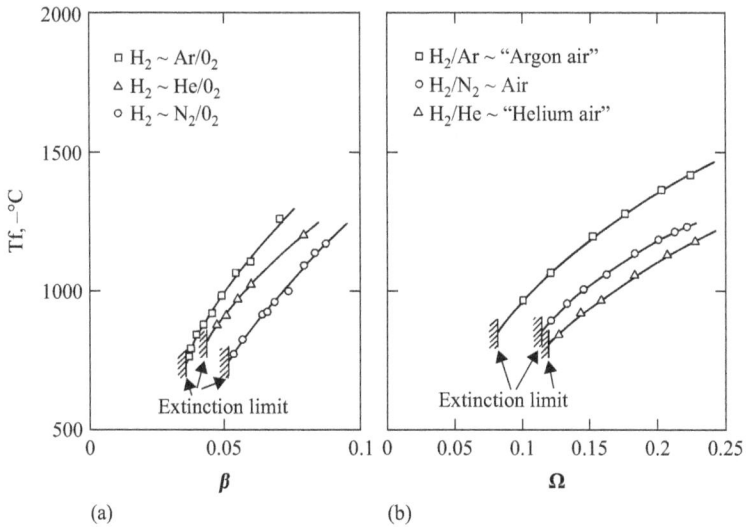

(a)   (b)

Figure 1.28. Measured flame temperatures as a function of (a) the oxygen concentration in the oxidizer β and (b) the fuel concentration in the ejected mixture Ω. (*Source*: Ishizuka, S. and Tsuji, H., Effects of Transport Properties and Flow Nonuniformity on the Temperature of Counterflow Diffusion Flames, Combustion Science and Technology 37(1984)171–191.[65] Used by permission of Taylor & Francis.)

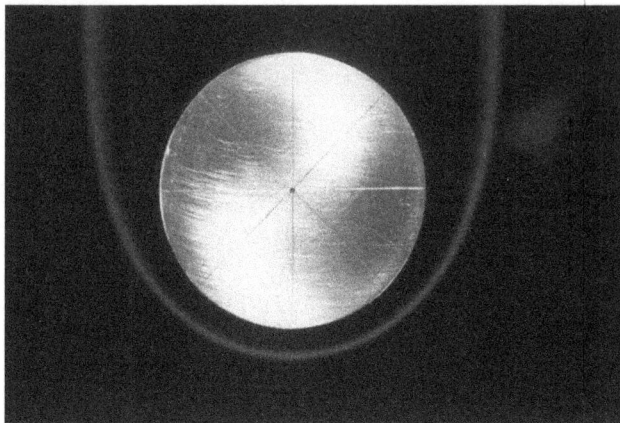

Figure 1.29. Hydrogen–air counterflow diffusion flame established in the forward stagnation region of a porous cylinder. (V_{air} = 1.59m/s, V_{wall} = 32.44 cm/s, 2V/R = 105 s^{-1}, $-f_w$ = 7.8).[66]

nitrogen, the temperature of argon-diluted flame is highest, that of helium-diluted flame is lower than, but relatively close to, that of argon-diluted flame, and the temperature of nitrogen-diluted flame is lowest (Fig.1.28(a)). This may be reasonable because the specific heats of monatomic molecule, argon and helium, are the same and lower than the specific heat of diatomic molecule, nitrogen. However, in a case when the hydrogen is diluted with these inert gases while "various airs (21% O_2 + 79% inert gas by volume)" are used as oxidizer, the temperature of helium-diluted flame is much decreased and becomes lowest, as shown in Fig.1.28(b). The observed variations of the counterflow diffusions flame temperature are considered to be strongly dependent on the transport properties of the mixtures and, in addition, the flow nonuniformity where the flame is located, besides the thermodynamic properties of the mixtures.

To understand the variations of the flame temperature, it is important to know where the flame is located in the nonuniform, divergent flow field. Figure 1.29 is the picture of a hydrogen–air counterflow diffusion flame established in the forward stagnation region of a porous cylinder.[66] When a pure fuel of the fuel mass fraction $\alpha_F = 1$ is injected into the air stream of the oxygen mass fraction $Y_{O\infty}$=0.233, the flame is established on the air side of the stagnation streamline. This is because the flame is established at a position where the mass fluxes of fuel and oxygen are stoichiometric, that is,

$$j \cdot \rho D_F \, grad Y_F = -\rho D_O \, grad Y_O, \qquad (1.12)$$

in which j is the number of grams of oxygen to burn 1 g of fuel, ρ is density, D_F and D_O are the mass diffusivities of fuel and oxygen, respectively, and Y_F and Y_O are the mass fractions of fuel and oxygen, respectively. Here, mass transfer by convection is ignored because the values of Y_F and Y_O are very small in the reaction zone. The value of j is 8 for hydrogen, which is greater than those of hydrocarbon fuels (j = 4 for methane, 3.636 for propane), and hence, the pure hydrogen flame is established far from the stagnation streamline in the air stream.

When the fuel is diluted and the mass fraction of fuel α_F is gradually decreased, however, the flame shifts toward the fuel side to maintain the relation of Eq.1.12. Figure 1.30 shows some pictures of counterflow diffusion flames, when a hydrogen–nitrogen mixture is injected into the air stream.[66] These photographs were taken from a direction perpendicular to the axis of the porous cylinder so that the cylinder surface was photographed as a flat surface.

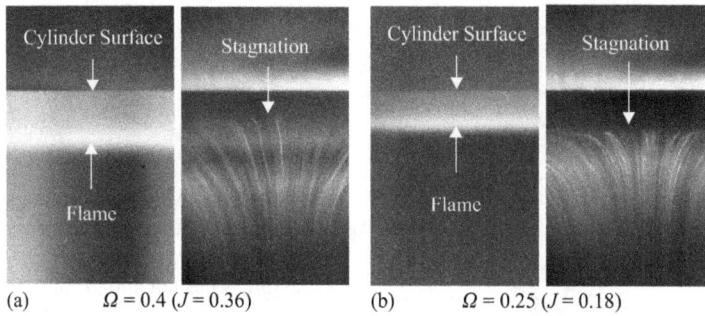

(a) $\Omega = 0.4$ $(J = 0.36)$ (b) $\Omega = 0.25$ $(J = 0.18)$

Figure 1.30. Direct photographs of counterflow diffusion flames of hydrogen/nitrogen mixtures in the air stream. These photographs were taken from a direction perpendicular to the axis of a porous cylinder so that the cylinder surface was taken to be flat; (a) $\Omega = 0.4$, (b) $\Omega = 0.25$, $-F_w = 2.0$, $2V/R = 50$ s^{-1}, left: non-seeded, right seeded with MgO particles in the air stream.[66]

Figure 1.30(a) shows the pictures when the hydrogen molar fraction Ω in the hydrogen–nitrogen mixture is 0.4, the nondimensional injection rate $-F_w = 2.0$, and the velocity gradient of the air stream $2V/R = 50$ s^{-1}, where V is the uniform air stream velocity and R ($= 3$ cm) is the radius of the porous cylinder. The diluted hydrogen flame is very weak in luminosity and hence, the left picture was taken with a long exposure time (4 sec). To determine the stagnation streamline, MgO particles of 5–10 μm were seeded into the air stream and illuminated with a high-pressure mercury vapor lamp. The right picture was taken for the seeded flow. The flame chemiluminescence was pictured only a little due to a short exposure time (1/8 sec). It is seen that the seeded particles pass through the flame zone and reach some position away from the cylinder surface. Then, the particles go out to disappear since they are out of a focused plane. This position can be considered as a stagnation (dividing) streamline. Thus, it is confirmed that the flame is located on the air side of the stagnation streamline.

When the hydrogen is further diluted, however, the flame is formed on the fuel side of the stagnation streamline. Figure 1.30(b) shows the pictures of the flame without and with seeded MgO particles, when the fuel molar fraction Ω is reduced to 0.25. Since the flame was very weak in luminosity, the left picture was taken with a very long exposure time (80 sec). On the other hand, the flame chemiluminescence was not photographed in the right picture because of a short exposure time (1/8 sec). However, by comparing these pictures, it has been found that the particles do not reach the flame; the flame is formed on the fuel side of the stagnation.

Figure 1.31 shows the variations of the locations of the flame and the stagnation point,[65] y_* and y_s, as a function of J, which is a product of the value of j for the fuel and the fuel mass fraction α_F in the injected hydrogen–nitrogen mixture, that is,

$$J = \alpha_F \cdot j, \tag{1.13}$$

At large values of J, the flame lies on the air side of the stagnation point for the three injection cases of $-F_w = 0.5$, 2.0, and 5.0. As the fuel concentration is decreased, the flame position y_* steeply decreases while the stagnation point y_s gradually decreases, and eventually, they

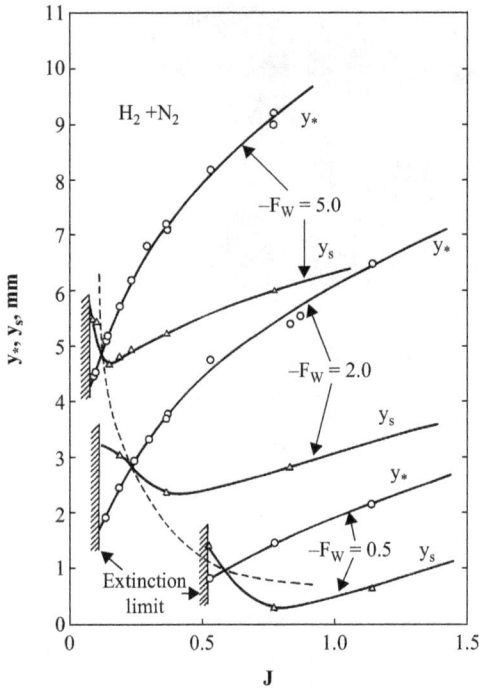

Figure 1.31. Measured locations of the center of the luminous flame zone y_* and the stagnation point y_s when a hydrogen/nitrogen mixture is ejected into the air stream (cylinder diameter: 6 cm; velocity gradient: 50 s^{-1}). (*Source*: Ishizuka, S. and Tsuji, H., Effects of Transport Properties and Flow Nonuniformity on the Temperature of Counterflow Diffusion Flames, Combustion Science and Technology 37(1984)171–191.[65] Used by permission of Taylor & Francis.)

are crossed and, finally, the flame is situated on the fuel side of the stagnation point in the fuel stream.

With increasing $-F_w$, the value of J at which the flame and the stagnation coincide is decreased. Its asymptotic value for large rates of injection is theoretically obtained as

$$J \to Y_{O\infty}\sqrt{\frac{S_F}{S_O}} \tag{1.14}$$

in which $Y_{O\infty}$ is the mass fraction of oxygen in the oxidizer stream, and S_F and S_O are the Schmidt numbers of fuel and oxygen, respectively.[65]

At large rates of blowing, the relative poison of the flame to the stagnation point can be evaluated by a parameter ϕ, which is given as

$$\phi \equiv \frac{J}{Y_{O\infty}\sqrt{S_F/S_O}} \tag{1.15}$$

Namely, $\phi > 1$ implies that the flame is formed on the oxidizer side of the stagnation point, whereas $\phi < 1$ implies that the flame is formed on the fuel side of the stagnation point.

It is interesting to note that when the flame is formed in the fuel stream, a nonuniform structure appears. Figure 1.32 shows such nonuniform, striped-pattern flames when hydrogen–nitrogen mixture is injected from the porous cylinder into an air stream.[66] When the injection velocity is small, striped-pattern flames of few millimeters are formed close to the cylinder

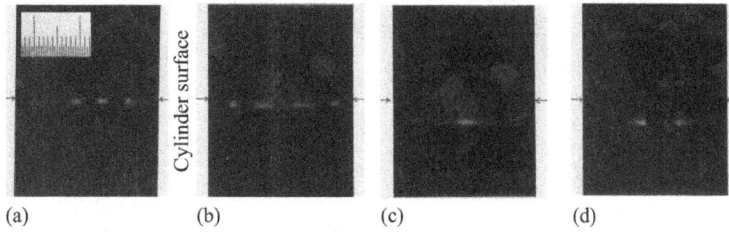

Figure 1.32. Striped-pattern flames ((a) 2V/R = 13 s^{-1}, v_w = 1.2 cm/s, Ω = 0.64, (b) 2V/R = 35 s^{-1}, v_w = 1.9 cm/s, Ω = 0.53, (c) 2V/R = 72 s^{-1}, v_w = 27 cm/s, Ω = 0.11, (d) 2V/R = 72 s^{-1}, v_w = 27 cm/s, Ω = 0.11).[66]

surface (Fig.1.32(a) and (b)). When the injection rate is large and the hydrogen concentration is decreased,[67] the flame is first deformed and the flame luminosity is intensified around the convex area toward the fuel side (Fig.1.32(c)), then, under the same condition of the velocity gradient 2V/R, the fuel injection velocity v_w and the fuel concentration Ω (molar fraction), the flame is extinguished at the middle (Fig.1.32(d)), in which the flame edges are intensified in burning due to the preferential diffusion of hydrogen.

It is also important to note that $\phi \gg 1$ ($\phi \ll 1$) implies that the flame is situated far from the stagnation point on the oxidant (fuel) side, and hence, the flame is situated in an almost one-dimensional uniform flow, whereas $\phi \cong 1$ implies that the flame is formed near the stagnation streamline, and hence, the flame is situated in a nonuniform flow.

Based on this knowledge, the flame temperature variations in Fig.1.28 can be understood as follows. In the case of Fig.1.28 (a), the oxygen is diluted with an inert gas for a pure fuel, and hence, the flames are formed far from the stagnation point in an almost uniform one-dimensional flow. Therefore, the flame temperature is mainly controlled by the thermodynamic properties of the mixture, resulting in a higher flame temperature for dilution with a monatomic molecule of argon or helium than that with a diatomic molecule, nitrogen. On the other hand, in the case of Fig.1.28 (b), the fuel is diluted with an inert gas for "various airs," and hence, the flames are situated in a nonuniform flow around the stagnation point. In such a nonuniform flow field, mass and heat transfer through a stream tube do occur, as illustrated in Fig.1.20. The flame temperatures are strongly dependent on the transport properties of the mixtures, especially the Lewis number of the fuel.

Figures 1.33 and 1.34 respectively show the calculated flame structure for the Lewis numbers of unity and 0.5, respectively, when the flame are situated close to the stagnation point.[65] In the case of unity Lewis number, the ratio of the flame temperature to the adiabatic flame temperature of the stoichiometric premixed flame ϕ is close to unity number (0.997) and the enthalpy H is uniformly distributed. On the other hand, in the case of Le = 0.5, the flame temperature becomes higher than the adiabatic flame temperature (ϕ = 1.38) and the enthalpy becomes higher around the flame sheet, at both sides of which the enthalpy becomes negative; the enthalpy profile becomes not uniform. Thus, it is theoretically found that the counterflow diffusion flame temperature is significantly influenced by the Lewis number, when the flame is situated in a nonuniform flow field. Therefore, it is reasonable to consider that in the case of Fig.1.28 (b), in which the flames seem to be situated in a nonuniform flow, the argon- and nitrogen-diluted flame temperatures become higher that the helium-diluted

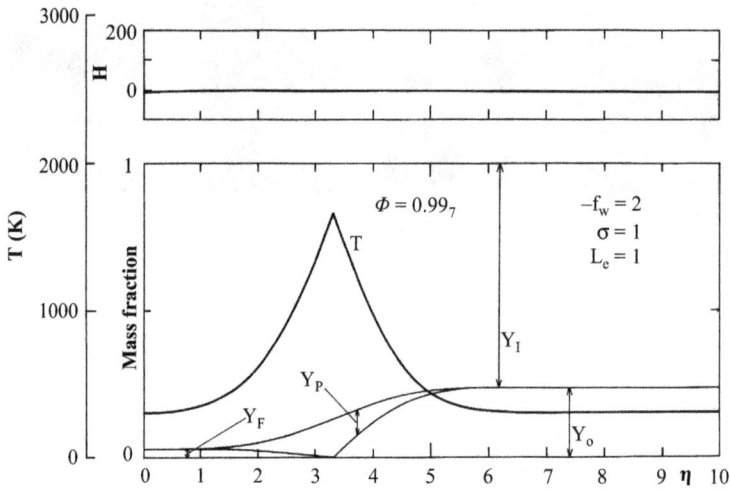

Figure 1.33. Structure of the diffusion flame in the case of Le = 1. (*Source*: Ishizuka, S. and Tsuji, H., Effects of Transport Properties and Flow Nonuniformity on the Temperature of Counterflow Diffusion Flames, Combustion Science and Technology 37(1984)171–191.[65] Used by permission of Taylor & Francis.)

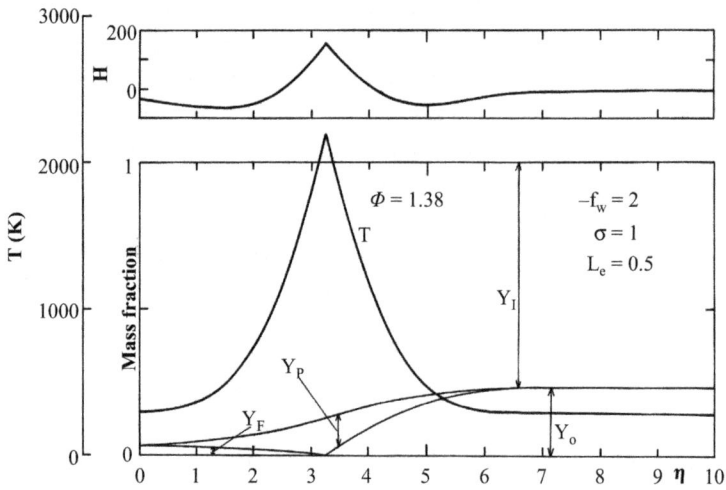

Figure 1.34. Structure of the diffusion flame in the case of Le = 0.5. (*Source*: Ishizuka, S. and Tsuji, H., Effects of Transport Properties and Flow Nonuniformity on the Temperature of Counterflow Diffusion Flames, Combustion Science and Technology 37(1984)171–191.[65] Used by permission of Taylor & Francis.)

flame temperature, because the Lewis numbers for argon and nitrogen are smaller than that for helium.

These peculiar properties of diffusion flames may appear in premixed tubular flames since the flame is situated in a stretched, nonuniform flow field. With regard to this, non-premixed tubular flame studies are indispensable for complete understanding of premixed tubular flame studies. Recently, by injecting a fuel from a porous rod of small diameter placed in the middle and by injecting air inward from the periphery, a tubular diffusion flame study has been started[68–71](see Fig.5.5). Using laser diagnostics and numerical simulations, the structure and characteristics of non-premixed tubular flames have been studied in detail together with those of premixed tubular flames.[53–56] The details of nonpremixed tubular flame characteristics are described by Robert W. Pitz in Chap.5.

1.4.2 MINIATURE LIQUID-FILM COMBUSTORS

Recently, an ingenious technique of liquid-film combustion has been proposed for a portable high power system, in which a liquid fuel is injected tangentially into a miniature-scale tube while a swirling air flow is introduced to make a very thin liquid film on the surface of the tube wall.[72–75] It has been revealed that, liquid-film combustion gains advantage over spray combustion when the burner diameter becomes smaller. For example, an elaborated analysis has shown that at atmospheric pressure, a 10-mm diameter combustor has a film surface area greater than that would occur for a droplet spray with a 10-μm Sauter mean radius.

Although combustion in the liquid-film combustor is heterogeneous and hence, much more complex than the gaseous tubular flame combustion, there are many similarities between the combustion in miniature liquid-film combustors and the premixed tubular flame combustion. First, the combustion is prevailed in a cylindrical, long space. Second, fuel is tangentially introduced into a tube and also air is tangentially introduced or a swirling air flow is used. There are, however, different aspects between the two. In the liquid-film combustors, the flame is anchored at a triple point located around the base of the combustor,[73] and the stable combustion is most under fuel-rich conditions which needs a secondary air injection to complete combustion within the tube.[75] Heat is conducted through a tube wall from the downstream hot gas region into the upstream liquid-film region for vaporizing the liquid fuel. Aluminum, bronze, soft steel, stainless steel, quartz, sapphire glass, and many kinds of wall material have been explored to vaporize the liquid fuel successfully. These second air injection and heat conduction through the tube indicate that the combustion is not similar in the axial direction, whereas similar solutions are assumed in most of the theoretical and computational tubular flame studies.

These dissimilar aspects, however, yield useful insights on practical tubular flame burners, especially rapidly mixed-type tubular flame combustion described later, because the fuel–air mixture composition is varied along with the axial distance in the burner. The heterogeneous processes in liquid-film combustors yield also useful insights to understand liquid-fueled tubular flame combustion in practical use, in which a liquid fuel is tangentially sprayed and, sometimes, the wall gets wet by the condensed liquid that forms a liquid film.

The details of observed characteristics similar to tubular flames in miniature liquid-film combustors are given by Derek Dun-Rankin in Chap.6.

1.5 PRACTICAL APPLICATION

Due to its thermal and aerodynamic advantages, tubular flame is robust in stability, and hence, a variety of tubular flame burners have been developed for practical use. In this section, prototype tubular flame burners and an inherently safe technique of rapidly mixed tubular flame combustion are briefly introduced.

1.5.1 PROTOTYPE TUBULAR FLAME BURNERS

Due to the thermal advantage and the aerodynamic advantage, tubular flames have technical merits for practical use. Up to now, a variety of tubular flame burners have been designed. The original one is a premixed tubular flame burner shown in Fig.1.35.[76] The tube diameter is 102.3 mm (4 inches) and the lengths of the injection section and combustion tube are 160 and 1000 mm, respectively. A premixed gas is injected tangentially through two rectangular slits of 3.7 mm wide and 95.4 mm long, which were placed at one closed end. The burned gas exits from the other open end.

Once ignited, a tubular flame is formed in the tube. Figure 1.36 shows the appearance of flames of propane–air mixtures of different equivalence ratios at the air flow rate $Q_{air} = 80$ m^3_N/h. The left pictures were taken through a quartz window equipped at the closed end plate, while the rights pictures were taken through a Pyrex tube of 1000 mm, which was used as a combustion tube. Note that the magnitudes of enlargement are different between the right and left pictures. It is seen that a tubular flame can be successfully stabilized in the burner. When the

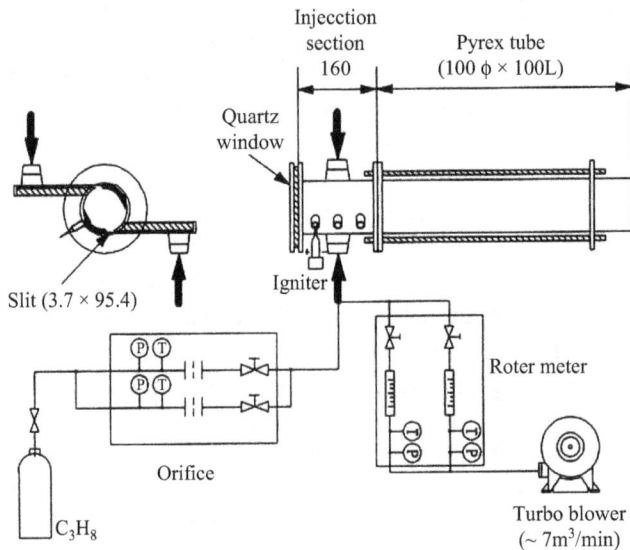

Figure 1.35. Proto-type of a tubular flame burner. (*Source*: Ishizuka, S., Hagiwara, R., Suzuki, M., Nakamura, A., and Hamaguchi, O., Combustion Characteristics of a Tubular Flame Burner, Transaction of JSME 65(1999)3845–3852.[76] Used by permission of JSME.)

mixture is stoichiometric (Fig.1.36 (a)), the flame is short; the blue flame does not extend to the Pyrex tube. The luminosity, which can be seen in the Pyrex tube, is due to thermal radiation of the hot burned gas. When the equivalence ratio is decreased to 0.7, the flame diameter decreases due to a decrease of the burning velocity and the flame extends to the Pyrex tube (Fig.8.36 (b)). The flame, which can be seen in the Pyrex tube, is a blue flame, and thermal radiation is not seen in the Pyrex tube. When the equivalence ratio is further reduced to 0.6 close to extinction (Fig.8.36 (c)), the flame is weakened in luminosity around the injection region, which is due to the stretch effect because the Lewis number of the lean propane–air mixtures is greater than unity, and eventually the flame blows off when the equivalence ratio is further reduced. For rich mixtures (Fig.8.36 (d)), a long tubular flame of blue-green luminosity is formed and extends to the Pyrex tube, followed with a thermally radiative zone downstream. The excess fuel is burned at the exit with the ambient air.

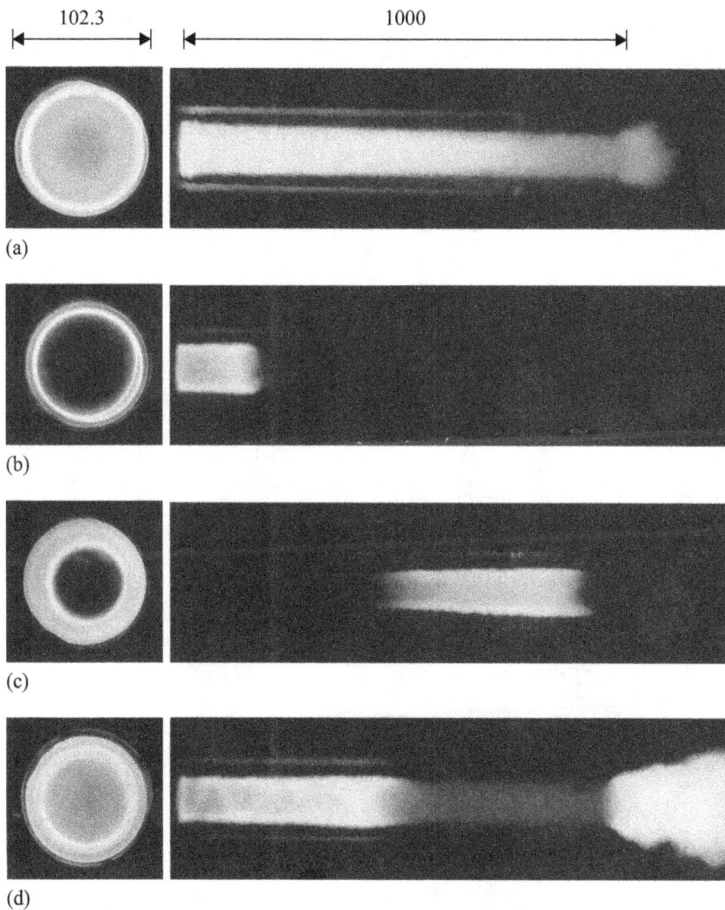

Figure 1.36. Appearance of flames (Q_{air} = 80 m^3/h(stp), equivalence ratio Φ (a) 1.0, (b) 0.7, (c) 0.6, and (d) 1.6). (*Source*: Ishizuka, S., Hagiwara, R., Suzuki, M., Nakamura, A., and Hamaguchi, O., Combustion Characteristics of a Tubular Flame Burner, Transaction of JSME 65(1999)3845–3852.[76] Used by permission of JSME.)

The technical merits of this burner are that (1) a laminar flame of a large area can be obtained easily by just injecting a combustible mixture tangentially into a tube and (2) a homogeneous burned gas can be obtained. Thus, this burner can be used, for example, to reduce the surface of a steel plate in steel works.

1.5.2 RAPIDLY MIXED TUBULAR FLAME COMBUSTION

However, premixed combustion is dangerous due to the occurrence of flame flashback. This becomes more serious as the burner size is increased. To overcome this problem, an inherently safe technique of rapidly mixed-type tubular flame combustion has been proposed. Figure 1.37 schematically shows this method together with the premixed-type combustion. Fuel and air are separately injected into a tube.[77]

To elucidate the rapidly mixed-type combustion, prototype burners of three kinds of inner diameters, 52, 75, and 102 mm, were made (Fig.1.38).[77] The burners have four tangential slits, through the two of which fuel is injected and through the other two of which air is injected. The burners also have four quartz windows at the periphery and one quartz window at the closed end. Figure 1.39 shows the flame appearance of premixed and rapidly mixed combustion.[77] It is seen that stable and uniform tubular flame can be obtained in the rapidly mixed combustion. It is experimentally found that if the swirl numbers are more than 5 and the injection velocities are higher than 20 m/s, very similar tubular flame to the premixed flame can be obtained in the rapidly mixed combustion.

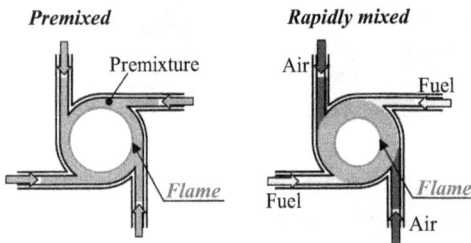

Figure 1.37. Concept of rapidly mixed combustion in a swirl-type tubular flame burner. (*Source*: Ishizuka, S., Motodamari, T., and Shimokuri, D., Rapidly-Mixed Combustion in a Tubular Flame Burner, Proceedings of the Combustion Institute 31(2007)1085–1092.[77] Used by permission of Elsevier.)

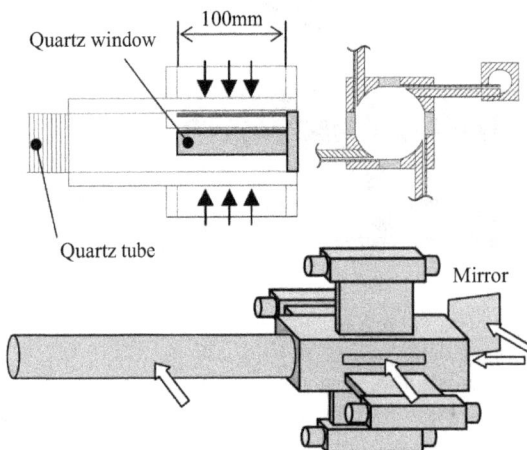

Figure 1.38. Swirl-type tubular flame burner with quartz windows. (*Source*: Ishizuka, S., Motodamari, T., and Shimokuri, D., Rapidly-Mixed Combustion in a Tubular Flame Burner, Proceedings of the Combustion Institute 31(2007)1085–1092.[77] Used by permission of Elsevier.)

Table 1.2 compares the conventional swirl combustor with the tubular flame burner. A major difference is to use volumetric, unit–volume combustion or superficial, unit–area combustion. In the former case, the combustion occurs under turbulent conditions, whereas the combustion is prevailed with a laminar flame in the latter. Thus, the former involves stochastic processes, which needs statistical models, whereas the latter only includes a deterministic process, hence, does not need any statistical modeling. As for mixing, the swirl combustor uses turbulence, for which a variety of flows such as strong swirl, opposed jets, cross flows and so on are used, whereas the rapidly mixed-type tubular flame combustion uses parallel mixing with molecular transport to suppress turbulence. In general, the swirl combustor aims at high-intensity, high-load combustion, whereas the tubular flame combustion aims at well-controlled combustion to reduce emissions of NOx and suspended particulate matter.

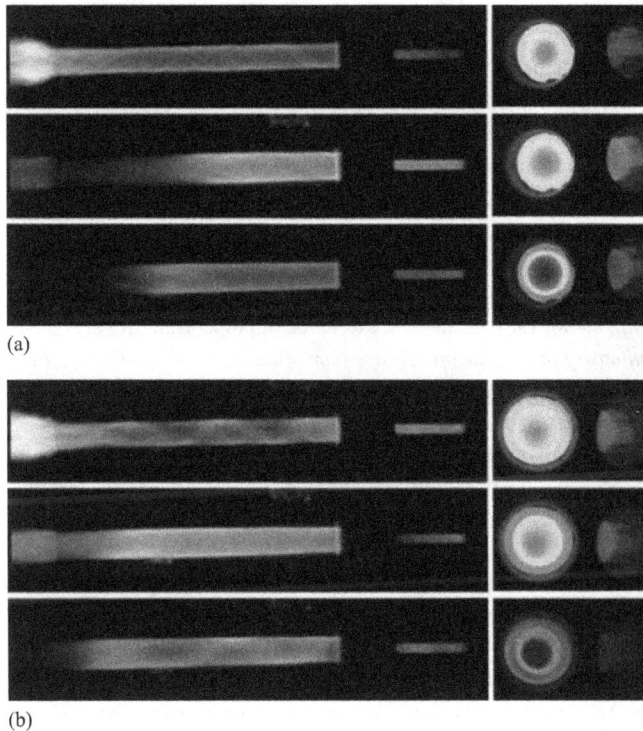

(a)

(b)

Figure 1.39. Appearance of flame in (a) premixed combustion; (b) rapidly mixed combustion; burner diameter, 52 mm; upper, $\Phi = 1.2$; middle, $\Phi = 1.0$; lower, $\Phi = 0.7$, $Q_{air} = 60$ m³/h; fuel methane. (*Source*: Ishizuka, S., Motodamari, T., and Shimokuri, D., Rapidly-Mixed Combustion in a Tubular Flame Burner, Proceedings of the Combustion Institute 31(2007)1085–1092.[77] Used by permission of Elsevier.)

Table 1.2. Comparison between the conventional swirl combustor and the tubular flame burner

Swirl Combustor	Tubular Flame Burner
Per unit reacting volume	Per unit flame area
Turbulent combustion	Laminar combustion
Stochastic process	Deterministic process
Turbulence swirl, opposed jets, cross flows	Molecular transport, parallel mixing
High-intensity combustion	Well-controlled combustion

Due to the technical merits, a variety of tubular flame burners have been designed. The smallest one is a microtubular flame burner, which is 2 mm in inner diameter. The largest one is a 2-MW gaseous and liquid-fueled burner with a 12-inch combustion tube. Small-scale applications are introduced by Daisuke Shimokuri in Chap.7, while large-scale applications will be presented by Satoru Ishizuka in Chap.8.

REFERENCES

1. Karlovitz, B., D. W. Denniston, Jr., D. H. Knapschaefer, and F. E. Wells. 1953. Studies on turbulent flames: A. Flame propagation across velocity gradients; B. Turbulence measurement in flames. In *Proceedings of the Fourth Symposium (International) on Combustion*, 613–20. Baltimore, MD: Williams and Wilkins. DOI: 10.1016/S0082-0784(53)80082-2.

2. Williams, F. A. 1985. *Combustion theory*. 2nd ed., 416. Redwood City, CA: Addison-Wesley.

3. Strehlow, R. A. and L. D. Savage. 1978. The concept of flame stretch. *Combustion and Flame* 31:209–11. DOI: 10.1016/0010-2180(78)90130-X.

4. Matalon, M. 1983. On flame stretch. *Combustion Science and Technology* 31:169–81. DOI: 10.1080/00102208308923638.

5. Law, C. K. 1989. Dynamics of stretched flames. In *Proceedings of the Twenty-Second Symposium (International) on Combustion*, 1381–1402. Pittsburgh, PA: The Combustion Institute. DOI: 10.1016/S0082-0784(89)80149-3.

6. Tennekes, H. 1983. Simple model for the small-scale structure of turbulence. *Physics of Fluids* 11:669–70. DOI: 10.1063/1.1691966.

7. Law, C. K., S. Ishizuka, and M. Mizomoto. 1981. Lean-limit extinction of propane/air mixtures in the stagnation-point flow. In *Proceedings of the Eighteenth Symposium (International) on Combustion*, 1791–8. Pittsburgh, PA: The Combustion Institute. DOI: 10.1016/S0082-0784(81)80184-1.

8. Ishizuka, S. 1993. Characteristics of tubular flames. *Progress in Energy and Combustion Science* 19:187–226. DOI: 10.1016/0360-1285(93)90015-7.

9. Ishizuka, S. and C. K. Law. 1982. An experimental study on extinction and stability of stretched premixed flames. In *Proceedings of the Nineteenth Symposium (International) on Combustion*, 327–35. Pittsburgh, PA: The Combustion Institute. DOI: 10.1016/S0082-0784(82)80204-X.

10. Tsuji, H. and I. Yamaoka. 1982. Structure and extinction of near-limit flames in a stagnation flow. In *Proceedings of the Nineteenth Symposium (International) on Combustion*, 1533–40. Pittsburgh, PA: The Combustion Institute. DOI: 10.1016/S0082-0784(82)80330-5.

11. Sato, J. 1982. Effects of Lewis number on extinction behavior of premixed flames in a stagnation flow. In *Proceedings of the Nineteenth Symposium (International) on Combustion*, 1541–8. Pittsburgh, PA: The Combustion Institute. DOI: 10.1016/S0082-0784(82)80331-7.

12. Law, C. K. and S. H. Chung. 1984. An invariant derivation of flame stretch. *Combustion and Flame* 55:123–5. DOI: 10.1016/0010-2180(84)90156-1.

13. Candel, S. M. and T. J. Poinsot. 1990. Flame stretch and the balance equation for the flame area. *Combustion Science and Technology* 70:1–15. DOI: 10.1080/00102209008951608.

14. Chomiak, J. 1977. Dissipation fluctuations and the structure and propagation of turbulent flames in premixed gases at high Reynolds numbers. In *Proceedings of the Sixteenth Symposium (International) on Combustion*, 1665–73. Pittsburgh, PA: The Combustion Institute. DOI: 10.1016/S0082-0784(77)80445-1

15. Tabaczynski, R. J., F. H. Trinker, and B. A. S. Shannon. 1980. Further refinement and validation of a turbulent flame propagation model for spark-ignition engines. *Combustion and Flame* 39:111–21. DOI: 10.1016/0010-2180(80)90011-5.

16. Klimov, A. M. and V. N. Lebedev. 1983. Limiting phenomena in turbulent combustion. *Fizika Goreniya i Vzryva* 19:7–9. DOI: 10.5082/83/1905-0540S07.50.

17. Batchlor, G. K. 1967. Analysis of the relative motion near a point (section 2.3). In *An introduction to fluid mechanics*, 79–84. Cambridge: Cambridge University Press.

18. Ishizuka, S. 1985. On the behavior of premixed flames in a rotating flow field: Establishment of tubular flames. In *Proceedings of the Twentieth Symposium (International) on Combustion*, 287–94. Pittsburgh, PA: The Combustion Institute. DOI: 10.1016/S0082-0784(85)80513-0.

19. Kobayashi, H. and M. Kitano. 1989. Extinction characteristics of a stretched cylindrical premixed flame. *Combustion and Flame* 76:285–95. DOI: 10.1016/0010-2180(89)90111-9.

20. Rayleigh, L. 1916. On the dynamics of revolving fluids. *Proceedings of the Royal Society of London* A93:148–54. DOI: 10.1098/rspa.1917.0010.

21. Ishizuka, S. 1989. An experimental study on extinction and stability of tubular flames. *Combustion and Flame* 75:367–79. DOI: 10.1016/0010-2180(89)90049-7.

22. Coward, H. F. and G. W. Jones. 1952. *Limits of Flammability of Gases and Vapors*, U.S. Bur. Mines, Bull. 503.

23. Zabetakis, M. G. 1965. *Flammability Characteristics of Combustible Gases and Vapors*, U.S. Bur. Mines, Bull. 627.

24. Ishizuka, S. 1991. Determination of flammability limits using a tubular flame geometry. *Journal of Loss Prevention in the Process Industry* 4:185–93. DOI: 10.1016/0950-4230(91)80035-S.

25. Saito, N., Y. Ogawa, S. Saso, C. Liao, and R. Sakei. 1996. Flame-extinguishing concentrations and peak concentrations of N_2, Ar, CO_2 and their mixtures for hydrocarbon fuels. *Fire Safety Journal* 27 (3):185–200. DOI: 10.1016/S0379-7112(96)00060-4.

26. Ogawa, Y., N. Saito, and C. Liao. 1998. Burner diameter and flammability limit measured by tubular flame burner. *Proceedings of the Combustion Institute* 27:3221–7. DOI: 10.1016/S0082-0784(98)80186-0.

27. Liao, C., N. Saito, S. Saso, and Y. Ogawa. 1996. Flammability limits of combustible gases and vapors measured by a tubular flame method. *Fire Safety Journal* 27 (1):49–68. DOI: 10.1016/S0379-7112(96)00021-5.

28. Kobayashi, H., M. Kitano, and Y. Otsuka. 1988. An analysis of a stretched cylindrical premixed flame. *Combustion Science and Technology* 57:17–36. DOI: 10.1080/00102208808923941.

29. Ishizuka, S., K. Mitoma, and D. Shimokuri. 2003. Extension of blow-off limits of tubular flame by strong rotation. *Proceedings of the Forty-First Symposium (Japanese) on Combustion*, 17–18 (in Japanese).

30. Beer, J. M. and C. A. Chigier. 1972. *Combustion Aerodynamics*, London: Applied Science Publication.

31. Sullivan, R. D. 1959. A two-cell vortex solution of the Navier-Stokes equations. *Journal of the Aerospace Sciences* 26:767–8.

32. Libby, P. A. and F. A. Williams. 1983. Strained premixed laminar flames under nonadiabatic conditions. *Combustion Science and Technology* 31:1–42. DOI: 10.1080/00102208308923629.

33. Mikolatis, D. W. 1984. The cylindrical stretched flame. *Combustion and Flame* 56:327–35. DOI: 10.1016/0010-2180(84)90066-X.

34. Tseng, L.-K., K. Abhishek, and J. P. Gore. 1995. An experimental realization of premixed methne/air cylindrical flames. *Combustion and Flame* 102:519–22. DOI: 10.1016/0010-2180(95)00122-M.

35. Yuan, S. W. and A. B. Finkelstein. 1956. Laminar pipe flow with injection and suction through a porous wall. *Transaction of ASME* 78:719–24.

36. Takeno, T. and S. Ishizuka. 1986. A tubular flame theory. *Combustion and Flame* 64:83–98. DOI: 10.1016/0010-2180(86)90100-8.

37. Takeno, T., M. Nishioka, and S. Ishizuka. 1986. A theoretical study on extinction of a tubular flame. *Combustion and Flame* 66:271–83. DOI: 10.1016/0010-2180(86)90140-9.

38. Takeno, T., S. Ishizuka, M. Nishioka, and J. D. Buckmaster. 1987. Extinction behavior of a tubular flame for small Lewis numbers Springer series in chemical physics. In *Complex chemical reaction systems*, vol. 47, 302–09. Berlin: Springer-Verlag.

39. Nishioka, M., T. Takeno, and S. Ishizuka. 1988. Effects of variable density on a tubular flame. *Combustion and Flame* 73:287–301 DOI: 10.1016/0010-2180(88)90024-7.

40. Libby, P. A., N. Peters, and F. A. Williams. 1989. Cylindrical premixed laminar flames. *Combustion and Flame* 75:265–80. DOI: 10.1016/0010-2180(89)90043-6.

41. Ju, Y., H. Matsumi, K. Takita, and G. Masuya. 1999. Combined effects of radiation, flame curvature, and stretch on the extinction and bifurcations of cylindrical CH_4/air premixed flame. *Combustion and Flame* 116:580–92. DOI: 10.1016/S0010-2180(98)00051-0.

42. Matthews, M. T., B. D. Dlugogorski, and E. M. Kennedy. 2006. The asymptotic structure of premixed tubular flames. *Combustion and Flame* 144:838–49. DOI: 10.1016/j.combustflame.2005.09.011.

43. Wang, P., J. A. Wehrmeyer, and R. W. Pitz. 2006. Stretch rate of tubular premixed flames. *Combustion and Flame* 145:401–14. DOI: 10.1016/j.combustflame.2005.09.015.

44. Kurdyumov, V. N. 2011. Diffusive-thermal instability of premixed tubular flames. *Combustion and Flame* 158:1718–26. DOI: 10.1016/j.combustflame.2011.01.012.

45. Smooke, M. D. and V. Giovangigli. 1991. Extinction of tubular premixed laminar flames with complex chemistry. In *Proceedings of the Twenty-Third Symposium (International) on Combustion*, 447–54. Pittsburgh, PA: The Combustion Institute. DOI: 10.1016/S0082-0784(06)80290-0.

46. Nishioka, M., K. Inagaki, S. Ishizuka, and T. Takeno. 1991. Effects of pressure on structure and extinction of tubular flame. *Combustion and Flame* 86:90–100. DOI: 10.1016/0010-2180(91)90058-J.

47. Yamamoto, K., T. Hirano, and S. Ishizuka. 1996. Effects of pressure diffusion on the characteristics of tubular flames. In *Proceedings of the Twenty-Sixth Symposium (International) on Combustion*, 1129–35. Pittsburgh, PA: The Combustion Institute. DOI: 10.1016/S0082-0784(96)80328-6.

48. Zhang, Y., S. Ishizuka, H. Zhu, and R. J. Kee. 2007. The effects of rotation on the characteristics of premixed propane/air swirling tubular flames. *Proceedings of the Combustion Institute* 31:1101–07. DOI: 10.1016/j.proci.2006.07.030.

49. Zhang, Y., S. Ishizuka, H. Zhu, and R. J. Kee. 2009. Effects of stretch and pressure on the characteristics of premixed swirling tubular methane-air flames. *Proceedings of the Combustion Institute* 32:1149–56. DOI: 10.1016/j.proci.2008.06.066.

50. Zhang, Y., J. Wu, and S. Ishizuka. 2009. Hydrogen addition effect on laminar burning velocity, flame temperature and flame stability of a planar and curved CH_4-H_2-air premixed flame. *International Journal of Hydrogen Energy* 34:519–27. DOI: 10.1016/j.ijhydene.2008.10.065.

51. Kee, R. J., A. M. Colclasure, H. Zhu, and Y. Zhang. 2008. Modeling tangential injection into ideal tubular flames. *Combustion and Flame* 152:114–24. DOI: 10.1016/j.combustflame.2007.07.019.

52. Sakai Y. and S. Ishizuka. 1991. Structures of tubular flames (structures of the tubular flames of lean methane/air mixtures in a rotating stretched-flow field). *JSME International Journal Series II* 34 (2):234–41.

53. Mosbacher, D. M., J. A. Wehrmeyer, R. W. Pitz, C.-J. Sung, and J. L. Byrd. 2002. Experimental and numerical investigation of premixed tubular flames. *Proceedings of the Combustion Institute* 29:1479–86. DOI: 10.1016/S1540-7489(02)80181-X.

54. Hu, S., P. Wang, and R. W. Pitz. 2009. A structural study of premixed tubular flames. *Proceedings of the Combustion Institute* 32:1133–40. DOI: 10.1016/j.proci.2008.06.183.

55. Wang, H. and R. W. Pitz. 2009. Extinction and cellular instability of premixed tubular flames. *Proceedings of the Combustion Institute* 32:1141–7. DOI: 10.1016/j.proci.2008.07.012.

56. Hall, C. A. and R. W. Pitz. 2013. A structural study of premixed hydrogen-air cellular tubular flames. *Proceedings of the Combustion Institute* 34:973–80. DOI: 10.1016/j.proci.2012.06.023.

57. Fendell, F. E. 1965. Ignition and extinction in combustion of initially unmixed reactants. *Journal of Fluid Mechanics* 21:281–303. DOI: 10.1017/S0022112065000186.

58. Ishizuka, S., K. Miyasaka, and C. K. Law. 1982. Effects of heat loss, preferential diffusion, and flame stretch on flame-front instability and extinction of propane/air mixtures. *Combustion and Flame* 45:293–308. DOI: 10.1016/0010-2180(82)90054-2.

59. Law, C. K., S. Ishizuka, and P. Cho. 1982. On the opening of premixed bunsen flame tips. *Combustion Science and Technology* 28:89–96. DOI: 10.1080/00102208208952545.

60. Mizomoto, M., Y. Asaka, S. Ikai, and C. K. Law. 1985. Effects of preferential diffusion on the burning intensity of curved flames. In *Proceedings of the Twentieth Symposium (International) on Combustion*, 1933–9. Pittsburgh, PA: The Combustion Institute. DOI: 10.1016/S0082-0784(85)80692-5.

61. Ishizuka, S. 1982. An experimental study on the opening of laminar diffusion flame tips. In *Proceedings of the Nineteenth Symposium (International) on Combustion*, 319–26. Pittsburgh, PA: The Combustion Institute. DOI: 10.1016/S0082-0784(82)80203-8.

62. Burke, S. P. and T. E. W. Schumann. 1928. Diffusion flames. *Industrial and Engineering Chemistry* 20:998–1004. (In *Proceedings of the Symposium on Combustion*, Vol. 1-2, 1948, pp.2-11. DOI: 10.1016/S1062-2888(65)80003-X.)

63. Ishizuka, S. 1983. Tip-opeing of laminar diffusion flame. *Proceedings of the Twenty-First Symposium (Japanese) on Combustion*, 287–9 (in Japanese).

64. Ishizuka, S. and Y. Sakai. 1988. Structure and tip-opening of laminar diffusion flames. In *Proceedings of the Twenty-first Symposium (International) on Combustion*, 1821–8. Pittsburgh, PA: The Combustion Institute. DOI: 10.1016/S0082-0784(88)80416-8.

65. Ishizuka, S. and H. Tsuji. 1984. Effects of transport properties and flow non-uniformity on the temperature of counterflow diffusion flames. *Combustion Science and Technology* 37:171–91. DOI: 10.1080/00102208408923752.

66. Ishizuka, S. 1979. *Combustion limits of diffusion flames Ph.D. thesis.* Tokyo: University of Tokyo, Department of Aeronautics. (in Japanese).

67. Ishizuka, S. and H. Tsuji. 1981. An experimental study of effect of inert gases on extinction of laminar diffusion flames. In *Proceedings of the Eighteenth Symposium (International) on Combustion*, 695–703. Pittsburgh, PA: The Combustion Institute. DOI: 10.1016/S0082-0784(81)80074-4.

68. Wang, P., S. Hu, and R.W. Pitz. 2007. Numerical investigation of the curvature effects on diffusion flames. *Proceedings of the Combustion Institute* 31 (1): 989–96 DOI: 10.1016/j.proci.2006.07.223.

69. Hu, S., P. Wang, R. W. Pitz, and M. D. Smooke. 2007. Experimental and numerical investigation of non-premixed tubular flames. *Proceedings of the Combustion Institute* 31(1):1093–9. DOI: 10.1016/j.proci.2006.08.058.

70. Hu, S. and R. W. Pitz. 2009. Structural study of non-premixed tubular hydrocarbon flames. *Combustion and Flame* 156 (1):51–61. DOI: 10.1016/j.combustflame.2008.07.017.

71. Hu, S., R. W. Pitz, and Y. Wang. 2009. Extinction and near-extinction instability of non-premixed tubular flames. *Combustion and Flame* 156 (1):90–8. DOI: 10.1016/j.combustflame.2008.09.004.

72. Sirignano, W. A., T. K. Pham, and D. Dunn-Rankin. 2002. Miniature-scale liquid-fuel-film combustor. *Proceedings of the Combustion Institute* 29:925–31. DOI: 10.1016/S1540-7489(02)80117-1.

73. Pham, T. K., D. Dunn-Rankin, and W. A. Sirignano. 2007. Flame structure in small-scale liquid film combustors. *Proceedings of the Combustion Institute* 31:3269–75. DOI: 10.1016/j.proci.2006.08.030.

74. Li, Y.-H., Y.-C. Chao, N. S. Amadé, and D. Dunn-Rankin. 2008. Progress in miniature liquid film combustors: Double chamber and central porous fuel inlet designs. *Experimental Thermal and Fluid Science* 32:1118-1131. DOI: 10.1016/j.expthermflusci.2008.01.005.

75. Mattioli, R., T. K. Pham, and D. Dunn-Rankin. 2009. Secondary air injection in miniature liquid fuel film combustors. *Proceedings of the Combustion Institute* 32:3091–8. DOI: 10.1016/j.proci.2008.06.174.

76. Ishizuka, S., R. Hagiwara, M. Suzuki, A. Nakamura, and O. 1999. Hamaguchi. Combustion characteristics of a tubular flame burner. *Transaction of JSME* 65:3845–52. (in Japanese).

77. Ishizuka, S., T. Motodamari, and D. Shimokuri. 2007. Rapidly-mixed combustion in a tubular flame burner. *Proceedings of the Combustion Institute* 31:1085–92. DOI: 10.1016/j.proci.2006.07.128.

CHAPTER 2

THEORY OF TUBULAR FLAMES

Tadao Takeno and Makihito Nishioka

2.1   INTRODUCTION

The problem of flame–flow interaction is one of our fundamental and also practical interests in combustion research,[1] and the behavior of flames in various flow fields has been studied experimentally by using a variety of burners. For example, the structure and propagation of the laminar premixed flame in a quasi one-dimensional (1D) flow have been studied by using the flat-flame burners,[2] while the structure and stability of the flame in a divergent flow have been studied by using stagnation flow or counterflow-type burners.[3] These are basically 1D flames established in 2D or 3D flows. Another example is the tubular flame. The flame was discovered by Prof. Ishizuka some 30 years ago and, in a series of the experimental study, the behavior of the flame has been studied by the group led by Prof. Ishizuka, using this swirl-type burner.[4]

One interesting characteristics of this flame is that it simultaneously experiences the stretch, along the longitudinal direction, and the curvature in the varying degree. On the other hand, the effects of stretch and curvature have extensively been studied so far,[5] but have not fully been understood yet. We believe that the tubular flame should present a good clue to understand the effects. Another motivation to study the tubular flame is to examine the possibility whether a minute tubular flame can actually be an element of turbulent combustion.[6] With this flame, we may find out whether the flame with the size as small as Kolmogorov microscale can be realized, and to what degree of the vortex stretching can sustain the flame.

Since then we have done a series of theoretical studies of the tubular flame. Our formulation is based on the similarity solution, which has been found so useful in the combustion theory. The solution is the simplest one among so many possible solutions, and yet is an exact solution. The flames stabilized in the stagnation flow or the counterflow burners have successfully been described in terms of the similarity solution. These flames have simple-plane 1D structures, in spite of the 2D or 3D flows. The flame structure can be described in terms of 1D ordinary differential equations, which makes it rather easy to predict the flame structure, with the detailed chemical kinetics and the exact transport properties. On the other hand, these flames are very stable to make a detailed measurement of flame structure through experiments.

Then the comparison between the predicted and the measured flame structure has brought us the very useful information, which has contributed greatly to the development of combustion science.

In this article we summarize what we have done so far in the theoretical study of the tubular flame. We believe, however, there still remain so many aspects to be explored theoretically. We hope this article is of some help to these efforts.

2.2 THEORETICAL FORMULATION

2.2.1 MODEL AND ASSUMPTIONS

Figure 2.1 shows the theoretical model adopted in our studies. The unburned mixture is injected diagonally into a cylindrical burner. In the actual experiment, the injection is made through one or two slits, but in this idealized model the mixture is injected through the whole periphery of porous wall with a uniform velocity.[7] The figure shows the axially symmetric flow system in an infinitely long tube with radius R. The cylindrical coordinate is used with the origin at the longitudinal and radial centers of the tube. x and r represent respectively the longitudinal and radial distance from the center, while α represents the azimuthal angle. u, v, and w represent respectively x, r, and α components of the velocity vector v. The unburned mixture is injected at a uniform velocity v_R $(0, -v_R, w_R)$. The following assumptions are adopted in the analysis.

1. The flame, as well as the flow, is steady and axially symmetric, and there are no external forces.
2. The bulk viscosity can be neglected.
3. The Soret and Dufour effects, as well as the pressure diffusion, can be neglected.
4. In the energy equation, the work done by pressure and the dissipation due to viscosity can be neglected. The radiative heat transfer can also be neglected.
5. The reacting mixture, as well as the burned gas, behaves like an ideal gas, and the thermodynamic pressure remains constant throughout the flow field.

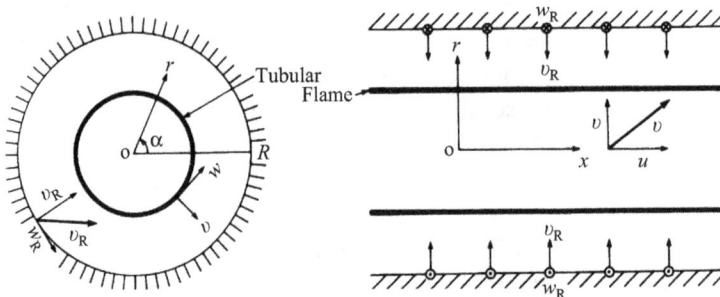

Figure 2.1. Axysymmetric flow in an infinitely long rotating porous tube with uniform injection.

2.2.2 FUNDAMENTAL EQUATIONS

The overall continuity equation is given by

$$\frac{\partial}{\partial x}(r\rho u)+\frac{\partial}{\partial r}(r\rho v)=0 \tag{2.1}$$

where ρ and p are density and pressure, respectively.

The momentum equations reduce to

$$\rho u\frac{\partial u}{\partial x}+\rho v\frac{\partial u}{\partial r}=-\frac{\partial p}{\partial x}+\frac{\partial}{\partial x}\left[\mu\left\{\frac{4}{3}\frac{\partial u}{\partial x}-\frac{2}{3}\frac{1}{r}\frac{\partial}{\partial r}(rv)\right\}\right]+\frac{1}{r}\frac{\partial}{\partial r}\left\{\mu r\left(\frac{\partial v}{\partial x}+\frac{\partial u}{\partial r}\right)\right\} \tag{2.2}$$

$$\rho u\frac{\partial v}{\partial x}+\rho v\frac{\partial v}{\partial r}-\frac{\rho w^2}{r}$$
$$=-\frac{\partial p}{\partial r}+\frac{\partial}{\partial r}\left[\mu\left\{2\frac{\partial v}{\partial r}-\frac{2}{3}\frac{\partial u}{\partial x}-\frac{2}{3}\frac{1}{r}\frac{\partial}{\partial r}(rv)\right\}\right]+\frac{\partial}{\partial x}\left[\mu r\left(\frac{\partial v}{\partial x}+\frac{\partial u}{\partial r}\right)\right]+\frac{2\mu}{r}\left(\frac{\partial v}{\partial r}-\frac{v}{r}\right) \tag{2.3}$$

$$\rho u\frac{\partial w}{\partial x}+\rho v\frac{\partial w}{\partial r}+\frac{\rho vw}{r}=\frac{\partial}{\partial x}\left(\mu\frac{\partial w}{\partial x}\right)+\frac{\partial}{\partial r}\left[\mu r\left(\frac{\partial w}{\partial r}-\frac{w}{r}\right)\right]+\frac{2\mu}{r}\left(\frac{\partial w}{\partial r}-\frac{w}{r}\right) \tag{2.4}$$

where μ is viscosity coefficient.

The continuity equation for each species is given by

$$\rho u\frac{\partial Y_i}{\partial x}+\rho v\frac{\partial Y_i}{\partial r}+\frac{\partial}{\partial x}\left(\rho Y_i V_{i,x}\right)+\frac{1}{r}\frac{\partial}{\partial r}\left(\rho Y_i V_{i,r}\right)+W_i\dot{\omega}_i=0\quad(i=1,...,N) \tag{2.5}$$

Y_i and $\dot{\omega}_i$ are mass fraction and mass production rate of species i, while $V_{i,x}$ and $V_{i,r}$ represent diffusion velocity of species i in the x and r direction, respectively.

The energy equation reduces to

$$\rho u c_p\frac{\partial T}{\partial x}+\rho v c_p\frac{\partial T}{\partial r}-\frac{\partial}{\partial x}\left(\kappa\frac{\partial T}{\partial x}\right)-\frac{1}{r}\frac{\partial}{\partial r}\left(\kappa r\frac{\partial T}{\partial x}\right)$$
$$+\rho\sum_{i=1}^{N}c_{p,i}Y_i V_{i,x}\frac{\partial T}{\partial x}+\rho\sum_{i=1}^{N}c_{p,i}Y_i V_{i,r}\frac{\partial T}{\partial r}+\sum_{i=1}^{N}h_i W_i\dot{\omega}_i=0 \tag{2.6}$$

In this equation, T is temperature and k and c_p are thermal diffusivity and specific heat of the mixture, respectively, and $c_{p,i}$ represents specific heat of species i. h_i is enthalpy and W_i is molecular weight of species i. In this equation, h_i is given in terms of heat of formation of species i, $h_{i,0}$ at standard temperature T^0 through

$$h_i=h_{i,0}+\int_{T^0}^{T}c_{pi}\,dT \tag{2.7}$$

$\dot{\omega}_i$ is the mole production rate of species i and is given by

$$\dot{\omega}_i = \sum_{m=1}^{M} \left(v''_{i,m} - v'_{i,m} \right) \left\{ k_{f,m} \prod_{n=1}^{N} \left(\frac{X_n \rho}{\overline{W}} \right)^{v'_{n,m}} - k_{b,m} \prod_{n=1}^{N} \left(\frac{X_n \rho}{\overline{W}} \right)^{v''_{n,m}} \right\} \quad (i = 1, ..., N) \quad (2.8)$$

where $v''_{i,m}$ and $v'_{i,m}$ are stoichiometric coefficients for species i (reactant) of mth reaction and of species i (product), respectively. $k_{f,m}$ and $k_{b,m}$ are specific reaction rate constants of kth forward and backward reaction. M is total number of the involved elementary reactions.

The equation of state is given by

$$p = \frac{\rho R^0 T}{\overline{W}} \tag{2.9}$$

where R^0 is the universal gas constant and $\overline{W}$ is the average molecular weight of the mixture, and the latter is given by

$$\overline{W} = \frac{1}{\displaystyle\sum_{i=1}^{N} \left(\frac{Y_i}{W_i} \right)} \tag{2.10}$$

The specific heat of the mixture is given by

$$c_p = \sum_{i=1}^{N} c_{p,i} X_i \tag{2.11}$$

where X_i is mole fraction of species i and is given by

$$X_i = \frac{\overline{W} Y_i}{W_i} \tag{2.12}$$

The appropriate boundary conditions for these equations are

$$r = 0: \quad \frac{\partial u}{\partial r} = v = w = 0, \quad \frac{\partial Y_i}{\partial r} = \frac{\partial T}{\partial r} = 0, \tag{2.13}$$

$$r = R: \quad u = 0, \quad v = -v_R, \quad w = w_R, \quad Y_i = Y_{i,R}, \quad T = T_R \tag{2.14}$$

$$x = 0: \quad u = 0, \quad \frac{\partial v}{\partial x} = \frac{\partial w}{\partial x} = 0, \quad \frac{\partial Y_i}{\partial x} = \frac{\partial T}{\partial x} = 0, \tag{2.15}$$

where suffix R denotes values at the wall. In the condition (2.14) we assume that the mixture composition is specified at the wall. This corresponds to the fact that in the experiment the unburned mixture of the specified composition is injected through slits.

2.3 SIMILARITY SOLUTION

2.3.1 INTRODUCTION

The fundamental equations described in the preceding section are very much complicated. It should be very hard to solve these equations simultaneously as they are. We shall try to find a way to make the problem simpler.

We remember that the experimentally observed tubular flame is the cylindrical shape in the strict sense. The cross sections at any locations are just a circle. The flame, in general, becomes visible because of some light-emitting species, produced in the chemical reactions of the flame, and the right circle indicates that the distribution of these species does not depend on the azimuthal angle. In addition, the tubular flame is a straight tube, which means that the flame radius does not depend on the axial location. Then the distribution of these light-emitting species does not depend on the axial position, neither. These observations suggest that the produced species in the chemical reactions of the flame, and even the reactions themselves, should depend only on the radial distance, being independent of the axial distance and of the azimuthal angle.

This is the so-called similarity solution and in the following section we describe this type of solution.

2.3.2 EQUATIONS TO BE SOLVED

Now we assume that the thermodynamic pressure remains constant in the axial direction. Then we adopt the similarity solution for u and v and introduce a stream function to satisfy the overall continuity equation.

$$\Psi(x,r) = xF(r) \tag{2.16}$$

where $F(r)$ is a function depending only on r. Now we have

$$r\rho v = -\frac{\partial \Psi}{\partial x} = -F, \qquad r\rho u = \frac{\partial \Psi}{\partial r} = x\frac{dF}{dr} \tag{2.17}$$

and the axial and radial velocity components are given by

$$u = \frac{x}{r\rho}\frac{dF}{dr}, \qquad v = -\frac{F}{r\rho} \tag{2.18}$$

Now we seek a similarity solution of the form in which Y_i, T, $\dot{\omega}_i$, and ρ depend only on radial distance r. Then we have

$$v = v(r), \qquad w = w(r), \qquad Y_i = Y_i(r), \qquad T = T(r), \qquad \rho = \rho(r) \tag{2.19}$$

It then follows that the transport properties (viscosity μ, thermal conductivity λ, and binary diffusion coefficient D_{ij}) are the functions only of r. Equation (2.18) suggests that the radial velocity v is a function only of r, while the axial velocity u increases linearly with x. The introduction of the above stream function makes it possible to reduce the compressible Navier–Stokes equations to a single ordinary differential equation[8]

$$\left(\frac{dF}{dr}\right)^2 - \rho r F \frac{d}{dr}\left(\frac{1}{\rho r}\frac{dF}{dr}\right) - \rho r\left[\mu r \frac{d}{dr}\left(\frac{1}{\rho r}\frac{dF}{dr}\right)\right] + \pi r^2 H = 0 \qquad (2.20)$$

where H is a constant representing the axial pressure gradient. The energy conservation equation is reduced to

$$-c_p \frac{F}{r}\frac{dT}{dr} - \frac{1}{r}\frac{d}{dr}\left(\lambda r \frac{dT}{dr}\right) + \rho \sum_{k=1}^{K} Y_k V_k c_{p,k}\frac{dT}{dr} + \sum_{k=1}^{K} h_k \dot{\omega}_k W_k = 0 \qquad (2.21)$$

where c_p and $c_{p,k}$ are specific heat at constant pressure of the mixture and kth species, respectively. V_k, W_k, and h_k are diffusion velocity, molecular weight, and specific enthalpy of kth species, respectively. $\dot{\omega}_k$ is the molar rate of production by chemical reactions of kth species per unit volume. The species conservation equation is reduced to

$$-\frac{F}{r}\frac{dY_k}{dr} + \frac{1}{r}\frac{d}{dr}\left(r\rho Y_k V_k\right) + \dot{\omega}_k W_k = 0, \qquad k = 1, 2, ..., K \qquad (2.22)$$

The equation of state is given by

$$\rho = \frac{p\bar{W}}{R^0 T} \qquad (2.23)$$

where p, $\bar{W}$, and R^0 are pressure, mean molecular weight and universal gas constant, respectively. The boundary conditions are

$$\begin{aligned} r = 0: \quad & F = \frac{dF}{dr} = \frac{dT}{dr} = \frac{dY_k}{dr} = 0, \\ r = R: \quad & F = -R\rho_R v_R, \quad \frac{dF}{dr} = 0, \quad T = T_R, \quad Y_k = Y_{k,R} \end{aligned} \qquad (2.24)$$

where suffix R indicates the value at the wall. The above system of equations and boundary conditions constitutes a two-point boundary value problem, in which H is to be determined as an eigenvalue. The derived solutions must be substituted into another momentum equation to obtain the circumferential velocity field. However, this velocity is not directly related to the flame behavior and hence will not be discussed hereafter.

2.4 SIMPLIFIED MODEL WITH ONE-STEP KINETICS AND SIMPLE TRANSPORT PROPERTIES

2.4.1 FORMULATION

The fundamental equations (2.20) through (2.24) are still very much complicated to analyze. We will introduce the simplified system in the following section, in which the chemical kinetics and the transport properties are replaced by the simpler model. In the beginning, we introduce the additional assumptions to simplify the chemical kinetics.

6. The mixture undergoes an overall one step chemical reaction described by

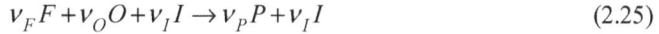

$$v_F F + v_O O + v_I I \rightarrow v_P P + v_I I \tag{2.25}$$

where v_i is the stoichiometric coefficient of species i, while F, O, I, and P represent fuel, oxygen, product, and inert gas, respectively. Now, we will be concerned only with these four species. Furthermore, we assume that the reaction order is unity with respect to fuel and oxygen.

Next, we introduce the following assumption to simplify the transport properties.

7. Viscosity coefficient μ, thermal conductivity $\tilde{\lambda}$, and diffusion coefficient D, which are common to all species, depend on temperature in the following manner.

$$\mu \propto T, \qquad \tilde{\lambda} \propto T, \qquad D \propto T^2 \tag{2.26}$$

It should be noticed that this assumption makes the three transport parameters—Lewis number Le, Prandtl number Pr, and Schmidt number Sc—remain constant throughout the flame.

$$\mathrm{Le} = \frac{\tilde{\lambda}}{\rho D c_p} = \frac{\tilde{\lambda}_R}{\rho_R D_R c_p}, \qquad \mathrm{Pr} = \frac{\mu_R c_p}{\tilde{\lambda}_R}, \qquad \mathrm{Sc} = \mathrm{Le} \times \mathrm{Pr} = \frac{\mu_R}{\rho_R D_R} \tag{2.27}$$

where c_p is assumed a constant. The suffix R indicates the value at the wall.

In this simplified system, momentum equation (2.20) remains the same. On the other hand, the continuity equations (2.22) for fuel, oxygen, product, and inert gas reduce to

$$\rho u \frac{\partial Y_F}{\partial x} + \rho v \frac{\partial Y_F}{\partial r} - \frac{\partial}{\partial x}\left(\rho D r \frac{\partial Y_F}{\partial x}\right) - \frac{1}{r}\frac{\partial}{\partial r}\left(\rho D r \frac{\partial Y_F}{\partial r}\right) = w_F \tag{2.28}$$

$$\rho u \frac{\partial Y_O}{\partial x} + \rho v \frac{\partial Y_O}{\partial r} - \frac{\partial}{\partial x}\left(\rho D r \frac{\partial Y_O}{\partial x}\right) - \frac{1}{r}\frac{\partial}{\partial r}\left(\rho D r \frac{\partial Y_O}{\partial r}\right) = w_O \tag{2.29}$$

$$\rho u \frac{\partial Y_P}{\partial x} + \rho v \frac{\partial Y_P}{\partial r} - \frac{\partial}{\partial x}\left(\rho D r \frac{\partial Y_P}{\partial x}\right) - \frac{1}{r}\frac{\partial}{\partial r}\left(\rho D r \frac{\partial Y_P}{\partial r}\right) = w_P \tag{2.30}$$

$$\rho u \frac{\partial Y_I}{\partial x} + \rho v \frac{\partial Y_I}{\partial r} - \frac{\partial}{\partial x}\left(\rho D r \frac{\partial Y_I}{\partial x}\right) - \frac{1}{r}\frac{\partial}{\partial r}\left(\rho D r \frac{\partial Y_I}{\partial r}\right) = w_I \tag{2.31}$$

Now the energy equation (2.21) reduces to

$$\rho u c_p \frac{\partial T}{\partial x} + \rho v c_p \frac{\partial T}{\partial r} - \frac{\partial}{\partial x}\left(\tilde{\lambda}\frac{\partial T}{\partial x}\right) - \frac{1}{r}\frac{\partial}{\partial r}\left(\tilde{\lambda} r \frac{\partial T}{\partial r}\right)$$
$$+ \left(h_F^0 w_F + h_O^0 w_O + h_P^0 w_P + h_I^0 w_I\right) = 0 \tag{2.32}$$

In the equation of state (2.9) the average molecular weight $\bar{W}$ reduces to

$$\bar{W} = \frac{1}{\dfrac{Y_F}{W_F} + \dfrac{Y_O}{W_O} + \dfrac{Y_P}{W_P} + \dfrac{Y_I}{W_I}} \tag{2.33}$$

Now we seek a similarity solution of the form in which Y_i, T, w_i, and ρ depend only on radial distance r. Then we have

$$v = v(r), \qquad w = w(r), \qquad Y_i = Y_i(r), \qquad T = T(r), \qquad \rho = \rho(r) \tag{2.34}$$

2.4.2 NONDIMENSIONAL SYSTEM

Now we introduce a nondimensional system to make the analysis simpler. The nondimensional space coordinates are given by

$$\zeta = \frac{x}{R}, \qquad \xi = \frac{r}{R} \tag{2.35}$$

The nondimensional stream function is given by

$$\psi(x,r) = 2\rho_R v_R x f(r) \tag{2.36}$$

and the nondimensional velocity components U, V, and h are given by

$$U = \frac{Ru}{2v_R x} = \frac{s}{\xi}\frac{df}{d\xi}, \qquad V = -\frac{v}{2v_R} = \frac{s}{\xi}f, \qquad h = \frac{w}{w_R} \tag{2.37}$$

where the nondimensional density ratio s, temperature θ pressure difference π are given by

$$s = \frac{\rho_R}{\rho}, \qquad \theta = \frac{T}{T_R}, \qquad \pi = \frac{2(p - p_R)}{\rho_R v_R^2} \tag{2.38}$$

where p_R is the pressure at the wall, which may change along the x direction. The assumption (2.7) and the fact that the thermodynamic pressure remains constant, and the similarity solution (2.19) makes the following three transport parameters defined by Eq. (2.27)—Lewis number Le, Prandtl number Pr, and Schmidt number Sc—constant throughout the flame.

$$\text{Le} = \frac{\tilde{\lambda}}{\rho D c_p} = \frac{\tilde{\lambda}_R}{\rho_R D_R c_p}, \qquad \text{Pr} = \frac{\mu_R c_p}{\tilde{\lambda}_R}, \qquad \text{Sc} = \text{Le} \times \text{Pr} = \frac{\mu_R}{\rho_R D_R} \tag{2.39}$$

The other nondimensional parameters which appear in the system are in the following. The injection Reynolds number λ represents the injection velocity of the unburned mixture, while the velocity ratio w_R/v_R represents the relative intensity of swirling flow. The reduced frequency factor B, activation energy E_n, and heat of combustion q describe the kinetic and thermodynamic properties of the mixture.

$$\lambda = \frac{\rho_R R v_R}{\mu_R}, \qquad B = \frac{v_F R^2 \rho_R^2 B_1 T^{\alpha_1}}{2\mu_R W_O}, \qquad E_n = \frac{E}{R^2 T_R},$$

$$q = \frac{\left(v_F W_F h_F^0 + v_O W_O h_O^0 - v_P W_P h_P^0\right)}{v_F W_F c_p T_R} \tag{2.40}$$

The equation of state (2.23) reduces to

$$s = \left(c Y_F + d\right)\theta \tag{2.41}$$

where constants c and d are given in terms of i, j, W_p, and ϕ as

$$c = \frac{\left\{\phi + (i+1)j\right\}\left\{1 + j\left(W_F/W_O\right) - (1+j)\left(W_F/W_I\right)\right\}}{\phi + j\left(W_F/W_O\right) + ij\left(W_F/W_I\right)},$$

$$d = \frac{j(1-\phi)\left(W_F/W_O\right) + (1+j)\phi\left(W_F/W_P\right) + ij\left(W_F/W_I\right)}{\phi + j\left(W_F/W_O\right) + ij\left(W_F/W_I\right)} \tag{2.42}$$

The constants i and j depend on properties of fuel, oxidizer, and inert gas, while the equivalence ratio ϕ specifies the mixture ratio.

$$i = \frac{W_I X_{I,R}}{W_O X_{O,R}}, \qquad j = \frac{v_O W_O}{v_F W_F}, \qquad \phi = \frac{v_O X_{F,R}}{v_F X_{O,R}} \tag{2.43}$$

Now the similarity solution yields the following equations and boundary conditions to be solved for the nondimensional radial velocity f and the density ratio $s = \rho_R/\rho$ as

$$2\lambda \frac{f}{\xi}\frac{d}{d\xi}\left(\frac{s}{\xi}\frac{df}{d\xi}\right) + \frac{1}{\xi}\frac{d}{d\xi}\left\{s\xi\frac{d}{d\xi}\left(\frac{s}{\xi}\frac{df}{d\xi}\right)\right\} - 2\lambda\frac{s}{\xi^2}\left(\frac{df}{d\xi}\right)^2 + 8\lambda k = 0 \tag{2.44}$$

$$2\lambda\left(\frac{f}{\xi}\frac{dh}{d\xi} + \frac{fh}{\xi^2}\right) + \frac{d}{d\xi}\left\{s\left(\frac{dh}{d\xi} - \frac{h}{\xi}\right)\right\} + \frac{2s}{\xi}\left(\frac{dh}{d\xi} - \frac{h}{\xi}\right) = 0 \tag{2.45}$$

$$\frac{1}{2\mathrm{PrLe}}\frac{s}{\xi}\frac{d}{d\xi}\left(s\xi\frac{dY_F}{d\xi}\right) + \lambda\frac{sf}{\xi}\frac{dY_F}{d\xi} + BY_F Y_O \exp\left(-\frac{E_n}{\theta}\right) = 0 \tag{2.46}$$

$$Y_O = Y_{O,R} + j\left(Y_F - Y_{F,R}\right), \qquad Y_P = (1+j)\left(Y_{F,R} - Y_F\right), \qquad Y_I = Y_{I,R} \tag{2.47}$$

$$\frac{1}{2\mathrm{Pr}}\frac{s}{\xi}\frac{d}{d\xi}\left(s\xi\frac{d\theta}{d\xi}\right) + \lambda\frac{sf}{\xi}\frac{d\theta}{d\xi} + qBY_F Y_O \exp\left(-\frac{E_n}{\theta}\right) = 0 \tag{2.48}$$

$$\xi = 0: \quad f(0) = f'(0) = h(0) = 0, \quad Y_F'(0) = \theta'(0) = 0 \tag{2.49}$$

$$\xi = 1: \quad f(1) = 1/2, \quad f'(1) = 0, \quad h(1) = 1, \quad Y_F(1) = Y_{F,R}, \quad \theta(1) = 1 \tag{2.50}$$

The above system of equations, without Eq. (2.45), constitutes a two-point boundary value problem, in which the unknowns are the boundary values $f''(0)$, $Y_F(0)$, $\theta(0)$, and the eigenvalue k. The obtained $f(\xi)$ and $s(\xi)$ are substituted into Eq. (2.45) to derive $h(\xi)$, by solving another two-point boundary value problem. The distribution of pressure difference $\pi(\zeta, \xi)$ can be obtained by integrating the following equations:

$$\frac{\partial \pi}{\partial \zeta} = -32k\zeta \tag{2.51}$$

$$\begin{aligned}
\frac{\partial \pi}{\partial \xi} = &-\frac{8f}{\xi}\frac{d}{d\xi}\left(\frac{sf}{\xi}\right) + 2\left(\frac{w_R}{v_R}\right)^2\frac{h^2}{s\xi} + \frac{8}{3\lambda}\frac{d}{d\xi}\left\{\frac{fs^2}{\xi^2} - \frac{s^2}{\xi} - 2s\frac{d}{d\xi}\left(\frac{sf}{\xi}\right)\right\} \\
&+ \frac{4s}{\lambda}\frac{d}{d\xi}\left(\frac{s}{\xi}\frac{df}{d\xi}\right) - \frac{8s}{\lambda\xi}\left\{\frac{d}{d\xi}\left(\frac{sf}{\xi}\right) - \frac{sf}{\xi^2}\right\}
\end{aligned} \tag{2.52}$$

2.4.3 INCOMPRESSIBLE FLOW SYSTEM

We further introduce additional assumptions to make the analysis easier. The assumption is the so-called incompressible flow assumption. The density remains constant throughout the flow field, in spite of the heat release due to chemical reactions, and of the resulting temperature increase.[7] The assumption decouples the momentum equations from the energy equation and the species conservation equations. The effects of the density change on the flame behavior will be studied later. We further assume that the transport properties are all common to each species and constant.

8. The flow is incompressible and the density remains constant throughout the whole flow field.
9. The transport properties are all common to each species and constant.

2.4.4 FLOW FIELD

Now s in Eqs (2.44) through (2.52) reduces to unity and the momentum equations (2.44) and (2.45) and the boundary conditions (2.49) and (2.50) become

$$\xi^2 f''' + (2\lambda f - 1)\xi f'' - (2\lambda f - 1)\xi f' - 2\lambda\xi f'^2 + 8\lambda k\xi^2 = 0 \tag{2.53}$$

$$f(0) = f'(0) = 0, \quad f(1) = 1/2, \quad f'(1) = 0 \tag{2.54}$$

$$\xi^2 h'' + (2\lambda f + 1)\xi h' + (2\lambda f - 1)h = 0 \tag{2.55}$$

$$h(0) = 0, \quad h(1) = 1 \tag{2.56}$$

Equations (2.53) and (2.54) for f constitute a two-point boundary value problem, in which a constant k is to be determined as an eigenvalue of the system for a given value of λ. The obtained f is substituted into Eqs (2.55) and (2.56) to derive h by solving another two-point boundary value problem. In this way, the whole flow field is completely described by specifying a single parameter λ, the injection Reynolds number.

2.4.5 CONCENTRATION AND TEMPERATURE FIELD

The concentration and temperature field is determined by solving the species conservation equation for each species and the energy equation. The equations for fuel concentration and temperature in this nondimensional system reduce to

$$\frac{1}{2\mathrm{PrLe}}\frac{1}{\xi}\frac{d}{d\xi}\left(\xi\frac{dY_F}{d\xi}\right)+\lambda\frac{f}{\xi}\frac{dY_F}{d\xi}-BY_FY_O\exp\left(-\frac{E_n}{\theta}\right)=0 \tag{2.57}$$

$$\frac{1}{2\mathrm{Pr}}\frac{1}{\xi}\frac{d}{d\xi}\left(\xi\frac{d\theta}{d\xi}\right)+\lambda\frac{f}{\xi}\frac{d\theta}{d\xi}-qBY_FY_O\exp\left(-\frac{E_n}{\theta}\right)=0 \tag{2.58}$$

The appropriate boundary conditions are

$$Y_F'(0)=\theta'(0)=0,\quad Y_F(1)=Y_{F,R},\quad \theta(1)=1 \tag{2.59}$$

In Eqs (2.57) and (2.58) oxygen mass fraction Y_O is given by

$$Y_O(\xi)=Y_{O,R}-j\left(Y_{F,R}-Y_F(\xi)\right) \tag{2.60}$$

where $j=v_OW_O/(v_FW_F)$ is gram of oxygen required to burn 1 g of fuel to effect complete combustion. The above system of equations and boundary conditions constitutes, again, a two-point boundary value problem. We may solve the system after the flow field has been solved for a given value of λ. It should be noticed that the circumferential flow field is not involved in this system.

We may assume that the unburned mixture injected through the porous wall contains no product, and in that case mass fractions at the wall can be expressed in terms of the equivalence ratio φ of the mixture as

$$Y_{F,R}=\frac{\varphi}{\varphi+j(i+1)},\qquad Y_{O,R}=\frac{j}{\varphi+j(i+1)},\qquad Y_{I,R}=\frac{ij}{\varphi+j(i+1)} \tag{2.61}$$

where the equivalence ratio φ and i are defined by

$$\varphi=\left(\frac{v_O}{v_F}\right)\left(\frac{X_{F,R}}{X_{O,R}}\right),\qquad i=\frac{W_IX_{I,R}}{W_OX_{O,R}} \tag{2.62}$$

In the above definitions X_i represents mole fraction of species i in the mixture.

The foregoing formulation indicates that the concentration and temperature field, and hence the flame behavior, can be described in terms of nine parameters (λ, Le, Pr, i, j, φ, B, E_n, q). Among these parameters five (Le, Pr, B, E_n, and q) should depend on φ, and we have to introduce some kind of theoretical model to describe the dependence. It must be remembered, here again, that the circumferential velocity does not affect the flame behavior.

2.4.6 SIMPLIFICATION FOR Le = 1

Now we introduce the following additional assumption that

10. Lewis number is equal to unity.

In this case, we can use the Shvab–Zeldovich function to represent Y_F and Y_O in terms of θ.

$$Y_F(\xi) = Y_{F,R} - \frac{1}{q}\big(\theta(\xi)-1\big), \qquad Y_O(\xi) = Y_{O,R} - \frac{1}{q}\big(\theta(\xi)-1\big) \qquad (2.63)$$

Therefore, we need not solve the oxygen conservation equation and the only equations to be solved are (2.53) and (2.58). Now we can make use of Eqs (2.61) and (2.63) to derive the attainable maximum flame temperature by putting $Y_F = 0$ for lean mixture as

$$\theta_a = 1 + \frac{\varphi q}{\varphi + j(i+1)} \qquad (2.64)$$

and by putting $Y_O = 0$ for rich mixture as

$$\theta_a = 1 + \frac{q}{\varphi + j(i+1)} \qquad (2.65)$$

2.4.7 RESULTS FOR SIMPLIFIED MODEL

2.4.7.1 Flow Field

The two-point boundary value problems described above were solved numerically by using the forward integration method of Runge–Kutta–Gill, with the generalized Newton–Raphson method to correct for the initial values. The whole calculation was made with an accuracy of more than eight significant figures.

Our main concern in the present study is the flame stability, or the flame behavior, when the injection Reynolds number λ is increased. We are mostly concerned, therefore, in the flow field for rather large values of λ (≥ 10). In the extreme case of infinitely large λ, Eq. (2.53) reduces to

$$\xi f f'' - f f' - \xi f'^2 + 4k\xi^2 = 0 \qquad (2.66)$$

This is the case of inviscid flow and the equation has the following analytical solution which satisfies the imposed boundary condition (2.54).[9]

$$f(\xi) = \frac{1}{2}\sin\left(\frac{\pi}{2}\xi^2\right) \tag{2.67}$$

Then the eigenvalue and $f''(0)$ for this incompressible flow solution reduce to

$$k = \frac{\pi^2}{16}, \qquad a \equiv \frac{1}{2}f''(0) = \frac{\pi}{4} \tag{2.68}$$

Figure 2.2 shows the numerically calculated k and a for finite values of λ. They decrease with λ approaching asymptotically the values given by (2.68) and can be expressed in terms of the second-order polynomials of $(1/\lambda)$ with an accuracy of less than 0.2% error for λ larger than 10.

$$k = \frac{\pi^2}{16} + \frac{1.3321}{\lambda} - \frac{0.15972}{\lambda^2}, \qquad a = \frac{\pi}{4} + \frac{0.84022}{\lambda} - \frac{2.3985}{\lambda^2}, \tag{2.69}$$

The flow field for finite values of λ can well be described in terms of the incompressible flow solution, provided λ is not too small. An example of the numerically calculated flow field is shown in Fig. 2.3 for the case $\lambda = 100$. The incompressible flow solution was plotted also in the figure for comparison, but the two curves were found so close that no perceptible difference could be observed. f increases monotonically from the axis toward the wall. The resultant longitudinal and radial velocities U and V are plotted as well. U becomes maximum on the axis giving the maximum value $2a$, while V increases toward the wall to attain a maximum just before the wall. The existence of this maximum brings out a curious flame behavior as will be described later. The maxima U_m and V_m shown in Fig. 2.3 are plotted against λ in Fig. 2.4. The radius ξ_m which gives the maximum V_m is plotted as well. As seen in the figure, they remain almost constant for a wide range of λ, and deviate slightly from the constants only when λ becomes smaller than 100. U_m and V_m increase, whereas ζ_m decreases, as λ is decreased. These constants can easily be derived from the incompressible flow solution as

$$U_m = \pi/2, \qquad V_m = 0.5334, \qquad \xi_m = 0.8614 \tag{2.70}$$

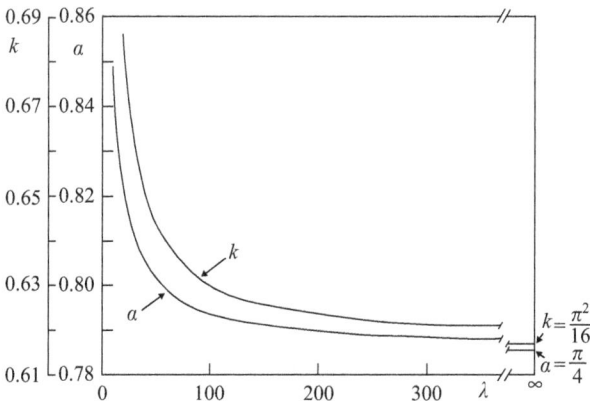

Figure 2.2. Calculated eigenvalues k and a plotted against the injection Reynolds number λ.

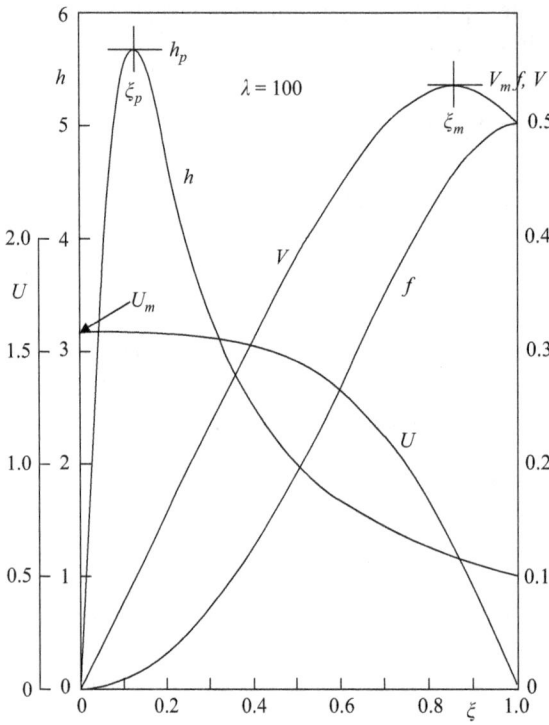

Figure 2.3. Flow field calculated for the injection Reynolds number 100.

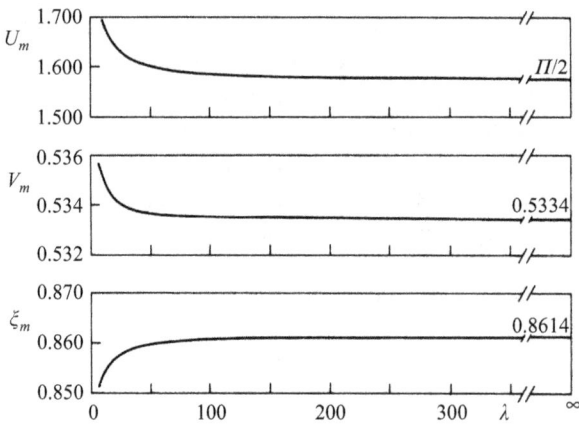

Figure 2.4. Maximum radial and longitudinal velocities V_m and U_m, and radius ξ_m, where radial velocity becomes maximum, plotted against the injection Reynolds number λ.

When λ is infinitely large the order of Eq. (2.55) is reduced to yield the first-order differential equation

$$\xi h' + h = 0 \tag{2.71}$$

and the solution which satisfies the boundary condition at $\xi = 1$ is given by

$$h(\xi) = 1/\xi \tag{2.72}$$

This solution describes the free vortex flow of constant circulation. When $\lambda = 0$, on the other hand, Eq. (2.55) reduces to

$$\xi^2 h'' + \xi h - h = 0 \tag{2.73}$$

The solution which satisfies the boundary condition on the axis is

$$h(\xi) = \xi \tag{2.74}$$

This solution describes the forced vortex flow of constant vorticity.

As mentioned before, we are interested in the flow field for rather large values of λ, and in that case the radial and longitudinal flow fields are well described by the incompressible flow solution (2.67). The circumferential flow field, on the other hand, has pronounced characteristics of the Rankin combined vortex. That is, in the central inner region the flow is described by the forced vortex flow (2.74), while in the outer region it is described by the free vortex flow (2.72). Fortunately, Eqs (2.55) and (2.56) can be solved analytically, provided f approaches 0 as $\xi \to 0$, to yield

$$h(\xi) = \frac{1}{\xi} \frac{\displaystyle\int_0^{\xi} \xi \exp\left[-\int_0^{\xi} \frac{2\lambda f}{\xi} d\xi\right] d\xi}{\displaystyle\int_0^{1} \xi \exp\left[-\int_0^{\xi} \frac{2\lambda f}{\xi} d\xi\right] d\xi} \tag{2.75}$$

When we study the flow characteristics for large values of λ we may replace f in the equation by the incompressible flow solution. Then we have the solution

$$h(\xi) = \frac{1}{\xi} \frac{\displaystyle\int_0^{\frac{\pi}{2}\xi^2} \xi \exp\left[-\frac{\lambda}{2} S_i(t)\right] dt}{\displaystyle\int_0^{\frac{\pi}{2}} \xi \exp\left[-\frac{\lambda}{2} S_i(t)\right] dt} \tag{2.76}$$

where

$$S_i(t) = \int_0^t \frac{\sin t}{t} dt = t - \frac{t^2}{3 \bullet 3!} + \frac{t^3}{5 \bullet 5!} - \dots \tag{2.77}$$

Therefore, if we adopt the following approximation, $1 \gg \pi^2/72$ and take the first term of the expansion, we obtain the following simple approximate solution:

$$h_a(\xi) = \frac{1}{\xi} \frac{1 - \exp\left[-(\pi/4)\lambda\xi^2\right]}{1 - \exp\left[-(\pi/4)\lambda\right]} \cong \frac{1 - \exp\left[-(\pi/4)\lambda\xi^2\right]}{\xi} \tag{2.78}$$

This solution holds for the whole range of ξ and then $b = h'(0)$ becomes

$$b = \frac{\pi\lambda}{4} \frac{1}{1-\exp\left[-(\pi/4)\lambda\right]} \cong \frac{\pi\lambda}{4} \tag{2.79}$$

where the last approximation in these equations may safely be adopted for λ larger than 10. The approximate solution (2.78) happens to coincide with the one derived by Burgers in his theory of turbulence[10] and retains the essential properties of the exact solution. It yields the approximate inner solution (2.74) in the very vicinity of $\xi = 0$, as well as the outer solution (2.72) for $\lambda\xi^2$ large enough. The profile of h calculated numerically is shown in Fig. 2.3 for the case $\lambda = 100$. The approximate solution (2.78) is also plotted, but it is so close to the numerical solution that they cannot be distinguished in the figure. As seen in the figure, the profile clearly exhibits the Rankine combined vortex characteristics. In Fig. 2.5 the numerically calculated b is plotted against λ as compared with the approximate value given by Eq. (2.79). The approximation is rather accurate, and although the deviation increases with a decrease in λ, it remains below 5% at $\lambda = 100$. As λ is increased, the inner forced vortex region shrinks, increasing the peak circumferential velocity. We may make use of Eq. (2.78) to derive the following approximate expressions for the peak position ξ_p and the peak value h_p as a function of λ

$$\xi_p = 1.1209\sqrt{\frac{4}{\pi\lambda}}, \qquad h_p = \frac{0.71533}{\xi_p} \tag{2.80}$$

ξ_p gives a measure for size of the core vortex flow region, and it is interesting to note that the size depends only on λ being independent of the circumferential injection velocity w_R. In Fig. 2.5 ξ_p and h_p are plotted against λ. The approximate values (2.80) are very accurate, and they almost coincide in the figure with the exact values calculated numerically.

2.4.7.2 Development of Flame

The mixture selected was methane and air in accordance with the experimental study.[11] The air was assumed to be composed of 79% N_2 and 21% O_2 by volume, and the calculated values

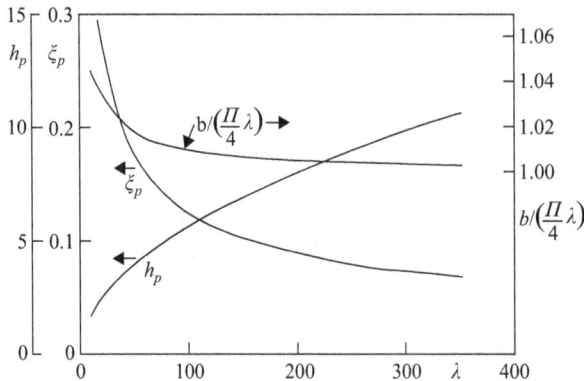

Figure 2.5. $b \equiv h'(0)$, maximum circumferential velocity h_p, and radius ξ_p, where it becomes maximum, plotted against the injection Reynolds number λ.

were $i = 3.2917, j = 4.0$. The activation energy E was assumed to be independent of the equivalence ratio φ and the value of 30.0 kcal/mol was adopted. The latter was obtained by comparing the experimental mass burning velocity with that predicted by the simplified flame theory.[12] The wall temperature T_R was 300 K and the reduced activation energy E_n was 50.3271. Lewis number, as well as Prandtl number, was fixed at unity, and the effects of the injection Reynolds number λ and the equivalence ratio φ on the flame behavior were studied. The reduced heat of combustion q was selected so that the maximum attainable temperature given by Eq. (2.64) coincided with the adiabatic flame temperature derived by the chemical equilibrium calculation. Thus for $\varphi = 1$, the adiabatic flame temperature $T_{ad} = 2225.82$ K and q became 116.618. This value was used when φ was changed around unity, while another value of 138.945, based on $T_{ad} = 1479.71$ K, was used when φ was changed around 0.5. Selection of the reduced frequency factor B was most arbitrary. Three values of $1.5 \times 10^8, 5.0 \times 10^8, 3.0 \times 10^9$ were used in accordance with the specific purpose of the calculation. The radial temperature distribution always exhibits a maximum on the axis and it decreases outward. We may rightly call this the maximum temperature $\theta_0 = \theta(0)$ as the flame temperature. The solid curve in Fig. 2.6 shows the flame temperature plotted against λ for fixed values of B and φ. As seen in the figure, the familiar response curve is obtained. For a given value of λ there exist three solutions simultaneously: the upper, the middle, and the lower. As λ is decreased along the upper solution, the flame temperature approaches asymptotically the limiting value θ_a given by Eq. (2.64) to effect complete combustion. It should be noticed, however, that θ_0 does not necessarily come to θ_a exactly at the limit of $\lambda = 0$. As λ is increased, on the other hand, it decreases gradually until λ reaches the critical value $\lambda_{cr} = 100.1$, above which there exist no solutions except the trivial lower solution. This corresponds to the flame blow-off. The temperature rising above unity of the lower solution is extremely small and tends to become zero as λ is increased without limit. The response curve obtained when the incompressible flow solution was used for the flow filed is shown as well, by the dotted curve. The deviation from the numerical solution is very small.

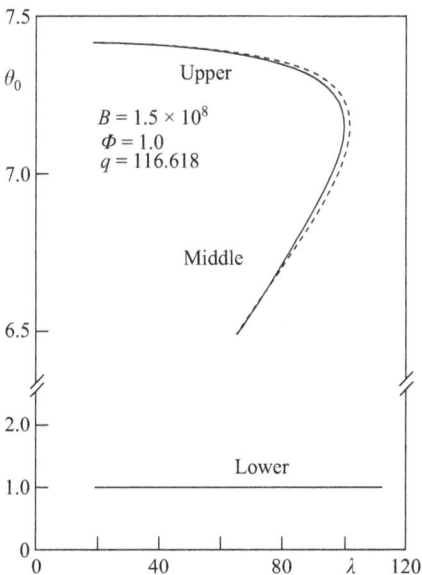

Figure 2.6. Flame temperature θ_0 plotted against injection Reynolds number λ.

As λ approaches λ_{cr}, the flame temperature of the upper solution approaches to that of the middle solution, and in Fig. 2.7 the flame structures of the two solutions are compared for $\lambda =$ 80. For both solutions the temperature and concentrations remain uniform in the extended outer region and the changes occur only in the small central region near the axis. Indicating that the flow is frozen in the most outer region and the reaction is concentrated in near the central axis. The reaction zone is situated more close to the axis for the middle solution. The local energy balance in the reaction region of the two solutions is compared in Fig. 2.8, in which Q_R, Q_C, and Q_D represent the contribution of the reaction, the convection, and the thermal conduction terms in the energy equation Eq. (2.58). As is in the normal flame case, it is mainly the conduction that balances the reaction in the reaction zone, while the conduction and the convection balance in the outer preheat zone. It is interesting to note that for the upper solution, the reaction zone is situated apart from the axis, whereas for the middle solution, a part of the reaction zone reaches the axis to produce an incomplete combustion. In the region very near the axis,

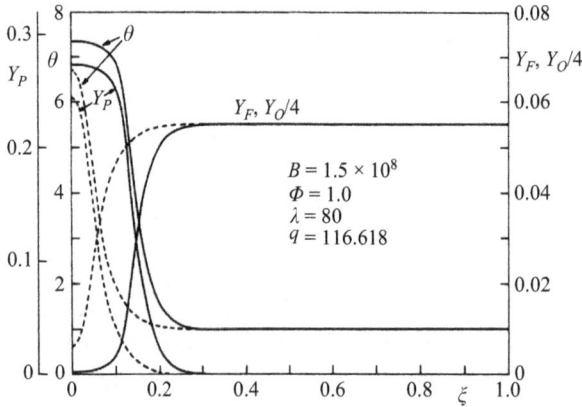

Figure 2.7. Temperature and concentration distributions for upper (solid curves) and middle (dotted curves) solutions.

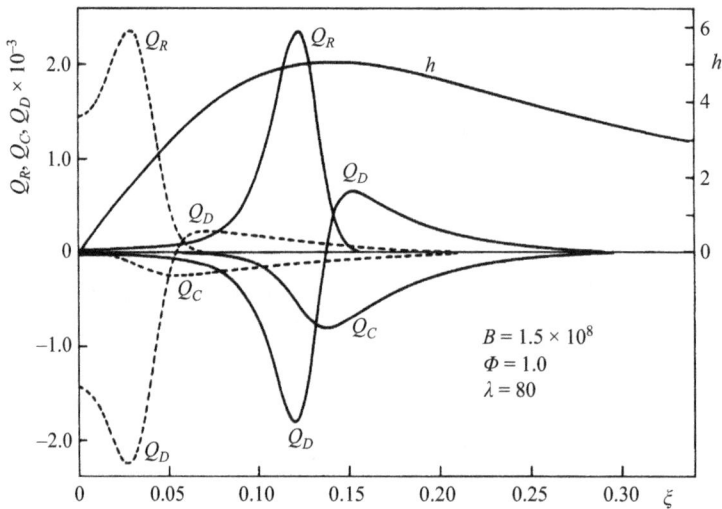

Figure 2.8. Local energy balance in flame for upper (solid curves) and middle (dotted curves) solutions.

the radial velocity V becomes very small, and hence the convective, as well as the conductive, term has to reduced, as seen in the preheat region of the middle solution. In the figure h is also shown. It can be seen that the both reaction zones are located, in this case, inside the forced vortex flow region.

2.4.7.3 Effects of Injection Reynolds Number

Figure 2.9 shows the variation of heat release rate profile Q_R when λ is changed along the response curve shown in Fig. 2.6. As λ is increased along the upper solution, the reaction zone moves toward the axis almost in parallel, keeping the similar profile. In the profile the maximum value can be deduced from the condition

$$\frac{dQ_R}{d\xi} = \frac{dQ_R}{d\theta} \frac{d\theta}{d\xi} = 0 \qquad (2.81)$$

We may make use of Eq. (2.63) to express Q_R in terms of θ, and we may deduce the temperature θ_R where Q_R becomes maximum, by using the condition $dQ_R/d\theta = 0$. Then the maximum value Q_R, as well as θ_R, is given as a function of (i, j, ϕ, q, E_n). When the flame temperature θ_0 is larger than θ_R, there should exist a position ξ_R in the flame where the temperature is just equal to θ_R to give this maximum value. Therefore, as λ is increased along the upper solution the position ξ_R, where Q_R becomes maximum, moves toward the axis, giving the identical maximum value, until λ reaches the critical value $\lambda_{cr} = 100.1$. The degree of incomplete combustion increases with λ leaving a considerable amount of reactant $Y_F(0) = Y_O(0)/4 = 0.002327$ at the critical state. Some 4% of the injected mixture remains unburned. As λ is decreased along the middle solution, on the other hand, θ_0 continues to decrease while $Y_F(0)$ continues to increase, until θ_0 reaches θ_R, where ξ_R comes to the axis. Then Q_R becomes maximum on the axis, where $d\theta/d\xi = 0$, and the maximum value is obtained by substituting θ_0 into $Q_R(\theta)$. For a further decrease in λ, ξ_R remains on the axis, whereas $Y_F(0)$ increases still further.

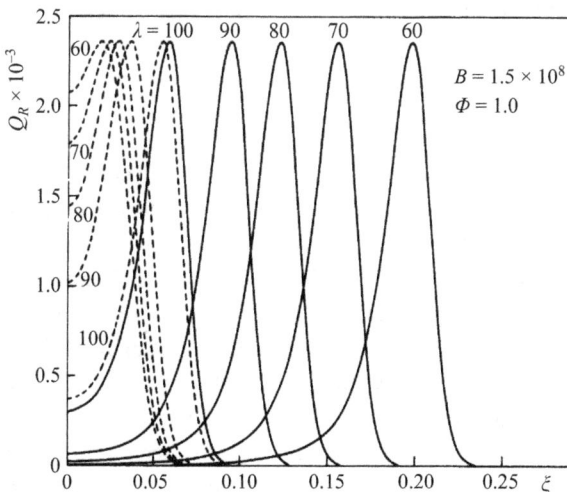

Figure 2.9. Variation with injection Reynolds number l of heat release rate profile for upper (solid curve) and middle (dotted curve) solutions.

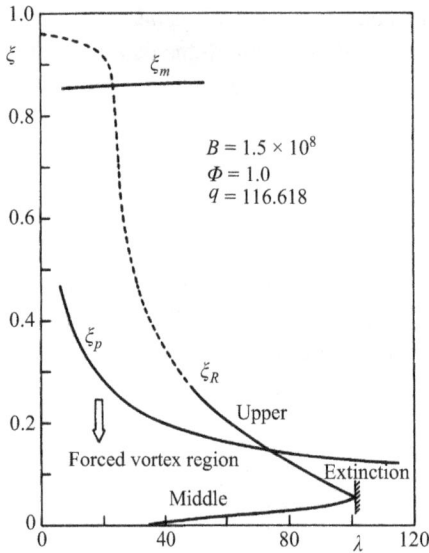

Figure 2.10. Radius ξ_R, where the heat release rate becomes maximum, and an outer boundary of core vortex region ξ_p, plotted against the injection Reynolds number.

Figure 2.10 shows the radius ξ_R, or the representative reaction zone position, plotted against λ. The solid curve represents the results obtained for the exact flow solution, while the dotted curve represents that obtained for the incompressible flow solution. The outer boundary of forced vortex region, which may be represented by ξ_p, is also shown to study the reaction zone position relative to the vortex. As seen in the figure, the reaction zone is located inside the vortex when λ is large. As λ is decreased along the upper solution, however, the reaction zone is shifted outside to be located in the free vortex region. As λ is decreased still further, the shift becomes increasingly rapid until ξ_R approaches ξ_m, where V becomes maximum. We may expect that the flame will flashback when we decrease λ beyond this, as now the radial velocity V decreases toward the wall as seen in Fig. 2.3. However, the flame continues to exist and ξ_R increases very slowly, approaching 0.96 ca. asymptotically as $\lambda \to 0$. In the latter limiting case, the flame with no convective term is established just apart from the wall. In this unusual flame the mixture supply, as well as the product removal, is provided by the diffusion process alone.

2.4.7.4 The Effects of Equivalence Ratio

The variation of the response curve when the equivalence ratio ϕ is changed is shown in Fig. 2.11. The maximum attainable temperature changes with ϕ according to Eq. (2.64), and as ϕ is decreased from unity, the curve shifts toward lower left. The critical injection Reynolds number λ_{cr} is decreased and the curve becomes gradually sharpened. Then the drop in the flame temperature from θ_a at λ_{cr} becomes smaller and smaller. Figure 2.12 compares the reaction zone profile for the different values of ϕ for a fixed value of λ. As ϕ is decreased, the total heat release in the flame decreases, and along the upper solution the reaction zone is shifted toward the axis until ϕ reaches a critical value ϕ_{cr}, below which no solutions except the trivial lower solution can be obtained. The lower part of Fig. 2.13 shows this behavior more clearly: here the variation with ϕ of ξ_R is shown. The upper part shows the variation with ϕ of the flame temperature θ_0.

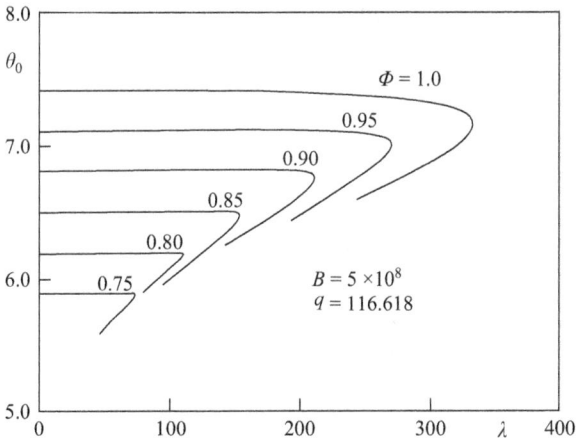

Figure 2.11. Variation with an equivalence ratio f of the response curve in flame temperature θ_0 – injection Reynolds number λ plane.

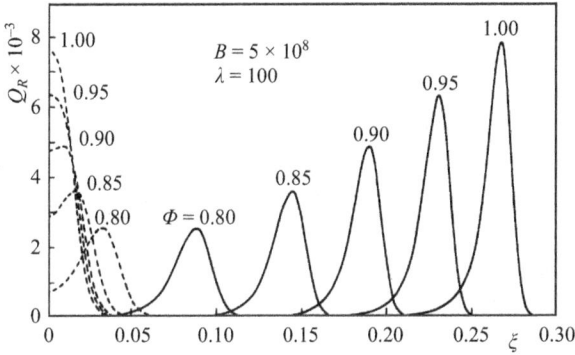

Figure 2.12. Variation with an equivalence ratio f of heat release rate profile for upper (solid curves) and middle (dotted curves) solutions.

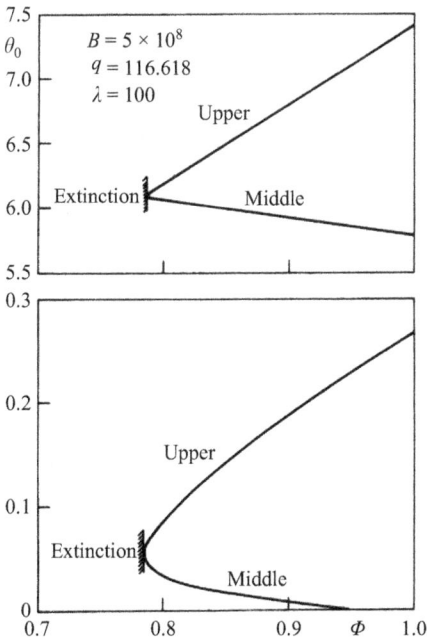

Figure 2.13. Flame temperature θ_0 and reaction zone position ξ_R equivalence ratio ϕ.

As ϕ is decreased, θ_0 for the upper solution decreases, while it increases for the middle solution, until ϕ reaches the critical value.

2.4.7.5 Flame Stability

Although the rigorous stability analysis of the multiple solutions has not yet been made, the following simple consideration seems to suggest that most of the upper solutions are stable to radial flow disturbances. As seen in Fig. 2.10, the reaction zone position ξ_R for the upper solution moves toward the axis with an increase in λ. If a momentary increase in the injection velocity v_R is given, the reaction zone moves toward the axis and the mixture velocity coming into the reaction zone is decreased except for the region very near the wall. Then the reaction zone will flashback to the original position. When the injection velocity is decreased, the opposite mechanism will operate to bring back the reaction zone to the original position. It may be considered, therefore, that the upper solution with ξ_R smaller than ξ_m should be stable to the flow disturbances. On the contrary, the upper solution with ξ_R larger than ξ_m becomes unstable. The flame established in the accelerating flow is unstable, naturally. Then the stability problem appears very simple. However, the situation is rather complicated when we note that the same argument reveals that the middle solution should be stable as well. In addition, the flame stability is closely related to the flow stability, and we will discuss this problem later. Instead, here we will advance a simplified argument so as to yield the theoretical extinction condition. In the experiment, the flame extinction was observed when fuel concentrations were decreased while keeping the mixture injection velocity constant, or when the injection velocity was increased for a fixed fuel concentration. The injection velocity is represented, in the present analysis, by the injection Reynolds number λ. Therefore the experimental operation to cause the extinction corresponds to decrease ϕ for a fixed value of λ (Fig. 2.13), or to increase λ for a fixed value of ϕ (Figs 2.6, 2.10, and 2.11) along the upper solution. Then it will be evident that the critical values ϕ_{cr} and λ_{cr} give the extinction condition, as far as the upper solution is stable.

Figure 2.14 shows the response curve calculated for this purpose. In accordance with the experimental stability diagram, the equivalence ratio of 0.5 was selected as the standard value and ϕ was varied around this value. The value of B was determined such that for $\phi = 0.5$ the extinction should occur at around $\lambda = 100$, which is the value observed in the experiment. As seen in the figure, the critical injection Reynolds number λ_{cr} decreases very rapidly with a decrease in ϕ. The deduced flame stability diagram in the $\phi - \lambda$ plane is shown in Fig. 2.15.

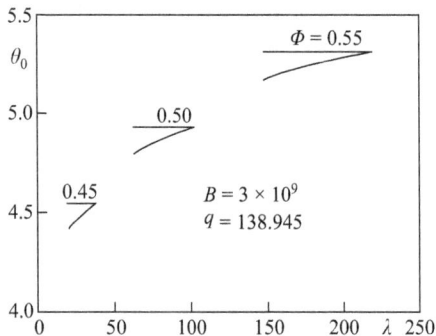

Figure 2.14. Response curve in flame temperature θ_0 – injection Reynolds number λ plane.

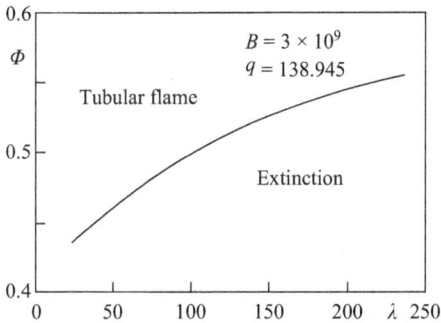

Figure 2.15. Predicted flame stability diagram in the equivalence ratio ϕ– injection Reynolds number λ plane.

It is to be compared with Fig. 2.4 of the experiment,[10] where the stability is given in the fuel concentration–injection velocity plane.

Finally, some remarks must be made on the physical mechanism that causes the extinction. At the extinction condition the calculated temperature distribution reveals that the gradient at the wall is substantially zero, indicating that there is eventually no heat loss to the wall. On the other hand, appreciable amounts of fuel and oxygen remain unreacted at this condition. Then we can say that the extinction is caused due to incomplete combustion.

2.4.8 DISCUSSIONS ON RESULTS FOR SIMPLIFIED MODEL

2.4.8.1 Comparison with Experiment

In the present analysis the hypothetical uniform introduction of the mixture into the tubular burner was assumed, and hence when we try to correlate the predicted flame behavior with those observed in the experiment,[10,11] we should restrict ourselves to the case when the flame is established apart from the wall. Under this restriction, we may safely say that the predicted effects of the injection Reynolds number λ and the equivalence ratio ϕ on the flame behavior correlate satisfactory with those observed for lean methane–air mixtures in the experiment. Figure 2.10 reveals that as λ is increased for a fixed value of ϕ, the flame size ξ_R of the upper solution becomes smaller and smaller until the flame is extinguished at a certain critical value, in agreement with the observation. Similarly, as ϕ is decreased for a fixed value of λ, the flame size becomes smaller, leading to the extinction at a certain critical value (Fig. 2.13). The almost linear decrease of the flame size with ϕ, as well as the critical flame size $\xi_R = 0.06$, correlates most satisfactory with the observation (Fig. 2.8 of[10] and Fig. 2.5 of[11]). The derived stability diagram in the $\phi - \lambda$ plane (Fig. 2.15) seems to correlate with the stability diagram for the tubular burner (Fig. 2.4 of[11]), although the increase in the critical fuel concentration with the injection velocity is not so apparent in the observation. This is presumably due to the rather narrow range of the injection velocity varied in the experiment. On the other hand, the predicted temperature distribution shown in Fig. 2.7 correlates satisfactorily with the observed distribution (Fig. 2.5 of[11]). These agreements suggest that the actual flame can well be described in terms of the axisymmetric similarity solution adopted in this study.

The distinguished flame behavior observed for lean hydrogen–air, as well as lean propane–air, mixtures could not be predicted in the present study. The hydrogen flame was found to merge into a rod at the extinction limit, whereas in the present study the flame causes the extinction at finite flame sizes. The hydrogen flame is known to be peculiar in that the Lewis number is considerably different from unity and the burning velocity is rather high. The behavior will be explained, qualitatively at least, by the Lewis number effect combined with the thin reaction zone width. This will be a subject of the future studies.

In order to make more detailed comparisons with the experiment to predict quantitative flame behavior, we have to develop, first of all, a theory which is based on the viscous flow solution with the density change. Furthermore, we have to introduce a full chemical kinetics with the exact transport properties, instead of the simplified model adopted in the present study. The involved numerical calculation should become very large. However, it is still within a range of possibility, and we have done this kind of calculations, and the result will be described later.

2.4.8.2   Characteristics of Tubular Flame

The tubular flame studied in the present study provides another example of the flame established in a nonuniform flow. The flow field can be solved exactly for the physically realistic boundary conditions. We may study the flame behavior in terms of the physically realistic parameters, which makes it possible to compare the theoretical prediction with that observed for the actual flame. In that sense the tubular flame has much the same characteristics with the flames stabilized in the stagnation flow.[3] In addition, the structure and stability of the flame, especially those of the flame stabilized near the axis, appear very similar to the stagnation flames. This should be a consequence of the fact that the both flow configurations happen to have the common character that the velocity normal to the flame decreases almost linearly toward the axis or the stagnation. However, there is some difference in the flame behavior, as well as in the flow characteristics. In the tubular flame, the flame behavior through the whole flow field, from the axis to the wall, can be studied, in contrast to the stagnation flame, where the study has to be restricted to a rather narrow region near the stagnation. This should be a consequence of the fact that the stagnation flow field has to be modified as it leaves from the stagnation. Then the more wide range of the flame behavior can be predicted for a single flow field, and we may elucidate flame characteristics, such as the stretch and the curvature, by comparing the predicted results with those observed in the experiment. In this respect the unusual flame behavior near the wall is very interesting in relation to the unique radial velocity profile. This unique flame behavior should be another subject of future study.

The most important feature of the flow field studied in the present study is that it is accompanied by the circumferential flow. The present study is based on the axisymmetric similarity solution, and the result has revealed that the flame behavior has nothing to do with this circumferential velocity. Figure 2.3 shows the Rankine combined vortex characteristics. The Rayleigh criterion predicts that the flow is stable in the core and is neutrally stable outside.[13] This is the reason that we have been so keen about the flame position relative to the core vortex. The present analysis is based on the constant density flow, and the actual density change produced by the flame may be favorable for the stability and may extend the stable flame region. Therefore, we think that the central flow field in the axisymmetric tubular flame should be very stable, and is most likely to be realized. This will explain why in the experiment the axisymmetric flame could easily be established in spite of the fact that the mixture introduction into the burner is far from axisymmetric. As soon as the mixture is injected through the several slits the local flow is induced to establish the axisymmetric flow field in the central region. In other words, we may say that the role of circumferential flow is to produce the axisymmetric flow in the central region against the nonsymmetric injection. In the present study the axisymmetric introduction is assumed and the axisymmetric flow field is established automatically and then the flame behavior becomes independent of the circumferential flow.

The discussions developed so far remain speculations. Further studies should be made to make these uncertainties clear.

2.5   EFFECTS OF VARIABLE DENSITY

2.5.1   MODEL AND ASSUMPTIONS

In the foregoing section, the tubular flame has been analyzed in terms of the simplified model with the overall one-step chemical kinetics and the simplified constant transport properties. In addition we have introduced the additional assumption that the flow is incompressible and the density remains constant. In the actual flame, however, the density will change throughout the flow field, since the temperature is greatly increased due to the heat release. This density change will produce the flow expansion, which may affect the flame behavior. The accelerated transport properties with the increased temperature may also affect the flame behavior. In this section we will study the effects of this temperature rise on the flame behavior in the simplified model.

In this section, we will remove the assumptions 1.8 and 1.9 of the previous section. However, as described before, the three transport parameters, that is, Lewis number Le, Prandtl number Pr, and Schmidt number Sc, instead of the transport properties themselves, remain constant. The other assumptions are basically identical with those adopted in the previous section. The same nondimensional system is adopted in this section either. Now the similarity solution yields the equations and boundary conditions to be solved for the nondimensional radial velocity f and the density ratio are Eqs (1.44) through (1.49) described in the previous section.

The flame experiences the stretch K in the axial direction.[14] The same expression with that of the incompressible flow can be applied to the present compressible flow case as well:

$$K = \frac{\delta}{S_u} \frac{\partial u}{\partial x} = \frac{2 v_R \kappa_R}{R^2 S_u^2} \lambda U = \frac{\beta^2 e^\beta}{2 BaLe} \left(\frac{\theta_a - 1}{\theta_a} \right)^2 \lambda U \qquad (2.82)$$

where S_u and δ represent the normal burning velocity and the representative flame thickness of the normal flame. $v_R \equiv \mu_R / \rho_R$ is the kinematic viscosity, while $\kappa_R \equiv \tilde{\lambda}_R / \rho_R c_p$ is the thermal diffusivity of the mixture at the wall. The nondimensional parameters in the above expression are the adiabatic flame temperature θ_a, Zeldovich number β, and a constant defined by[14]

$$\theta_a = \frac{T_{ba}}{T_R}, \qquad \beta = \frac{E_n}{\theta_a}, \qquad a = \frac{j|1-\varphi|}{\varphi + j(i+1)} \qquad (2.83)$$

This expression reveals that for the similarity solution the flame stretch becomes independent of the axial distance. It depends on the radial distance through U and is proportional to the product of the injection Reynolds number and the nondimensional axial velocity.

2.5.2 COMPARISON WITH INCOMPRESSIBLE SOLUTIONS

In the following section, the obtained flow and flame characteristics are compared with those of the incompressible flow solution for the case when Le = 1. Figure 2.16 shows the response curve of the flame temperature θ_0, which is the maximum temperature on the axis, plotted against the injection Reynolds number λ. For both the incompressible and compressible flow solutions, there exist simultaneously three – the upper, the middle, and the lower – solutions for a given value of λ smaller than a critical value λ_e. The upper solution has been considered to correspond to the actual flame observed in the experiment.[4,11,15] As λ is increased along the upper solution, θ_0 decreases gradually and the flame causes extinction at the critical injection Reynolds number λ_e. As seen in the figure, λ_e is some 30% smaller for the compressible flow, indicating that the flame will cause the extinction at 30% smaller injection velocity than that predicted by the incompressible flow solution. It is also noted that the decrease of the flame temperature as the extinction is approached is a little bit larger for the compressible flow solution.

Figure 2.17 shows the comparison of the flame structure of the upper solution for the case when $\lambda = 60$. The profiles of the temperature θ, the fuel mass fraction Y_F, and the heat release rate Q_R are compared. ξ_R designates the position where Q_R becomes the maximum for the compressible flow solution. As seen in the figure, the location of the reaction zone is shifted outside in the compressible flow. Furthermore, the reaction zone width is largely extended accompanied by a considerable decrease in the temperature and concentration gradients. This can be attributed to the accelerated transport processes at the increased temperature combined with the accelerated radial velocity described below. An examination of the structures revealed that the former contribution is more significant. The extended reaction zone provides an increase in the unburned fuel concentration on the axis. Figure 2.18 compares the unburned fuel mass fraction $Y_{F,0}$ remaining on the axis plotted against λ. As λ is increased along the upper solution, $Y_{F,0}$ increases very rapidly to produce the unburned fuel on the axis. The increase in the vicinity

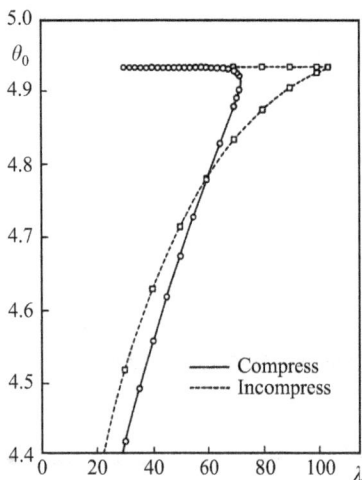

Figure 2.16. Flame temperature plotted against injection Reynolds number λ: Comparison between compressible and incompressible flow solutions.

of λ_e is especially large for the compressible flow solution, which will explain the gradual decrease of the flame temperature observed in Fig. 2.16. Figure 2.19 compares the velocity distributions for $\lambda = 60$ and reveals a remarkable effect of the flow expansion on the flow field. The axial velocity U begins to be accelerated appreciably at the leading edge of preheat zone and takes a maximum value on the axis. The radial velocity V starts to increase after the mixture is injected from the wall reaching a small maximum just apart from the wall and then starts to decrease toward axis, and then starts to increase again at the edge of preheat zone. And then reaches another maximum at around ξ_R; then decreases again toward axis almost linearly. These velocity profiles are quite similar to those observed for the plane stagnation flame.[16] On the other hand, the circumferential velocity h is decelerated considerably because of the increased viscosity and decreased density of the burned gas, giving about half maximum velocity. Furthermore, the position of the maximum ξ_R is shifted outside to extend the forced vortex region in the core of the Rankine combined vortex. As seen in the figure, the compressible flow makes the combined vortex characteristics less apparent.

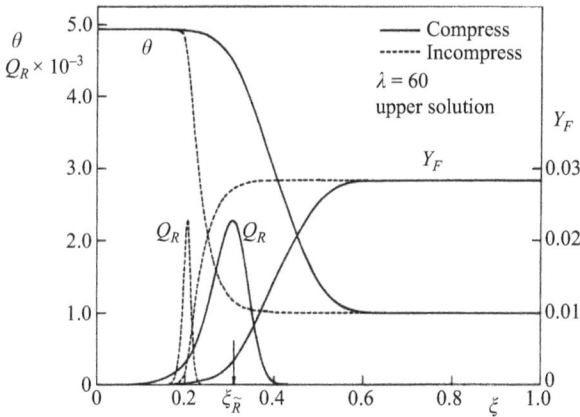

Figure 2.17. Temperature, fuel mass fraction, and heat release rate distributions: Comparison between compressible and incompressible flow solutions.

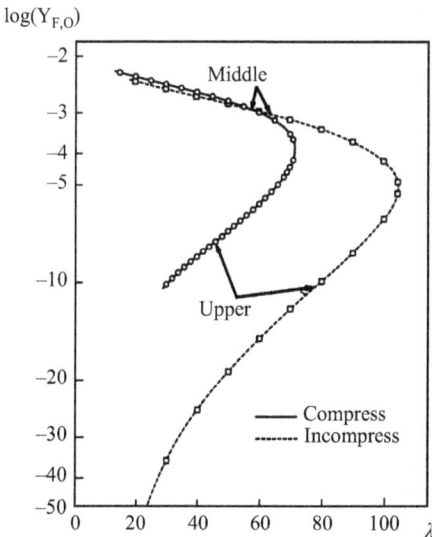

Figure 2.18. Fuel mass fraction remaining on the axis plotted against injection Reynolds number λ: Comparison between compressible and incompressible flow solutions.

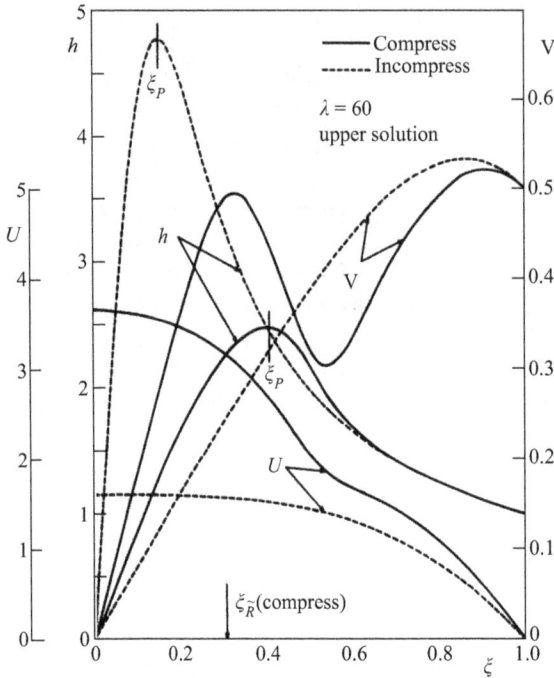

Figure 2.19. Flow field in the tubular flame: Comparison between compressible and incompressible flow solutions.

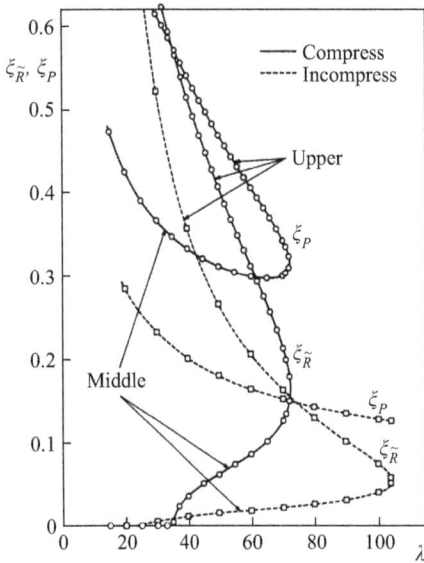

Figure 2.20. Representative reaction zone position ξ_R and the outer boundary of core vortex region ξ_p plotted against injection Reynolds number λ: Comparison between compressible and incompressible flows.

In Fig. 2.20, the outer boundary of forced vortex region, represented by ξ_p and the representative reaction zone position ξ_R are plotted against λ. As seen in the figure, ξ_p is largely separated each other for the upper and the middle solutions in the compressible flow. This may be a consequence of the fact that now the temperature distribution is considerably different, and the flow filed is closely coupled with the temperature. In the compressible flow both the outer boundary and the reaction are shifted outside, but the shift of the outer boundary is larger. As a

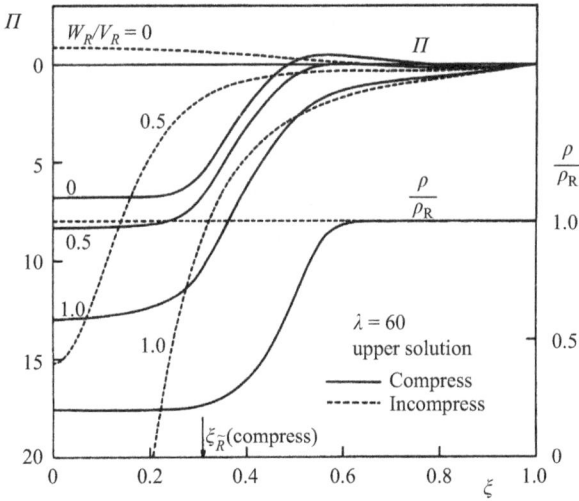

Figure 2.21. Radial distribution of density and pressure: Comparison between compressible and incompressible flows.

consequence, for the compressible flow the reaction zone is always situated outside the forced vortex region for any λ larger than 35. It should be noticed that the extinction flame radius becomes less than half of the incompressible flow.

Figure 2.21 compares the radial pressure distribution for three velocity ratios $w_R/v_R = 0$, 0.5, and 1.0. The density distribution is also shown. In the incompressible flow, the density remains constant, whereas in the compressible flow it begins to decrease at the edge of the preheat zone and takes a minimum value on the axis. When there is no rotating flow, the pressure first decreases near the wall because of the small increase in the radial velocity shown in Fig. 2.19. Then it increases monotonically toward the axis for the incompressible flow, in accordance with the monotonic decrease in the radial velocity. In the compressible flow, on the other hand, it takes a gradual maximum somewhere around the edge of preheat zone, and then decreases toward the axis to accelerate the flow. When the rotating flow is introduced, the centrifugal force contributes to produce the pressure drop in the core, and the drop is very rapid for the incompressible flow. The lighter density and the decelerated rotating velocity of the burned gas suppress this rapid pressure decrease in the compressible flow.

2.5.3 EFFECTS OF INJECTION VELOCITY

Figures 2.22 through 2.26 show effects of the injection Reynolds number λ, or the injection velocity, on the flame structure and the flow field of the upper solution for the case when Le = 1. Figure 2.22 shows the change in the temperature and heat release rate profiles. As λ is increased, the flame is shifted almost parallel toward the axis, keeping the similar reaction rate profile. The behavior is very similar to that observed for the incompressible flow shown in Fig. 2.9, except for the extended and outward shifted reaction zone position mentioned before.

Figures 2.23 and 2.24 show the changes in the axial and radial velocity, respectively. In the ordinate λU and λV, instead of U and V, are taken, since v_R is contained in the definition of the latter. In the compressible flow, the flame and the flow are closely coupled, and the flow expansion due to the heat release accelerates the flow in accordance with the flame movement shown

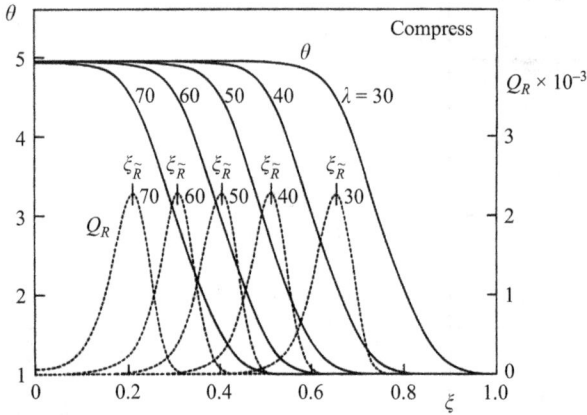

Figure 2.22. Variation with the injection Reynolds number of temperature and heat release rate profiles.

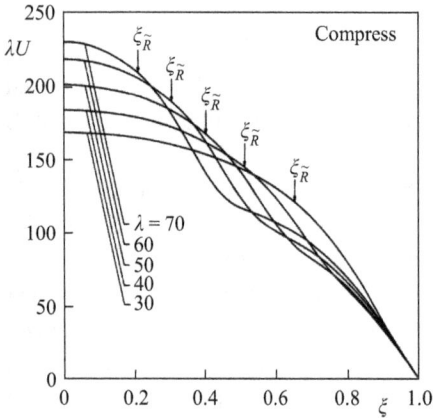

Figure 2.23. Variation with injection Reynolds number of an axial velocity profile.

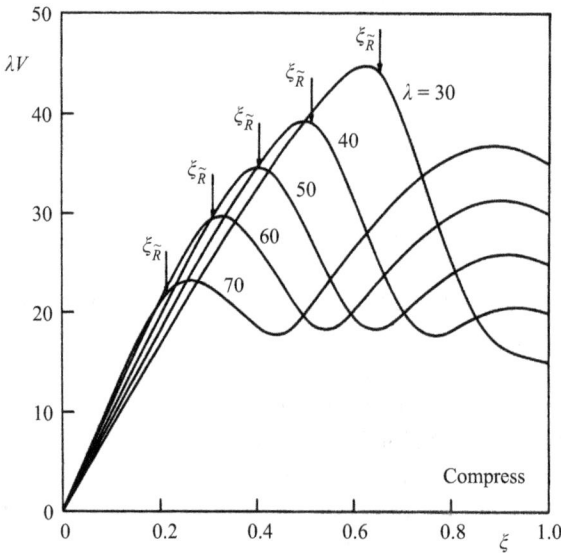

Figure 2.24. Variation with injection Reynolds number of a radial velocity profile.

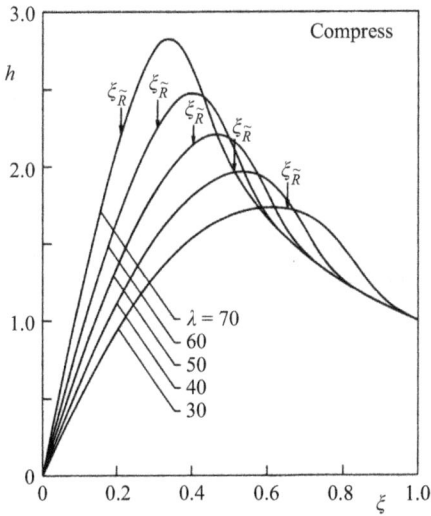

Figure 2.25. Variation with injection Reynolds number of a circumferential velocity profile.

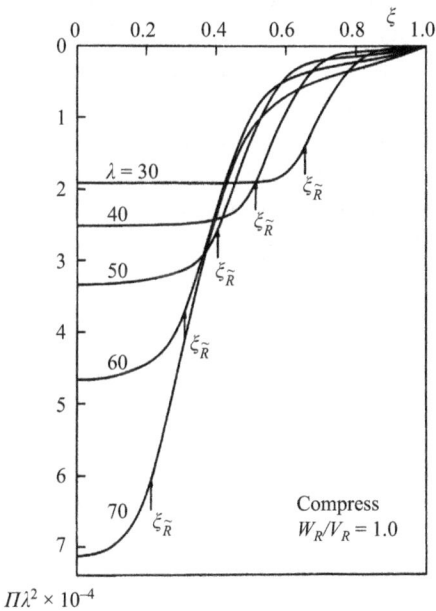

Figure 2.26. Variation with injection Reynolds number of a radial pressure profile.

in Fig. 2.22. When λ is small and the edge of preheat zone is situated near the wall, the flow is accelerated mainly in the radial direction, since the flow in the axial direction is restricted at the wall. This produces a large peak of the radial velocity around the reaction zone, whereas the axial velocity increases monotonically toward the axis. As λ is increased and the edge is shifted inside, the axial acceleration becomes important as compared to the radial acceleration, since the latter is so restricted as to give zero radial velocity on the axis. The peak radial velocity becomes smaller, whereas the maximum axial velocity on the axis becomes larger. It should be noticed that λU shown in Fig. 2.23 is directly proportional to the flame stretch given by Eq. (2.76). As seen in the figure, the stretch at the reaction zone position increases with λ.

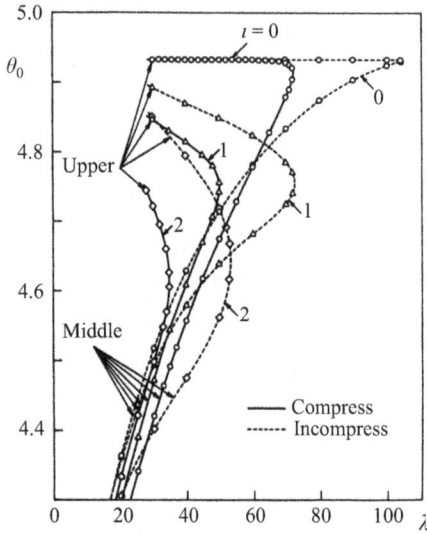

Figure 2.27. Flame temperature plotted against injection Reynolds number for Le ≥ 1.

Figure 2.25 shows the changes in the circumferential velocity distribution. When λ is small, the temperature increase near the wall produces an increase in viscosity in this region, which will contribute to destroy the free vortex flow characteristics of the Rankine combined vortex. When λ is large, on the other hand, the increase of viscosity in the core region, where the viscosity plays a role originally, hardly affects the overall flow characteristics to keep the combined vortex flow. The change in the pressure distribution is seen in Fig. 2.26 for the case when the velocity ratio is kept at unity. The low pressure region in the core becomes narrower as λ is increased and the flame is shifted inside. The accompanying increase in the centrifugal force provides a large pressure drop in the core. It should be noticed that the reaction zone is always situated at the position where the pressure gradient is rather large.

2.5.4 EFFECTS OF LEWIS NUMBER

Now we will study the effects of Lewis number on the flame behavior for the case when the deviation from unity is order of $1/\beta$, where β is Zeldovich number defined in Eq. (2.83). We will introduce Lewis number parameter l by

$$\frac{1}{\text{Le}} = 1 - \frac{l}{\beta} \qquad (2.84)$$

and we will adopt numerical values -2, -1, 0, 1, and 2 for l to study the effects of Lewis number. The corresponding values of Lewis number are 0.836112, 0.910742, 1.0, 1.108655, and 1.243802. Figure 2.27 shows the flame temperature θ_0 plotted against λ for three values of Le ≥ 1. The flame temperature decreases from the adiabatic flame temperature with λ along the upper solution for $l > 0$. This is due to an increase in the flame stretch, defined by Eq. (2.82). The behavior for respective values of l, and their dependence on l, is similar to those of the incompressible flow. However, the decrease of the flame temperature is much larger. That is, the stretch effect is stronger in the compressible flow for the same value of λ. The extinction Reynolds numbers λ_e are some 30% smaller than those of the incompressible flow for these Lewis numbers. The

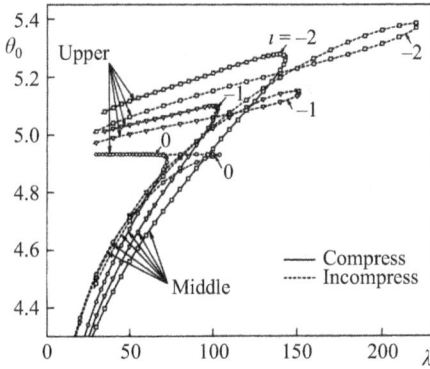

Figure 2.28. Flame temperature plotted against injection Reynolds number for Le ≤ 1.

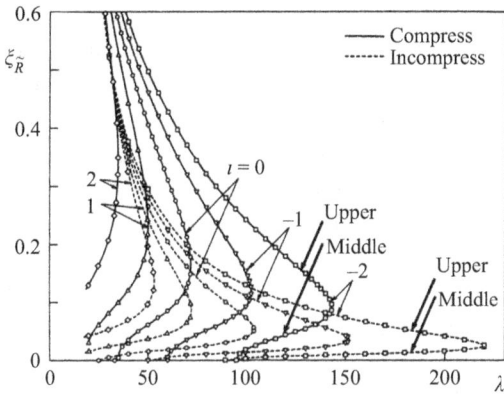

Figure 2.29. Reaction zone position plotted against injection Reynolds number for five different Lewis numbers.

response curves for the case when Le ≤ 1 are compared in Fig. 2.28. When Le is <1, the flame temperature increases with λ, or with the stretch. The increase is monotonic, and no loop is generated as in the incompressible flow case. In the latter case, the two opposing effects of a decrease in λ on the flame temperature along the middle solution will produce the loop: the increase in the stretch and the increase of the unburned mixture. In the compressible flow, the latter effect is always predominant to give the monotonic decrease of flame temperature as λ is decreased along the middle solution. The increase of flame temperature along the upper solution is larger than that of the incompressible flow, suggesting the stronger effect of stretch here again. The critical injection Reynolds number is some 30–40% smaller than that of the incompressible flow. In Fig. 2.29 the reaction zone position ξ_R is plotted against λ for the five values of l. The general trend is similar to that of the incompressible flow. However, the compressible flow predicts the smaller extinction Reynolds number, and the reaction zone position of the upper solution is shifted outside, except for the case $l = 2$, resulting in a larger extinction flame radius.

2.5.5 DISCUSSIONS ON THE EFFECTS OF VARIABLE DENSITY

2.5.5.1 Effects of Stretch on Flame Temperature

The foregoing results have revealed that the flame and the flow field are strongly coupled in the compressible flow. The flame structure and stability are affected by the flow, whereas the flow field is strongly affected by the flame. Figures 2.27 and 2.28 show that the flame will cause the extinction at 30–40% smaller injection velocity than that predicted for the incompressible flow. This can be attributed to the increase in the flame stretch caused by the flow expansion. That is, in the compressible flow the actual stretch operating on the flame is increased for the same injection velocity because of the axial flow acceleration. Equation (2.82) and Fig. 2.19 indicate that the stretch is increased some two times. The ambiguity will arise because the stretch depends on the radial position, and in order to make a quantitative assessment of the stretch effect we have to locate the radial position where we should evaluate the stretch. Although there should be several possibilities, we adopt here the position where heat release rate becomes maximum, that is, ξ_R. Now we can evaluate the stretch by substituting into Eq. (2.82) the value of U at ξ_R.

Figure 2.30 shows the response curve of the upper solution: the flame temperature plotted against the stretch defined in this way, instead of λ of Fig. 2.28. The results for the incompressible flow and for the asymptotic analysis described in the next section are shown as well for comparison. Please note that ξ_R for the incompressible flow is situated inside that of the compressible flow (Fig. 2.29). It is interesting to note that the correlation between the incompressible flow and the asymptotic analysis is appreciably improved by taking the stretch in the abscissa instead of the injection Reynolds number.[14] The reason for this improvement requires some delicate discussion on the velocity gradient adopted in the potential flow. However, we are mostly concerned here with the variable density effect and shall concentrate on the comparison between the compressible and the incompressible flow. When we compare the response curves in this figure with those shown in Figs 2.27 and 2.28, we note two important distinctions. First is the critical flame stretch K_e the flame causes the extinction. K_e of the compressible flow is always larger than that of the incompressible flow, in contrast to λ_e, which is smaller for the compressible flow. The increase of K due to the flow expansion is so large as to bring about this reversal. Second is the effect of the stretch on the flame temperature of the upper solution. For the same value of K the compressible flow predicts the lower flame temperature Le < 1 and the higher one for Le > 1. This is very opposite to the case when the comparison is made for the same value of λ. This reversal is another consequence of the increase in K due to the flow expansion. In the following sections we shall discuss the physical reason for these reversals.

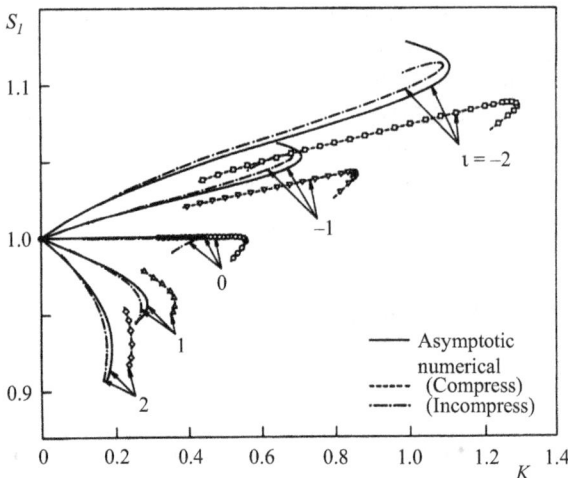

Figure 2.30. Flame temperature plotted against flame stretch for upper solution: Comparison between asymptotic analysis and compressible and incompressible numerical solutions.

The response curves shown in Fig. 2.30 suggest that the stretch is less effective in the compressible flow for the same value of K. This should be closely related to the changes in the flame structure shown in Fig. 2.17. The reaction zone width is considerably extended in the compressible flow because of the accelerated transport processes. The extended reaction zone will reduce the average velocity gradient over the reaction zone, and thus will eventually weaken the stretch which actually operates on the flame. In other words, for a given increase in flame area the stay time of a reacting gas particle in the reaction zone is increased, in spite of the accelerated normal flow velocity, which should contribute to reduce the stretch effect. Then the deviation of the flame temperature from the adiabatic flame temperature should become smaller for the compressible flow for the same values of K. The flame, therefore, can sustain reaction stably up to a larger value K giving the higher extinction stretch K_e.5

2.5.5.2 Mass Burning Velocity

Another important flame property to be discussed is the burning velocity. In the stretched flame, in general, the velocity normal too the flame changes with distance and hence the burning velocity comes to depend on the position where we take the flame front.[1,17] In our study of the incompressible flow the burning velocity is defined at the position where the heat release rate becomes maximum.[14] The definition is consistent with that of the flame stretch introduced in this section, and we may adopt this definition in the compressible flow as well. However, as explained in Figs 2.23 and 2.24, the radial velocity profile is a consequence of the complicated 3D flow acceleration under the given restrictions, and we may doubt that this definition would give a certain physically significant quantity. On the other hand, when the mass flux ρv is plotted against ξ, it remains almost linear through the reaction zone for all values of λ. Therefore, we will adopt here the mass flux at ξ_R as the mass burning velocity.

Figure 2.31 shows the response curve of the mass burning velocity divided by the mass burning velocity of the normal flame plotted against the flame stretch for five values of Lewis number parameters l. As seen in the figure, the agreement between the asymptotic analysis and the incompressible flow is largely improved in this response curve here again, by taking K as the abscissa instead of λ. Although these two solutions predict somewhat a smaller extinction

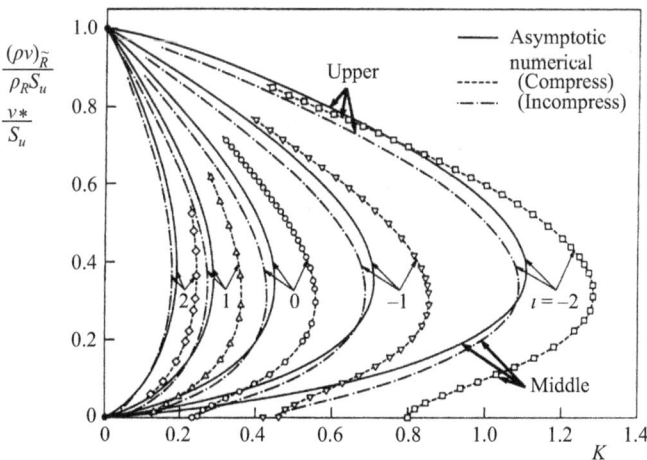

Figure 2.31. Reduced mass burning velocity plotted against flame stretch: Comparison between asymptotic analysis and compressible and incompressible numerical solutions.

stretch than that of the compressible flow, the curve derived by the asymptotic analysis predicts the correct general trend. In fact, we may safely say that that the qualitative behavior, especially the dependence on Lewis number, can well be predicted in terms of the asymptotic analysis based on the incompressible flow.

The above agreement is rather surprising when we consider the remarkable change induced in the flow field. As discussed before, the increase of the stretch due to the flow expansion is offset by the increase of the reaction zone width to bring about a better agreement. Although this is a result for the specific mixture and the apparent agreement may be coincidental, the above explanation seems quite general. In view of the similar agreement observed for the plane stagnation flame,[18] we may hope that the agreement should be a universal one, suggesting the usefulness of the asymptotic analysis.

2.6 ASYMPTOTIC ANALYSIS

2.6.1 MODEL AND ASSUMPTIONS

In the foregoing section, we have studied the tubular flame in terms of the similarity solution and the simplified model with the overall one-step chemical kinetics and the simplified transport properties. The results have shown that the similarity solution can predict the observed extinction behavior correctly. Now we think that we have understood the physics and chemistry of the tubular flame. On the other hand, we are concerned mostly in the extinction behavior, and we want to develop a simple analysis to predict the extinction. This is the main target of this section, and we will find out the parameter which governs the extinction behavior. To develop the asymptotic analysis, we will adopt the following assumptions described before

6. The mixture undergoes an overall one-step chemical reaction described by

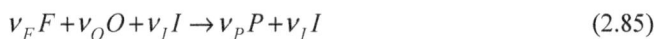

$$v_F F + v_O O + v_I I \rightarrow v_P P + v_I I \tag{2.85}$$

7. The flow is incompressible and the density remains constant through the whole flow field.
8. The transport properties are all common to each species and constant.

In addition, we further assume that

9. The chemical reactions proceed in an infinitesimally thin reaction surface.

The analysis is based on the flame surface model with the incompressible and nonviscous flow field, given by Eq. (2.62).

2.6.2 NONDIMENSIONAL SYSTEM

We will adopt the same nondimensional system as before. The flow field is given by Eq. (2.62). Now the equations and boundary conditions to be solved are Eqs (2.57) through (2.59). The mass fractions at the wall are given in terms of φ, i, j as

$$Y_{F,R} = \frac{\varphi}{\varphi + j(i+1)}, \qquad Y_{O,R} = \frac{j}{\varphi + j(i+1)} \tag{2.86}$$

while the adiabatic flame temperature θ_a is given by

$$\theta_a = 1 + qY_{F,R} \quad (0 \le \varphi \le 1), \qquad \theta_a = 1 + \frac{q}{j}Y_{O,R} \quad (1 \le \varphi) \tag{2.87}$$

The total enthalpy ψ is

$$\psi = \theta + qY_F \quad (0 \le \varphi \le 1), \qquad \psi = 1 + \frac{q}{j}Y_O \quad (1 \le \varphi) \tag{2.88}$$

The flame experience the flame stretch K in the axial direction given by Eq. (2.82). The equation reveals that K is proportional to U, as well as to λ, which means that in the flow field under consideration, it becomes maximum on the axis and decreases outward, reaching zero at the wall.

2.6.3 ASYMPTOTIC ANALYSIS

Now the two-point boundary value problem Eqs (2.44) through (2.50) will be solved by asymptotic analysis for the case when $1 - \varphi = O(1)$, that is, for lean or rich mixtures.

$$\beta = \frac{E_n}{\theta_a}, \qquad \frac{1}{Le} = 1 - \frac{l}{\beta} \tag{2.89}$$

2.6.3.1 Outer Solutions

θ and ψ are expanded as

$$\theta(\xi) = \theta_0(\xi) - \frac{1}{\beta}\theta_1(\xi), \qquad \psi(\xi) = \psi_0(\xi) - \frac{1}{\beta}\psi_1(\xi) \tag{2.90}$$

The equations and boundary conditions for $\theta_0(\xi)$, $\psi_0(\xi)$, and $\psi_1(\xi)$ reduce to

$$\frac{d}{d\xi}\left(\xi\frac{d\theta_0}{d\xi}\right) + 2\Pr\lambda f\frac{d\theta_0}{d\xi} = 0, \qquad \theta_0'(0) = 0, \qquad \theta_0(1) = 1 \tag{2.91}$$

$$\frac{d}{d\xi}\left(\xi\frac{d\psi_0}{d\xi}\right) + 2\Pr\lambda f\frac{d\psi_0}{d\xi} = 0, \qquad \psi_0'(0) = 0, \qquad \psi_0(1) = \theta_a \tag{2.92}$$

$$\frac{d}{d\xi}\left(\xi\frac{d\psi_1}{d\xi}\right) + 2\Pr\lambda f\frac{d\psi_1}{d\xi} = 0, \quad \psi_1'(0) = -l\frac{d}{d\xi}\left(\xi\frac{d\theta_0}{d\xi}\right), \quad \psi_1'(1) = \psi_1(1) = 0 \tag{2.93}$$

The solution for this system is divided into two regions by the flame surface position ξ_* :

$$0 \le \xi \le \xi_* : \quad \psi_0(\xi) = \theta_0(\xi) = \theta_a = \text{constant}, \qquad \psi_1(\xi) = \psi_f = \text{constant} \qquad (2.94)$$

$$\xi_* \le \xi \le 1 : \quad \psi_0(\xi) = \theta_a = \text{constant} \qquad (2.95)$$

$$\theta_0(\xi) = \theta_a - (\theta_a - 1) I(\xi) / I(1) \qquad (2.96)$$

$$\psi_1(\xi) = \psi_f \left\{ 1 - I(\xi) / I(1) \right\} + AI(\xi) \int_\xi^1 \frac{f}{\xi} d\xi$$
$$+ A \left\{ \int_{\xi_*}^\xi \frac{fI(\xi)}{\xi} d\xi - \frac{I(\xi)}{I(1)} \int_{\xi_*}^1 \frac{fI(\xi)}{\xi} d\xi \right\} \qquad (2.97)$$

where

$$I(\xi) = \int_{\xi_*}^\xi \frac{1}{\xi} \exp \left[-2\Pr\lambda \int_{\xi_*}^\xi \frac{f}{\xi} d\xi \right] d\xi, \quad I(\xi) = \int_{\xi_*}^1 \frac{1}{\xi} \exp \left[-2\Pr\lambda \int_{\xi_*}^\xi \frac{f}{\xi} d\xi \right] d\xi \quad (2.98)$$

and a constant A is given by

$$A = 2\Pr\lambda l (\theta_a - 1) / I(1) \qquad (2.99)$$

2.6.3.2 Inner Solutions

The extended space coordinate η is introduced by

$$\eta = \beta(\xi - \xi_*) \qquad (2.100)$$

Then θ and ψ are expanded as

$$\theta(\eta) = \theta_a - \frac{1}{\beta} t_1(\eta), \qquad \psi(\eta) = \theta_a + \frac{1}{\beta} \phi_1(\eta) \qquad (2.101)$$

The equations for $t_1(\eta)$ and $\phi_1(\eta)$ reduce to

$$\frac{d^2}{d\eta^2}(t_1 + \phi_1) - \Lambda(t_1 + \phi_1) \exp \left(-\frac{t_1}{\theta_a} \right) = 0 \qquad (2.102)$$

$$\frac{d^2 t_1}{d\eta^2} - \Lambda(t_1 + \phi_1) \exp \left(-\frac{t_1}{\theta_a} \right) = 0 \qquad (2.103)$$

where constant Λ and a are given by

$$\Lambda = 2\Pr Ba/(\beta^2 e^\beta), \qquad a = \left|Y_{O,R} - jY_{F,R}\right| = \frac{j|1-\varphi|}{\varphi + j(i+1)} \tag{2.104}$$

The boundary conditions are

$$\eta \to -\infty: \quad t_1 + \phi_1 \to 0, \quad \phi_1 \to \psi_f, \quad \frac{d}{d\eta}(t_1 + \phi_1) \to 0, \quad \frac{d\phi_1}{d\eta} \to 0 \tag{2.105}$$

Then we get the solution as

$$\phi_1(\eta) = \psi_f = \text{constant}$$

$$\left(\frac{dt_1}{d\eta}\right)^2 = 2\Lambda\theta_a^2 \exp\left(\frac{\psi_f}{\theta_a}\right)\left\{1 - \frac{(1+\psi_f+\theta_a)}{\theta_a}\exp\left(-\frac{t_1+\psi_f}{\theta_a}\right)\right\} \tag{2.106}$$

Then we have

$$\left(\frac{dt_1}{d\eta}\right)_{\eta \to +\infty} = \sqrt{2\Lambda}\,\theta_a \exp\left(\frac{\psi_f}{2\theta_a}\right) \tag{2.107}$$

2.6.3.3 Matching Conditions

The first matching condition at the flame surface ξ_* is derived through Eqs (2.90), (2.100), and (2.101) as

$$\left(\frac{d\theta_0}{d\xi}\right)_{\xi_*+} = -\left(\frac{dt_1}{d\eta}\right)_{\eta \to +\infty} \tag{2.108}$$

The integration of Eq. (2.93) over the reaction zone width yields the second condition:

$$\left[\frac{d\psi_1}{d\xi}\right]^+ = -l\left[\frac{d\theta_0}{d\xi}\right]^+ \tag{2.109}$$

Then Eqs (2.96), (2.98), (2.107), and (2.108) can be used to yield

$$\left(\frac{\theta_a}{\theta_a-1}\right)\sqrt{2\Lambda}\exp\left(\frac{\psi_f}{2\theta_a}\right) = \frac{1}{\xi_* I(1)} \tag{2.110}$$

On the other hand, Eqs (2.96), (2.97), (2.99), and (2.109) yield

$$1 + \frac{\psi_f}{2\theta_a\gamma} = \frac{2\Pr\lambda}{I(1)}\int_{\xi_*}^1 \frac{f}{\xi}\int_\xi^1 \frac{1}{\xi}\exp\left[-2\Pr\lambda\int_\xi^\xi \frac{f}{\xi}d\xi\right]d\xi d\xi \tag{2.111}$$

where a constant γ is defined as

$$\gamma = l\left(\theta_a - 1\right)/\left(2\theta_a\right) \tag{2.112}$$

We can make use of Eqs (2.101) and (2.105) to obtain the flame temperature θ_* as

$$\theta_* = \theta_a + \psi_f/\beta \tag{2.113}$$

Reference to Eqs (2.88), (2.90), (2.94), and (2.91) reveals that the second term represents the contribution of the excess enthalpy produced by the unbalance in heat and mass diffusion rate. Then Eqs (2.110) and (2.112) constitute the simultaneous equations for determining the flame surface position ξ_* and θ_*.

2.6.4 APPROXIMATE SOLUTIONS

In the extinction problem under study we are concerned with rather large values of λ (> 10), and in that case the flow field can be described accurately by the simple nonviscous solution given below. In addition, we are concerned with small values of ξ and may adopt the following approximation.

$$f = \frac{1}{2}\sin\left(\frac{\pi}{2}\xi^2\right) \cong \frac{\pi}{4}\xi^2 \tag{2.114}$$

This velocity gives the linear radial velocity profile. Then we introduce a new space coordinate-z by

$$z = \frac{\pi \operatorname{Pr}\lambda}{4}\xi^2 \tag{2.115}$$

When Eqs (2.114) and (2.115) are substituted into Eq. (2.98), we obtain

$$I(1) = \exp\left(z_*\right)J\left(z_*\right)/2, \qquad z_* = \frac{\pi \operatorname{Pr}\lambda}{4}\xi_*^2 \tag{2.116}$$

where

$$J\left(z_*\right) \equiv \int_{z_*}^{\operatorname{Pr}\lambda/4}\frac{e^{-z}}{z}dz = \int_{z_*}^{\infty}\frac{e^{-z}}{z}dz - \int_{\operatorname{Pr}\lambda/4}^{\infty}\frac{e^{-z}}{z}dz \tag{2.117}$$

Since λ is large enough, we may adopt the following approximation

$$J\left(z_*\right) \cong \int_{z_*}^{\infty}\frac{e^{-z}}{z}dz = -E_i\left(-z_*\right), \qquad \exp\left(-z_*\right) - \exp\left(-\frac{\pi \operatorname{Pr}\lambda}{4}\right) \cong \exp\left(-z_*\right) \tag{2.118}$$

where $E_i(-z_*)$ is the exponential integral. Then Eqs (2.110) and (2.111) reduce to

$$\left(\frac{\theta_a}{\theta_a-1}\right)\sqrt{\frac{2\Lambda}{\pi\,\mathrm{Pr}\,\lambda}}\exp\left(\frac{\psi_f}{2\theta_a}\right)=\frac{\exp(-z_*)}{\sqrt{z_*\{-E_i(-z_*)\}}}\equiv F(z_*),$$

$$1+\frac{\psi_f}{2\theta_a\gamma}=-z_*+\frac{\exp(-z_*)}{\{-E_i(-z_*)\}}\equiv G(z_*)$$

(2.119)

where F and G are functions of z_* alone. Figure 2.32 shows F and G plotted against z_*. F has a minimum at $z_* = z_c$, while G increases monotonically with z_*. The calculation shows

$$z_c = 0.258951, \qquad F_c = 1.49144, \qquad G_c = 0.5.$$

(2.120)

2.6.5   RESPONSE CURVES

Now Eq. (2.114) can be used to express q_* etc., in terms of z_*, and the various response curves can be obtained. Eqs (2.115) and (2.116) yield

$$\theta_*(z_*) = \theta_a - \frac{2\theta_a\gamma}{\beta}(1-G)$$

(2.121)

while Eqs (2.115) and (2.119) give

$$\lambda(z_*) = \frac{2\Lambda}{\pi\,\mathrm{Pr}}\left(\frac{\theta_a-1}{\theta_a}\right)^2\frac{1}{F^2}\exp\left[-2\gamma(1-G)\right]$$

(2.122)

Equations (2.115) and (2.122) give

$$\xi_*(z_*) = \sqrt{\frac{2}{\Lambda}}\left(\frac{\theta_a-1}{\theta_a}\right)\sqrt{z_*}\,F\exp\left[\gamma(1-G)\right]$$

(2.123)

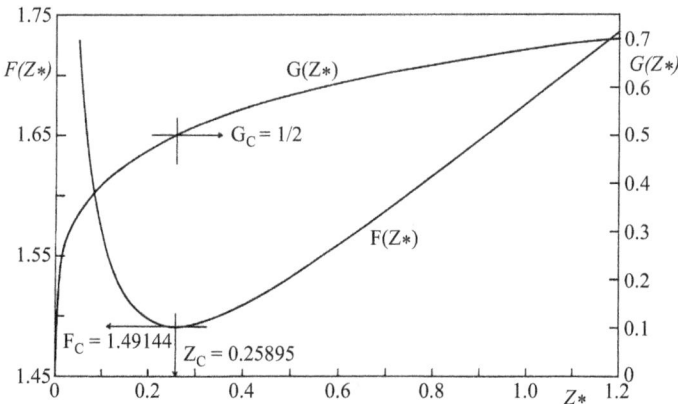

Figure 2.32. Functions $F(z_*)$ and $G(z_*)$.

The radial velocity v_* at the flame surface ξ_* will represent the apparent burning velocity of the flame (note the flow is incompressible). If the asymptotic solution for the planar normal flame of Le = 1 is used, S_u can be expressed in terms of the present nondimensional system as

$$S_u = \frac{\kappa}{R}\left(\frac{\theta_a}{\theta_a - 1}\right)\sqrt{2\Lambda} \qquad (2.124)$$

Then we make use of Eqs. (2.37), (2.114), (2.115), (2.122), and (2.124) to obtain

$$-\frac{v_*}{S_u} = \frac{\sqrt{z_*}}{F}\exp\left[-2\gamma(1-G)\right] \qquad (2.125)$$

The linear radial velocity approximation Eq. (2.114) yields the constant axial velocity

$$U = \frac{1}{\xi}\frac{df}{d\xi} \cong \frac{\pi}{2} \qquad (2.126)$$

and substitution of this into Eq. (2.82) yields

$$K(z_*) = \frac{1}{F^2}\exp\left[-2\gamma(1-G)\right] \qquad (2.127)$$

where Eqs (2.122) and (2.124) are used. Now the above equations make it possible to describe the response curves $\xi_* \sim \lambda$, $\theta_* \sim \lambda$, $-v_*/S_u \sim \lambda$, and $K \sim \lambda$. Equations (2.125) and (2.127) reveal that the response curve $-v_*/S_u \sim K$ contains only one parameter γ, which represents the effects of Le.

2.6.6 EXTINCTION CONDITIONS

In the response curve $\xi_* \sim \lambda$, the critical condition for the flame extinction is given by

$$\frac{d\lambda}{d\xi_*} = \frac{d\lambda}{dz_*}\frac{dz_*}{d\xi_*} = 0 \qquad (2.128)$$

which gives

$$\frac{dF}{dz_*} - \gamma F\frac{dG}{dz_*} = 0 \qquad (2.129)$$

where Eq. (2.122) has been used. Substitution of Eq. (2.119) yields

$$\left\{E_i(-z_*)\exp(z_*)\right\}^2\left(\gamma z_* - z_* - \frac{1}{2}\right) - E_i(-z_*)\exp(z_*)(\gamma z_* + 1) - \gamma = 0 \qquad (2.130)$$

This equation can be solved numerically to give the extinction values z_e, F_e, and G_e as functions of γ. Then the following extinction values are given as functions of γ alone.

$$\left(\frac{\theta_e - \theta_a}{\theta_a}\right)\beta = -2\gamma(1 - G_e) \tag{2.131}$$

$$\frac{\lambda_e}{\lambda_{e0}} = \left(\frac{F_c}{F_e}\right)^2 \exp\left[-2\gamma(1 - G_e)\right] \tag{2.132}$$

$$\frac{\xi_e}{\xi_{e0}} = \sqrt{\frac{z_e}{z_c}}\left(\frac{F_e}{F_c}\right)\exp\left[\gamma(1 - G_e)\right] \tag{2.133}$$

$$-\frac{v_e}{S_u} = \frac{\sqrt{z_e}}{F_e}\exp\left[-\gamma(1 - G_e)\right] \tag{2.134}$$

$$K_e = \frac{1}{F_e^2}\exp\left[-2\gamma(1 - G_e)\right] \tag{2.135}$$

where λ_{e0} and ξ_{e0} are the extinction values for $\gamma = 0$ and given by

$$\lambda_{e0} = \frac{2\Lambda}{\pi\,\mathrm{Pr}}\left(\frac{\theta_a}{\theta_a - 1}\right)^2\frac{1}{F_c^2} \tag{2.136}$$

$$\xi_{e0} = \sqrt{\frac{2}{\Lambda}}\left(\frac{\theta_a - 1}{\theta_a}\right)\sqrt{z_c}\,F_c \tag{2.137}$$

and z_e, F_e, G_e are given by Eq. (2.120). The above results reveal clearly that it is the parameter γ which governs the extinction behavior of the flame. The solid curves in Figs 2.33 through 2.35 show the results obtained by Eqs (2.131) through (2.135). In these figures, the open circles

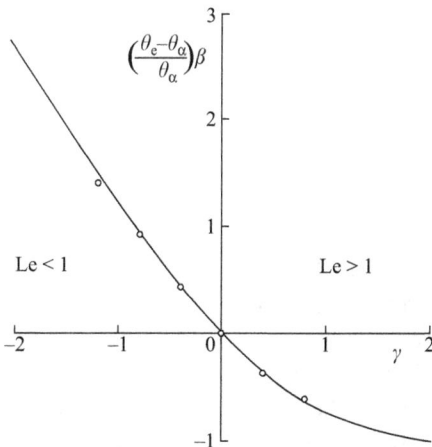

Figure 2.33. Deviation of flame temperature at extinction from the adiabatic flame temperature plotted against γ. The solid curve represents the results of asymptotic analysis, whereas open circles show the numerical solution.

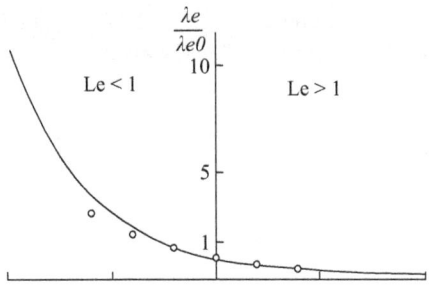

Figure 2.34. Relative Reynolds number and flame radius at extinction plotted against γ. The solid curve represents the results of asymptotic analysis, while open circles show the numerical solution.

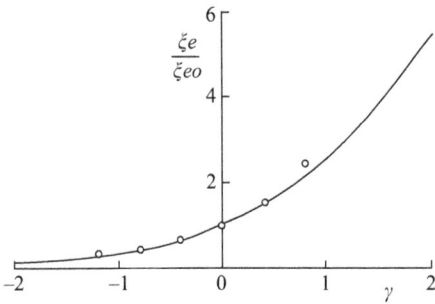

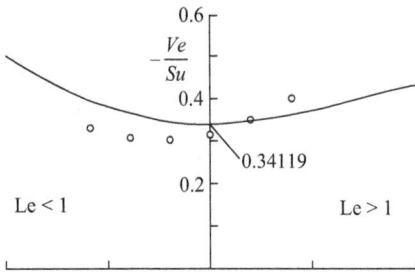

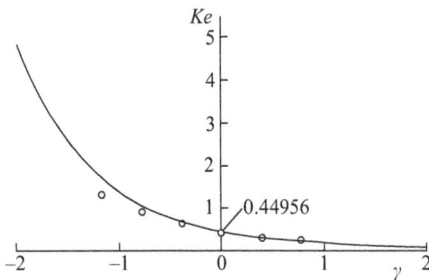

Figure 2.35. Apparent burning velocity and flame stretch at extinction plotted against γ. The solid curve represents the results of asymptotic analysis, while open circles show the numerical solution.

represent the results of the numerical solution as described later. As seen in Fig. 2.33, the extinction flame temperature increases above the adiabatic flame temperature when Le is smaller than unity, while it decreases below the adiabatic flame temperature when Le is larger than unity. As a result, in Fig. 2.34, the extinction Reynolds number increases while the extinction flame radius decreases with a decrease in Le. Figure 2.35 shows that the apparent burning velocity $-v_e$ at the extinction becomes minimal when Le = 1, and increases with an increase or a decrease in

Le. The variation, however, is rather small and it remains below the normal burning velocity. The extinction flame stretch K_e, on the other hand, increases rapidly with a decrease in Le as expected from the behavior of the extinction flame temperature shown in Fig. 2.33.

2.6.7 NUMERICAL EXAMPLE

In the numerical calculation Eqs (2.57) through (2.60) were solved by using the forward integration method of Runge–Kutta–Gill, with the generalized Newton–Raphson method to correct for the initial values. The non-viscous solution was used for the flow field. The whole calculation was performed with an accuracy of more than 12 significant figures. The mixture selected was methane and air and $i = 3.2917$, $j = 4.0$, in accordance with Section 2.3. The adopted numerical values for the nondimensional parameters were

$$\text{Pr} = 1.0, \quad B = 3.0 \times 10^9, \quad E_n = 50.3271, \quad \varphi = 0.5, \quad q = 138.945$$

and the derived values were

$$\theta_a = 4.93237, \quad \beta = 10.2034, \quad a = 0.113207, \quad \Lambda = 241.6761$$

The value of l, defined in Eq. (2.89), was varied and its effects on the response curves were studied.

The results of the calculation are shown in four response curves in Figs 2.27–2.30 as compared with the results of the asymptotic analysis. In these figures, the solid curves with the extinction points marked by fine vertical lines represent the results of the asymptotic analysis described in this section, while the dotted curves with the data point represent of the numerical solution. Figure 2.36 shows the reaction zone position ξ_R, as compared with the flame surface position ξ_* of the asymptotic analysis, plotted against injection Reynolds number λ for five values of l. ξ_R is defined as the position where the heat release rate becomes maximum in the finite reaction zone width. For a given value of λ there exist simultaneously three—the upper, the middle, and the lower—numerical solutions, whereas the trivial lower solution is not predicted

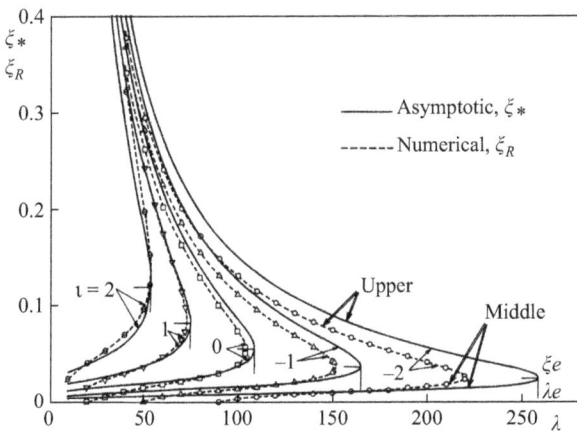

Figure 2.36. Reaction zone position plotted against Reynolds number λ.

by the asymptotic analysis based on the flame surface model of infinite reaction rate. As mentioned in Section 2.3, the upper solution is the one with less incomplete combustion, and has been considered to correspond with the actual flame observed in the experiment. The general trend predicted by the asymptotic analysis agrees well with that of the numerical solution. The response curve moves toward the lower right with a decrease in l, resulting in the larger extinction Reynolds number λ_e and the smaller extinction flame radius ξ_e. For fixed values of l and λ, however, the asymptotic analysis predicts a somewhat larger flame size for the upper solution leading to the larger value of extinction flame radius λ_e except for the case when $l = 2$. In other words, the response curve for ξ_* is shifted toward the larger value of λ as compared with the numerical solution. The shift becomes small with an increase in l. For the middle solution, ξ_R, decreases with a decrease in λ, but it remains finite, however small it may be. On the other hand, ξ_R reaches the axis at a certain value of λ, and this value increases with a decrease in l. It is interesting to note that although the asymptotic analysis predicts the larger value for λ_e, the predicted extinction flame radius ξ_e is very close to that of the numerical solution. Another point is that the asymptotic analysis predicts the finite extinction flame radius for any value of l. This contrasts sharply with the result for the plane flames in a stagnation flow, in which the extinction flame size becomes zero for l smaller than a certain value.[19–21]

Figure 2.37 compared the flame temperature θ_0, which is the maximum temperature on the axis, with the flame surface temperature θ_*. Only the results for $l \geq 0$ are shown. The agreement of the asymptotic analysis with the numerical solution is not as good as in Fig. 2.36. When $l = 0$ the asymptotic analysis predicts only one constant curve $\theta_* = \theta_a$, and the flame surface temperature remains the adiabatic flame temperature for any increase in λ until λ_e, where the extinction occurs suddenly. On the other hand, θ_0 for the upper solution remains slightly smaller than θ_a as λ is increased until the extinction, but there exists the middle solution and θ_0 decreases very rapidly with a decrease in λ. When $l > 0$, both θ_* and θ_0 for the upper solution decrease very rapidly in a similar fashion from θ_a with λ until the extinction. For the middle solution they decrease with a decrease in λ, but the decrease is faster for θ_0.

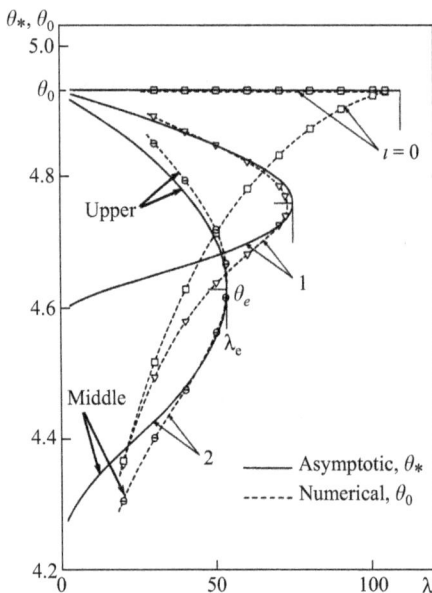

Figure 2.37. Flame temperature plotted against Reynolds number λ for $l \geq 0$.

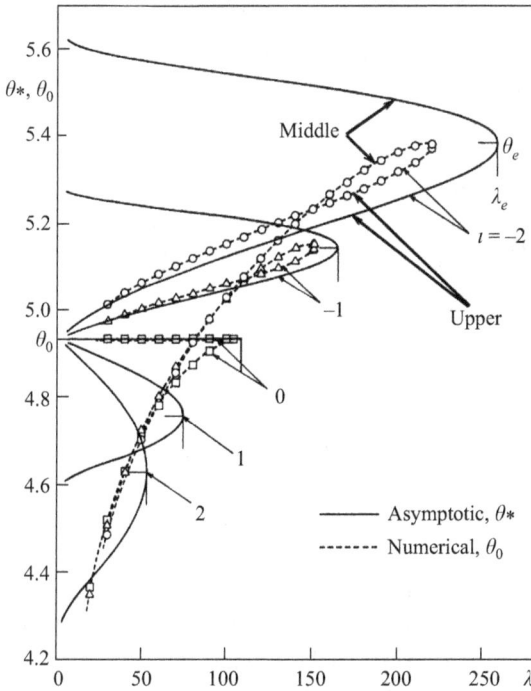

Figure 2.38. Flame temperature plotted against Reynolds number λ for $l \le 0$.

The situation is rather complicated for the case when $l < 0$, shown in Fig. 2.38. For respective values of l it is the lower part of the θ_* curves predicted by the asymptotic analysis, which corresponds to the upper solution, and the upper part corresponds to the middle solution. That is, the middle solution is accompanied by the higher flame temperature. The unbalance between heat and mass diffusion rates produces the excess enthalpy that causes a flame temperature higher than the adiabatic flame temperature, and the effect is more pronounced for the middle solution, in which the flame surface is located closer to the axis. For the upper solution θ_* increases from θ_a with λ until λ_e is reached, and then it increases along the middle solution with a decrease in λ approaching the maximum deviation given by the following equation.

$$\left(\theta_* - \theta_a\right)_{max} = -\frac{l\left(\theta_a - 1\right)}{\beta} = -\frac{2\theta_a \gamma}{\beta} \tag{2.138}$$

On the other hand, the numerical solution shows that θ_0 for the upper solution is higher than θ_*, and it increases with λ in the same way, as θ_*. However, θ_0 for the middle solution does not increase as θ_*, but increases with a decrease in λ. As a consequence of this, a loop is generated in the response curve, which will be discussed later in detail. The decrease becomes faster as λ is decreased, following a common curve for all values of l. It is very interesting to note, here again, that the extinction flame temperatures are very close to the value of θ_e predicted by the asymptotic analysis, in spite of the discrepancy in the extinction Reynolds number.

Figure 2.39 compares the apparent burning velocities. In the numerical solution $\tilde{v}_R$ was defined as the velocity at the position ξ_R. The asymptotic analysis predicts the correct general trend, although the predicted burning velocity for the upper solution is too large. For all

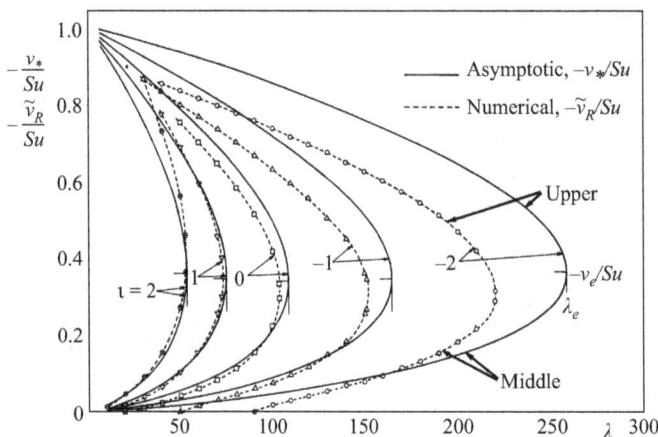

Figure 2.39. Apparent burning velocity plotted against Reynolds number λ.

values of l, $-v_*/S_u$ for the upper solution approaches unity, while that for the middle solution approaches zero, as λ is decreased to zero. An examination of the solution behavior at large and small values of z_* reveals this clearly, although the linear velocity approximation Eq. (2.114), as well as the approximation Eq (2.113), becomes questionable at these extreme values. Anyway, the apparent burning velocity is always finite. It must be remembered that in the plane stagnation flame, the apparent burning velocity becomes zero for certain values of l and λ.[18–20]

2.6.8 DISCUSSIONS

2.6.8.1 Comparison Between Asymptotic Analysis and Numerical Calculation

The foregoing results indicate that the general extinction behavior predicted by the asymptotic analysis agrees well with that of the numerical solution. In particular, the agreement of the extinction values, except for the Reynolds number, is most satisfactory as can be seen in Figs 2.33 through 2.35. It can be seen intuitively that the flame surface model, with the concentrated reaction zone of infinitely fast chemical reactions, overestimates the reaction rate, and hence it is quite natural that the predicted extinction Reynolds number λ_e becomes too large. The same reasoning will explain why the asymptotic analysis predicts the response curves which are shifted toward the larger value of λ, except for $l = 2$, as shown in Figs 2.36 through 2.39.

On the other hand, the discrepancy in the prediction appears most remarkably in the response curve of the flame temperature when $l = 2$ (Fig. 2.38). This is the first paper, to the authors' knowledge, that reports that a loop can be produced in the response curve of the flame temperature of the numerical solution. The loop is closely related to the reaction zone width. In the flame surface model of zero reaction zone width and complete combustion, the change in the flame temperature is entirely due to the excess enthalpy produced by the unbalance in heat and mass diffusion rates. When $l = 0$, therefore, the flame surface temperature remains the adiabatic flame temperature for any changes in λ, as can be seen in Fig. 2.37. Even for this case, on the other hand, the flame temperature of the numerical solution decreases slightly below the adiabatic flame temperature as λ is increased along the upper solution, whereas it decreases very rapidly when λ is increased along the middle solution. This decrease is closely correlated

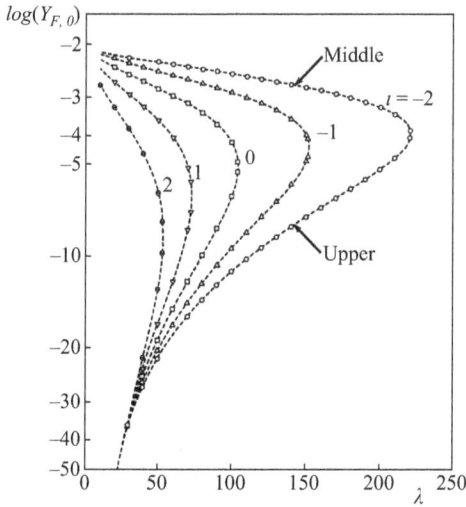

Figure 2.40. Unburned fuel mass fraction remaining on the axis plotted against Reynolds number λ.

with the behavior of the reaction zone position shown in Fig. 2.36 and it becomes larger as the reaction zone approaches the axis.

As mentioned before, the finite reaction zone width produces the unburned reactant on the axis, and this increases as the reaction zone approaches the axis. This can be seen clearly in Fig. 2.40, which shows the fuel mass fraction $Y_{F,0} \equiv Y_F(0)$ remaining on the axis plotted against λ. The upper part, of course, corresponds to the middle solution, while the lower part corresponds to the upper solution. As λ is decreased along the upper solution, $Y_{F,0}$ decreases very rapidly. Note the scale of the ordinate. The amount of the remaining unburned fuel becomes extremely small (less than 10^{-50}) as λ approaches 0, which causes the complete combustion to yield the adiabatic flame temperature, as shown in Fig. 2.38. When λ is decreased along the middle solution, on the other hand, the shift of reaction zone toward the axis and the finite reaction zone width produces a rapid increase in $Y_{F,0}$. Then the incomplete combustion is accelerated, and as a consequence of this a large drop in the flame temperature is produced.

When $l > 0$, the reaction zone is located far from the axis and the above explained effect of incomplete combustion is not significant, except for the rapid decrease in the flame temperature of the middle solution. In addition, the effect is the same as that of the excess enthalpy and the discrepancy between the numerical and analytical predictions is not so pronounced. On the other hand, the effect becomes significant when $l < 0$. As λ is increased along the middle solution, the reaction zone position closer to the axis makes the effect so powerful as to overcome the counter effect of the excess enthalpy, forcing the flame temperature to decrease, and as a result of this the loop is generated. Then we may say that it is the incomplete combustion due to the finite reaction zone width that is responsible for the loop.

Another problem to be discussed is the extinction flame radius ξ_e. The asymptotic analysis, predicts the finite value ξ_e, which can be made small without limit by decreasing l, or the Lewis number. It appears that we can make the tubular flame as small as we want provided we can decrease Le. However, this holds true only for the flame surface model. In the numerical solution the flame has a finite reaction zone width. What will happen if the flame radius is decreased to this width? Figure 2.41 shows a part of the response curve ξ_R calculated especially for the

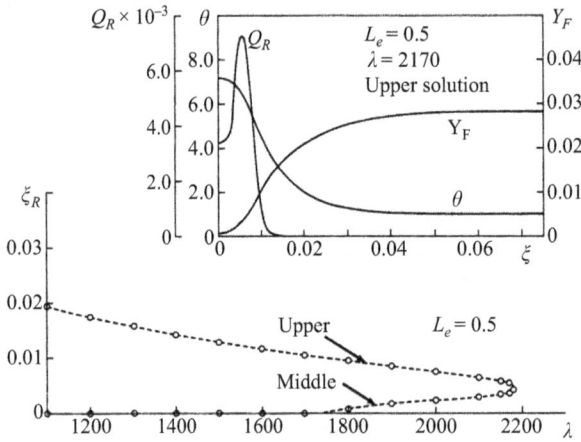

Figure 2.41. Reaction zone position plotted against Reynolds number and the flame structure at extinction when Lewis number is very small.

small value of Le = 0.5 in order to answer this question. The flame structure at the extinction condition is also shown. The flame causes the extinction when the flame radius becomes as small as the reaction zone width. As a matter of fact, the tubular flame cannot be made smaller than the reaction zone width. A minute tubular flame cannot be an element of turbulent combustion if the Kolmogorov microscale is the same order as, or is smaller than, the reaction zone width.

2.6.8.2 Extinction Behavior

Figures 2.33 through 2.35 show that the extinction values predicted by the asymptotic analysis agree very well with those of the numerical solution, although some deviations are observed in the $-v_e/S_u$ curve. Furthermore, the predicted dependence of extinction flame radius on Le shown in Fig. 2.34, is in full accord with the experimental observation that the flame diameter at extinction is smaller for a lean hydrogen–air flame than that for a lean methane–air flame, whereas it is larger for a lean propane–air flame than that for a lean methane–air flame.[22] The Lewis numbers of the limiting reactant in the flames for $\varphi = 0.5$ were calculated to be 0.409, 0.809, and 1.427, respectively, for the hydrogen, the methane, and the propane flames. Then Fig. 2.34 suggests that the size of extinction flame radius should be largest for the propane flame and smallest for the hydrogen flame. This agreement provides another example which reveals the usefulness of the asymptotic analysis. Similar agreement has been obtained for the effect of Lewis number of limiting component on the extinction behavior of the flame stabilized in plane stagnation flows.[19,22–24,47,48]

The extinction values predicted by the analysis depend on γ alone, and the expressions given in Eqs (2.131) through (2.135), are very general ones in the sense that they hold for any set of values of the involved nine parameters (λ, Le, Pr, i, j, φ, B, E_n, q). Then we may say that γ is the most important parameter governing the extinction behavior of the flame. It should be noticed that γ was deduced naturally in the course of the analysis. It can be rewritten as

$$\gamma = \frac{\beta}{2} \frac{(\theta_a - 1)}{\theta_a} \frac{(\text{Le} - 1)}{\text{Le}} = \frac{E(T_{ad} - T_R)(\text{Le} - 1)}{2R^0 T_{ad}^2 \text{Le}} \qquad (2.139)$$

The parameter contains all three properties of the mixture in a simple form: the activation energy (kinetic), Lewis number (transport), and the adiabatic flame temperature (thermodynamic). It is interesting to note that Lewis number affects the extinction through a combination of the other two properties, and not by itself. Incidentally, γ is equal to J multiplied by $(\theta_a^{-1})/2\theta_a$, which is introduced by Williams,[1] to explain the effects of a Lewis number differing from unity on the extinction of the plane stagnation flame.

The expression given by Eq. (2.135) for the extinction flame stretch K_e is very simple and depends on γ alone. On the other hand, the extinction Reynolds number λ_e is proportional to K_e, and the proportional constant contains Λ through λ_{e0} of Eq. (2.136), which introduces the additional complicated dependence of λ_e on the kinetic and thermodynamic properties of the mixture. Although λ is the important parameter in specifying the flow field, and hence is inevitably involved in the energy and species conservation equations governing the flame behavior, it is not λ_e but K_e which is more relevant to describe the extinction. This is a natural consequence of the fact that K is the reciprocal of the appropriate Damkohler number which actually governs the extinction behavior.[1] The present analysis shows clearly how the proportional constant, which contains the representative reaction time, is eliminated so as to yield the numerical values for K_e, which depends only on γ. The tubular flame is forced to be extinguished for the stretch exceeding this value. Then the minute tubular flame in turbulent combustion cannot sustain combustion when the vortex stretching becomes stronger than this value.

One characteristic of the tubular flame is that it is subjected to the curvature in addition to the stretch. This will explain the observed differences in the extinction behavior, predicted by the asymptotic analysis, from that of the plane stagnation flame. It has been well known in thermal explosion theory that it is harder to cause a spontaneous ignition in the cylindrical configuration, and easier to cause an extinction, than in the plane configuration, because of the increased cooling rate through the surface.[25,26] This effect becomes significant as the flame radius is decreased, and hence the tubular flame with radius smaller than some critical value cannot be established, leading to the finite extinction flame radius; this is in contrast to the plane flame, which can be made small without limit.

2.6.9   SOME CONCLUDING REMARKS

The present theoretical study on extinction of a tubular flame has brought about the following concluding remarks.

1. The general extinction behavior can well be predicted by asymptotic analysis based on the flame surface model. The predicted critical values at extinction of the flame temperature, the flame radius, the apparent burning velocity, and the flame stretch agree very well with those of the exact numerical solution, although the critical injection Reynolds number is somewhat larger.
2. The predicted effects of Lewis number on the critical flame radius are found to agree qualitatively with experimental observation.
3. In the numerical solution, when the Lewis number is smaller than unity, a loop appears in the response curve of the flame temperature plotted against the Reynolds number. The increase in incomplete combustion as the finite width reaction zone approaches the axis is responsible for this loop.

4. The tubular flame cannot be made smaller than the reaction zone width, and therefore a minute tubular flame cannot be an element of turbulent combustion if the Kolmogorov microscale is the same order as, or is smaller than the reaction zone width.

5. The relevant parameter describing the extinction condition is the flame stretch, and the flame is extinguished when it exceeds the critical value. The value is a function of a single parameter γ, given by Eq. (2.139), which contains kinetic, transport, and thermodynamic properties of the mixture in a simple form.

6. The extinction behavior of the tubular flame predicted by the asymptotic analysis is different from that of the plane flame stabilized in stagnation flows due to the effect of curvature, which provides the increased cooling, as compared with the plane flame.

2.7 NUMERICAL STUDY WITH FULL KINETICS AND EXACT TRANSPORT PROPERTIES

2.7.1 INTRODUCTION

In the foregoing section, we have studied about the tubular flame in terms of the similarity solution and the simplified model with some kind of assumptions. The results have shown that the similarity solution can predict correctly the observed flame behavior. The most important simplification made so far is the overall one-step kinetics. The chemical reactions preceding the flame are represented in terms of the single chemical reaction Eq. (2.20). In addition, some simplifications about the transport properties, such as the assumptions 6.7 or 6.9 in Section 2.6, have been introduced. These assumptions are very useful and help us understand the respective roles played by the concurrent multiple processes involved and about the flame qualitatively.

However, if we want to understand the flame quantitatively through more precise comparisons with the experimental observations, we have to develop a more rigorous theory, in which the detailed chemical kinetics and the realistic transport properties are taken into account. Some attempts have been made to combine the similarity solution with a detailed chemical reaction mechanism, and the results obtained so far are promising.[27] Furthermore, the subsequent calculation on the flame extinction behavior by Smooke and Giovangigli[28] agrees excellently with the observations made by Kobayashi and Kitano.[29] The latter have studied experimentally the tubular flame established in a nonrotating flow field, and have made an extensive measurement of the extinction flame radius and the extinction injection velocity, in wide range of equivalence ratio for methane–air and propane–air mixtures. This agreement between theory and experiment is very encouraging, and we are now confident that the similarity solution with the detailed kinetics can correctly describe what actually happens in the tubular flame.

One interesting problem with the tubular flame is how small we can make the flame. As mentioned above, the flame radius cannot be made smaller than the reaction zone width. However, the width can be decreased by increasing pressure, and hence we may expect to produce a very small flame at an elevated pressure. This is important in view of the fact that some turbulent combustion models in engines presume the existence of a coherent small-scale turbulent structure, containing highly dissipative vortex tubes of the diameter of Kolmogorov microscale.[30] The molecular processes, such as the transport processes and chemical reactions, should be concentrated in these narrow vortex tubes. At the atmospheric pressure, the observed limiting size of the tubular flame for hydrogen fuels is around 1 mm,[29,31] which is much larger than the typical Kolmogorov microscale of 0.01–0.04 mm in engines.[30] However, the pressure in engine combustion is higher, and at high pressure there is a chance that the limiting flame size will become the same order as of Kolmogorov scale. We would then be convinced of the assertion that the reacting vortex tube of Kolmogorov scale can sustain combustion, even if momentarily, before the extinction due to the increasing vortex stretching. The objective of the present section is to study numerically the effects of pressure on the structure and extinction of the tubular flame. The approach is based on the similarity solution with the detailed chemical reaction mechanism and the realistic transport properties. For the numerical calculation we have developed a computer code, which is a modified version of CHEMKIN program developed by Kee et al. for the normal 1D flame.[32] The latter is generally accepted to be among the most accurate and efficient programs.

2.7.2   MODEL AND EQUATIONS

The theoretical model is the same as that of the previous section and shown in Fig. 2.1. The cylindrical coordinate is used with the origin on the axis: x and r represent, respectively, axial and radial distance from the origin, while α represents azimuth angle. u, v, and w represent respectively the x, r, and α components of the velocity vector v. A combustible mixture is injected through a rotating porous wall with a uniform velocity v_R $(0, -v_R, w_R)$. The swirling mixture flows through the flame, and the produced burned gas flows outward along the axis. Now we are going to study the flame behavior in terms of the similarity solution described so far. The assumptions 2.1 through 2.5 adopted in 2.2.1 will be adopted here again and are described in the following.

1. The flame, as well as the flow, is steady and axisymmetric, and there are no external forces.
2. The reacting mixture behaves like an ideal gas.
3. The pressure diffusion can be neglected.
4. The thermodynamic pressure remains constant throughout the flow field.
5. In the energy equation, work done by pressure and viscous dissipation can be neglected.

Now the equations and boundary conditions to be solved are Eqs (2.20) through (2.24).

2.7.3   REACTION MECHANISM AND TRANSPORT PROPERTIES

The mixture studied was stoichiometric methane–air mixture. The calculation was performed for the case when the mixture at room temperature ($T_R = 300$ K) is injected into a tube of radius $R = 0.85$ cm. For a given pressure, the injection velocity (v_R) was increased from 100 cm/s until the flame is caused to be extinguished at the extinction velocity. The flame structure and the extinction limits were determined for the pressure ranging from 1 to 8 atm. The reaction mechanism adopted was so-called C1 chemistry, recommended by Kee et al.[32] In a numerical study, the burning velocity of methane–air mixture at elevated pressures calculated with this reaction mechanism has been found to agree well with the experimental data.[33,34] The mechanism involves 18 species and 58 elementary reactions, shown in Table 2.1 with their rate constants. The effects of pressure on the thermal dissociation and recombination reactions have been discussed by Gardiner and Troe,[35] and for reaction No. 1 in the table this pressure dependence was taken into account by the Lindemann form. The adopted limiting rate constants[36] are also shown in Table 2.1. The necessary thermodynamic and transport properties were obtained by the CHEMKIN database.[37–39] The solution procedure was similar to that developed for

Table 2.1. Adopted reaction mechanism and rate constants in the form $k_f = AT^\beta \exp\left(-\dfrac{E}{R^0 T}\right)$

No.	Reaction	A	β	E (cal/mol)
1.	$CH_3 + H + M = CH_4 + M$			
	High-pressure limit	6.00E + 16	−1.0	0
	Low–pressure limit	8.00E + 26	−3.0	0
2.	$CH_4 + O_2 = CH_3 + HO_2$	7.90E + 13	0.0	56000
3.	$CH_4 + H = CH_3 + H_2$	2.20E + 4	3.0	8750
4.	$CH_4 + O = CH_3 + OH$	1.60E + 6	2.36	7400
5.	$CH_4 + OH = CH_3 + H_2O$	1.60E + 6	2.1	2460
6.	$CH_3 + O = CH_2O + H$	6.80E + 13	0.0	0
7.	$CH_3 + OH = CH_2O + H_2$	1.00E + 12	0.0	0
8.	$CH_3 + OH = CH_2 + H_2O$	1.50E + 13	0.0	5000
9.	$CH_3 + H = CH_2 + H_2$	9.00E + 13	0.0	15100
10.	$CH_2 + H = CH + H_2$	1.40E + 19	−2.0	0
11.	$CH_2 + OH = CH_2O + H$	2.50E + 13	0.0	0
12.	$CH_2 + OH = CH + H_2O$	4.50E + 13	0.0	3000
13.	$CH + O_2 = HCO + O$	3.30E + 13	0.0	0
14.	$CH + O = CO + H$	5.70E + 13	0.0	0
15.	$CH + OH = HCO + H$	3.00E + 13	0.0	0
16.	$CH + CO_2 = HCO + CO$	3.40E + 12	0.0	690
17.	$CH_2 + CO_2 = CH_2O + CO$	1.10E + 11	0.0	1000
18.	$CH_2 + O = CO + H + H$	3.00E + 13	0.0	0
19.	$CH_2 + O = CO + H_2$	5.00E + 13	0.0	0
20.	$CH_2 + O_2 = CO_2 + H + H$	1.60E + 12	0.0	1000
21.	$CH_2 + O_2 = CH_2O + O$	5.00E + 13	0.0	9000
22.	$CH_2 + O_2 = CO_2 + H_2$	6.90E + 11	0.0	500
23.	$CH_2 + O_2 = CO + H_2O$	1.90E + 10	0.0	−1000
24.	$CH_2 + O_2 = CO + OH + H$	8.60E + 10	0.0	−500
25.	$CH_2 + O_2 = HCO + OH$	4.30E + 10	0.0	−500

(Continued)

Table 2.1. (*Continued*)

No.	Reaction	A	β	E (cal/mol)
26.	$CH_2O + OH = HCO + H_2O$	3.43E + 9	1.18	–447
27.	$CH_2O + H = HCO + H_2$	2.19E + 8	1.77	3000
28.	$CH_2O + M = HCO + H + M$	3.31E + 16	0.0	81000
29.	$CH_2O + O = HCO + OH$	1.81E + 13	0.0	3082
30.	$HCO + OH = CO + H_2O$	5.00E + 12	0.0	0
31.	$HCO + M = H + CO + M$	1.60E + 14	0.0	14700
32.	$HCO + H = CO + H_2$	4.00E + 13	0.0	0
33.	$HCO + O = CO_2 + H$	1.00E + 13	0.0	0
34.	$HCO + O_2 = HO_2 + CO$	3.30E + 13	–0.4	0
35.	$CO + O + M = CO_2 + M$	3.20E + 13	0.0	–4200
36.	$CO + OH = CO_2 + H$	1.51E + 7	1.3	–758
37.	$CO + O_2 = CO_2 + O$	1.60E + 13	0.0	41000
38.	$HO_2 + CO = CO_2 + OH$	5.80E + 13	0.0	22934
39.	$H_2 + O_2 = 2OH$	1.70E + 13	0.0	47780
40.	$OH + H_2 = H_2O + H$	1.17E + 9	1.3	3626
41.	$H + O_2 = OH + O$	5.13E + 16	–0.816	16507
42.	$O + H_2 = OH + H$	1.80E + 10	1.0	8826
43.	$H + O_2 + M = HO_2 + M^a$	3.61E + 17	–0.72	0
44.	$OH + HO_2 = H_2O + O_2$	7.50E + 12	0.0	0
45.	$H + HO_2 = 2OH$	1.40E + 14	0.0	1073
46.	$O + HO_2 = O_2 + OH$	1.40E + 13	0.0	1073
47.	$2OH = O + H_2O$	6.00E + 8	1.3	0
48.	$H + H + M = H_2 + M$	1.00E + 18	–1.0	0
49.	$H + H + H_2 = H_2 + H_2$	9.20E + 16	–0.6	0
50.	$H + H + H_2O = H_2 + H_2O$	6.00E + 19	–1.25	0
51.	$H + H + CO_2 = H_2 + CO_2$	5.49E + 20	–2.0	0
52.	$H + OH + M = H_2O + M^b$	1.60E + 22	–2.0	0
53.	$H + O + M = OH + M^c$	6.20E + 16	–0.6	0
54.	$H + HO_2 = H_2 + O_2$	1.25E + 13	0.0	0
55.	$HO_2 + HO_2 = H_2O_2 + O_2$	2.00E + 12	0.0	0
56.	$H_2O_2 + M = OH + OH + M$	1.30E + 17	0.0	45500
57.	$H_2O_2 + H = HO_2 + H_2$	1.60E + 12	0.0	3800
58.	$H_2O_2 + OH = H_2O + HO_2$	1.00E + 13	0.0	1800

[a] Third-body efficiencies: $k_{43}(H_2O) = 18.6k_{43}(Ar)$, $k_{43}(CO_2) = 4.2k_{43}(Ar)$, $k_{43}(H_2) = 2.86k_{43}(Ar)$, $k_{43}(CO) = 2.11k_{43}(Ar)$, $k_{43}(N_2) = 1.26k_{43}(Ar)$.
[b] Third-body efficiencies: $k_{52}(H_2O) = 5k_{52}(Ar)$
[c] Third-body efficiencies: $k_{53}(H_2O) = 5k_{53}(Ar)$

the calculation of the normal 1D flames.[32] The central difference formula was used for the convective terms. The adaptive placement of the mesh points to form the finer meshes was done in such a way to minimize the total number of mesh points needed to represent the solution accurately. No special techniques, such as the arc–length method[28,36] were used to predict the extinction limits. It was found, however, that with increasing pressure the convergence of the solution becomes progressively more difficult. Then we had to obtain a very accurate solution for a given injection velocity, so as to obtain an accurate initial estimate for the next injection velocity. Therefore the number of mesh points was increased up to around 185. In addition, the increment in the injection velocity had to be very small in order to obtain the convergence.

2.7.4   RESULTS AND DISCUSSIONS

2.7.4.1   The 1D Normal Flame

To confirm the validity of the adopted reaction mechanism and thermochemical and transport properties at elevated pressures, the calculation was done on the normal 1D flame for a stoichiometric methane–air mixture at room temperature (300 K). The results show that the calculated burning velocity decreases with pressure, and the present calculation and the similar numerical calculation,[34] based on the same reaction mechanism, predict rather stronger pressure dependence compared with experimental data.[40–42] We are not certain if the discrepancy can be explained in terms of the adopted reaction mechanism. However, the discrepancy is within the scattered range of the old experimental data[43] and the calculation predicts the correct general trend. We think that the adopted mechanism, as well as the solution procedure, works well at elevated pressures.

2.7.4.2   Flame Structure

Figure 2.42 compares the calculated flame structure at three representative pressures $p = 1, 3$, and 8 atm for the same injection velocity of $V_R = 200$cm/s. The flame is weakly strained and is stable for this injection velocity. Figure 2.42(a) shows the radial temperature and main species distributions, while Figs 2.42(b)–(d) show those of intermediate and radical species distributions. As pressure is increased, the flame is shifted toward the axis with the decreased reaction zone thickness. In the tubular flame, the radial velocity of the mixture coming into the flame decreases toward the axis, and hence the flame is shifted in this direction when the burning velocity is decreased. The normal burning velocity of the mixture decreases with pressure, and hence the flame is shifted inward as pressure is increased. On the other hand, since the diffusion coefficients of the binary mixture are inversely proportional to pressure, the diffusion rates of each species are decelerated at the elevated pressures. The thermal diffusivity is decreased as well due to the increase in density. These are the main cause of the reduction in the reaction zone thickness and of the steeper concentration and temperature gradients. In addition, the accelerated reaction rates at the elevated pressures increase the flame temperature T_0, which is the maximum temperature on the axis. The concentrations of species remaining on the axis are decreased due to the decelerated diffusion rates. As is seen in the figure, however, the chemical structure does not change significantly at the elevated pressures.

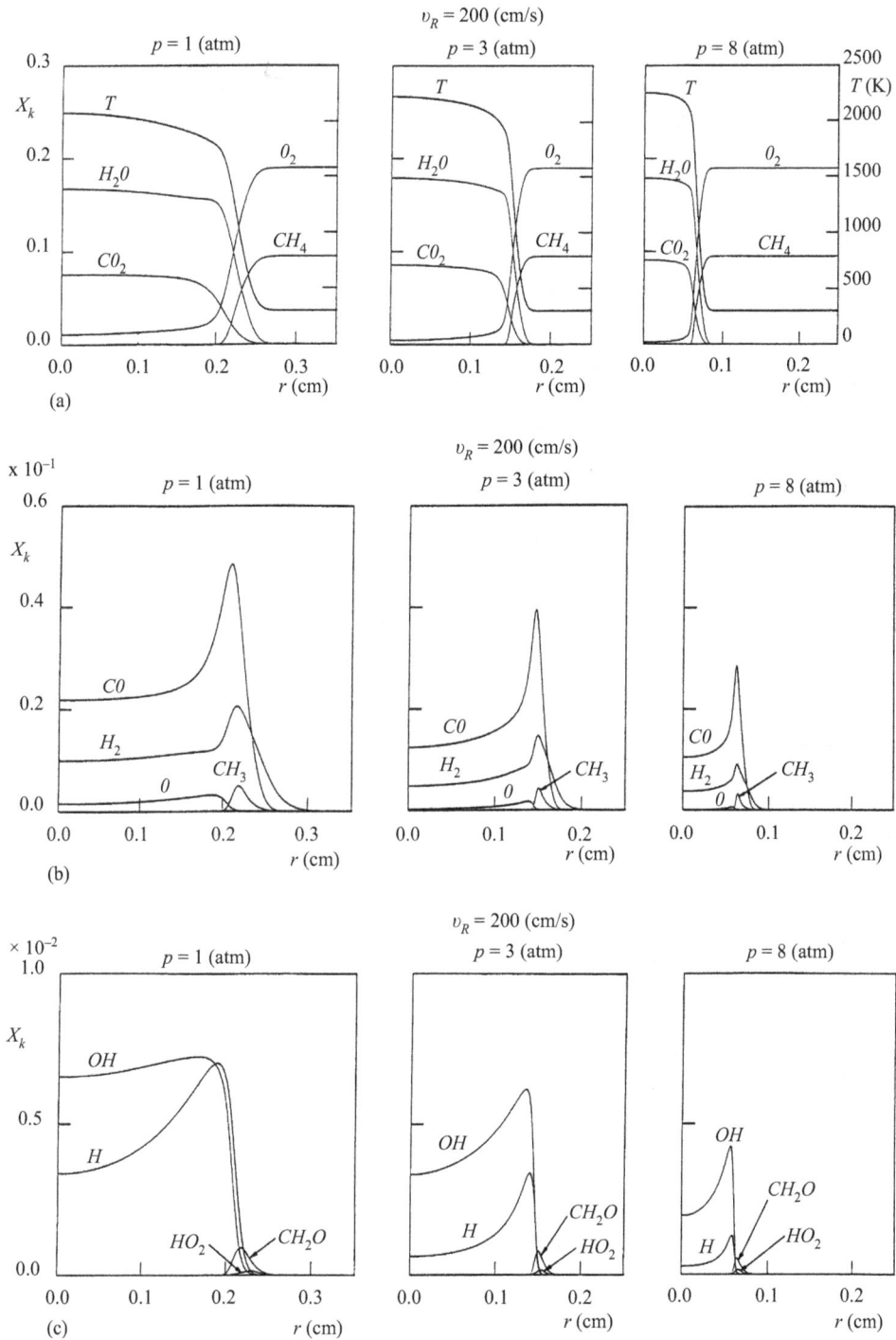

Figure 2.42. Comparison of flame structure for three representative pressures: (a) temperature and main species distributions, (b), (c), (d) intermediate and radical species distributions. (*Continued*)

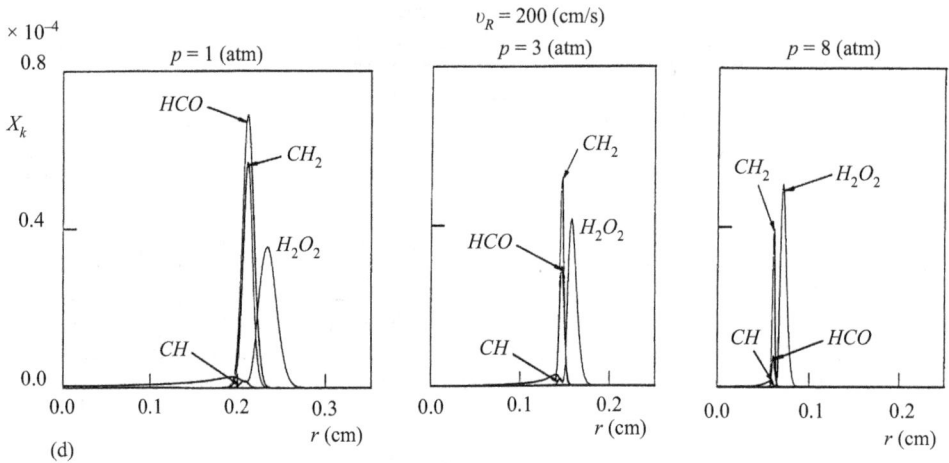

Figure 2.42. (*Continued*)

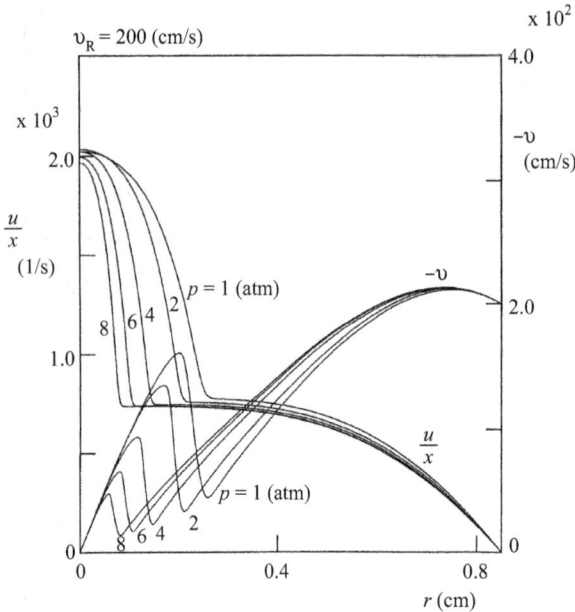

Figure 2.43. Effects of pressure on axial and radial velocity distributions.

Figure 2.43 shows the effects of pressure on the flow field. At atmospheric pressure the flame is situated a little bit apart from the axis, and the flow expansion due to heat release causes the acceleration in the radial, as well as the axial direction. This produces the peculiar radial velocity distribution with a large peak around the reaction zone. As pressure is increased, however, the flame is shifted inwards and the peak value is decreased because the restriction of zero velocity on the axis comes to play.[44] These changes in the flow field for the increased pressure are gradual and continuous, in accordance with the flame shift and the reduced reaction zone thickness. No abrupt changes occur in the flow field. Figure 2.44 shows the effects of

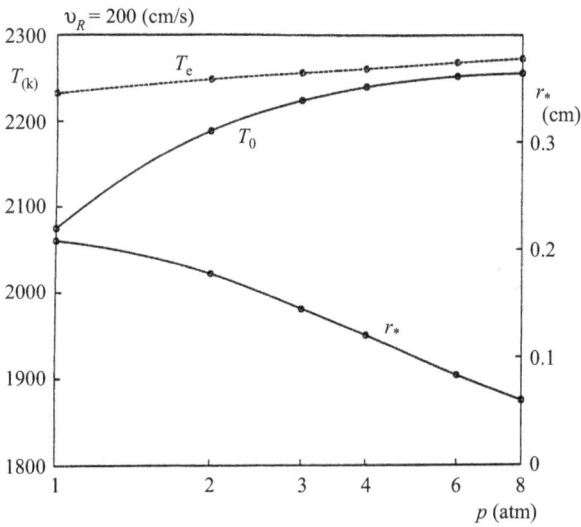

Figure 2.44. Pressure dependence of flame temperature and of flame radius: $v_R = 200$ cm/s.

accelerated reaction rates at elevated pressures. The flame temperature T_0, on the axis, is plotted against pressure as compared with the equilibrium flame temperature T_e, for the case when the injection velocity is 200 cm/s. The equilibrium values are calculated by the conventional equilibrium calculation for the respective pressures. For the same residence time in the flow field, the flame temperature approaches the equilibrium value as pressure is increased. The change in the flame radius r_*, defined as the position of maximum CH concentration, is shown as well in Fig. 2.44. As mentioned before, the flame becomes smaller with an increase in pressure due to the decrease in the burning velocity.

2.7.4.3 Extinction Limits

Figure 2.45 shows the flame temperature T_0 on the axis as a function of the average residence time R/v_R with pressure as a parameter. The equilibrium flame temperatures for the respective pressures are noted on the right ordinate. For a given pressure, as the residence is decreased the flame temperature decreases until at a critical residence time the flame is extinguished. The decrease in flame temperature is larger for lower pressures, and at elevated pressures the extinction occurs with a small drop in flame temperature. Moreover, it is very interesting to note that although the critical residence time decreases with pressure at first, it increases for higher pressures, yielding the minimum critical residence time at around 3 atm. The change in the flame radius with the residence time is shown in Fig. 2.46. For a given pressure, the flame radius decreases with a decrease in the residence time until the flame is extinguished at the critical residence time. The decrease is slower for higher pressure. On the other hand, the extinction flame radius decreases monotonically with pressure. However, the reduction becomes progressively smaller with pressure, and it appears that the reduction will not be significant for pressures higher than 8 atm.

The flame temperature on the axis T_0 and the flame radius $r_{*,e}$ at the extinction are plotted as a function of pressure in Fig. 2.47. The equilibrium flame temperature T_e is plotted as well.

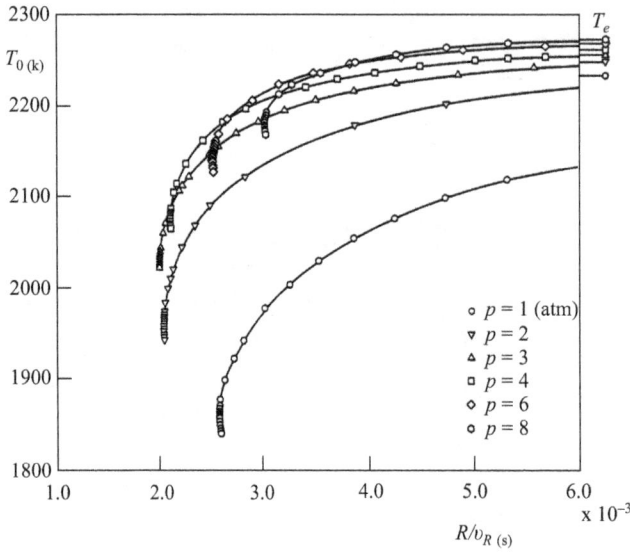

Figure 2.45. Dependence of flame temperature on residence time.

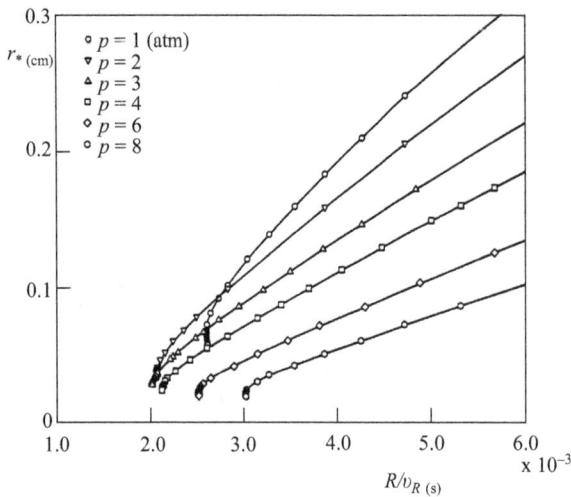

Figure 2.46. Dependence of flame radius on residence time.

As pressure is increased, the extinction flame temperature increases, approaching the equilibrium flame temperature, and the extinction occurs at relatively high flame temperatures. Further increases in pressure will extinguish the flame at a temperature that is close to the equilibrium flame temperature. This means that at elevated pressures the extinction will occur even if the residence time is long enough for the reaction to proceed. On the other hand, the extinction flame radius decreases with pressure, but the decreasing rate becomes very slow for the pressures higher than 4 atm, approaching asymptotically the minimum flame radius of around 0.2 mm. The flame with a radius smaller than this value cannot be established for the pressure up to 8 atm. It should be noted that this minimum flame radius is one order of magnitude larger than the Kolmogorov microscale encountered in engines.

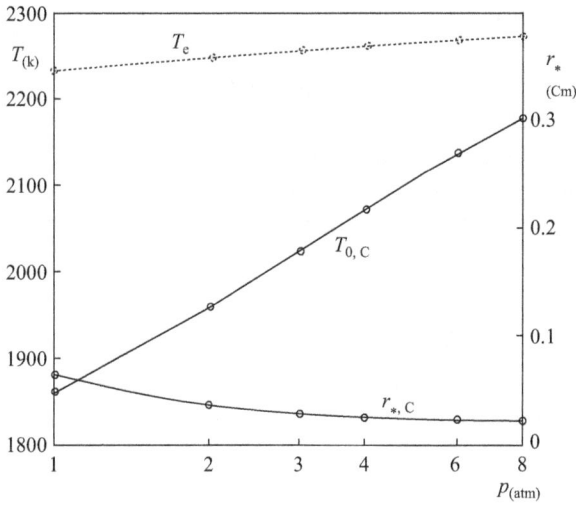

Figure 2.47. Dependences of extinction flame temperature and extinction flame radius on pressure.

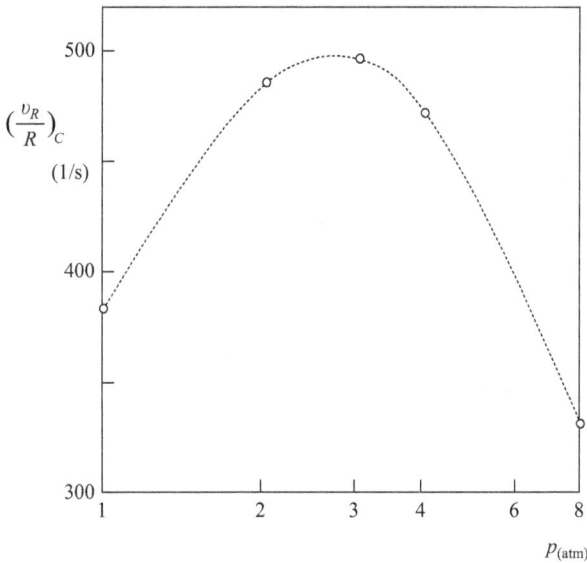

Figure 2.48. Dependence of critical strain rate on pressure.

Figure 2.48 shows the inverse of residence time at the extinction, or the critical strain rate, plotted against pressure. The critical strain rate increases with pressure at first, reaches a maximum at around 2.7 atm, and then decreases for higher pressure. This behavior is very interesting, since as noted above, the pressure seems to have little effect on the chemical structure, suggesting that the main reaction mechanism does not change significantly with pressure. It appears that the elevated pressure affects the flame behavior only through a decrease in the burning velocity and the accelerated reaction rates to approach the equilibrium flow. In addition, the flame has similar chemical structure even at the extinction limit, similar to the behavior of the planar flame.[35] In view of these behaviors, the extinction seems to be governed by the minimum flame radius rather than changes in the reaction mechanism. This is presumably due to the effects of curvature, since the latter influence should become dominant as the flame radius becomes small. In this connection, it should be remembered that the asymptotic analysis predicts that this effect can extinguish the flame before the reaction zone reaches the axis.[13,14,45] The asymptotic analysis is based on the flame surface model, and the elevated pressure is favorable for the model because of the accelerated reaction rates and decelerated transport rates. There is a fair chance that the extinction behavior can be explained in terms of the simplified asymptotic analysis, which can predict the correct pressure dependence of the burning velocity.

2.7.5 CONCLUDING REMARKS

The numerical study on the effects of pressure on the tubular flame behavior has yielded the following concluding remarks.

1. The flame radius decreases with pressure because of the decrease in the burning velocity, while the reaction zone width is decreased due to the decelerated transport rates at elevated pressures. The accelerated reaction rates, on the other hand, increase the flame temperature approaching the equilibrium flame temperature.
2. The extinction flame temperature increases, whereas the extinction flame radius decreases, with pressure. However, the latter decrease becomes very slow at higher pressures, approaching the minimum extinction flame radius of around 0.2 mm. The flame with a radius smaller than this value cannot be established for pressures ranging from 1 to 8 atm. It is rather difficult to realize the steady flame with sizes as small as Kolmogorov microscale.
3. The critical strain rate at the extinction limit increases with pressure at first and reaches a maximum at around 2.7 atm, but decreases for higher pressures. It appears that this behavior is closely related to the effects of curvature on the extinction, which can be explained in terms of the simplified asymptotic analysis.

2.8 FINAL CONCLUSIONS

In the foregoing section, we have studied the tubular flame in terms of the similarity solution. We believe that we have elucidated important aspects of the tubular flame. However, we still have not yet understood the role played by the circumferential flow. The present study has clarified that the flame behavior is independent of this flow. In spite of this, we believe that the flow should play a certain role in the flame behavior. At present we think of this problem in the following way.

Figures 2.21 and 2.27 show the radial density and pressure distributions. In the compressible flow solution, the density decreases inward and takes a minimum on the axis. The pressure as well decreases very rapidly toward the axis. This type of flow should be stable to produce a very stable tubular flame. In the actual experiment, the unburned mixture is injected tangentially into the cylindrical burner through one or two slits. However, in the theoretical model shown in Fig. 2.1, the introduction is assumed uniform through the whole periphery. In the experiment, there should be some transition stage, in which the biased flow becomes the uniform axisymmetric flow. We think that the density and pressure distributions described above will contribute to reduce the transition stage, and to produce the uniform axisymmetric flow field very quickly. This is the contribution of the circumferential flow.

The next problem is if the minute tubular flame can be an element of turbulent combustion. As pointed out in Section 2.6, the stable tubular flame cannot be as small as the Kolmogorov microscale. Then we think that the transient tubular flame may appear in the turbulent combustion, but the steady tubular flame cannot be an element of turbulent combustion.

As mentioned repeatedly in the foregoing section, the tubular flame simultaneously experiences the stretch, along the longitudinal direction, and the curvature in the varying degree. Although we have not yet made clear these effects on the flame behavior, we still believe that the flame should give a clue to understand the effects, and this problem should be the subject of next study.

REFERENCES

1. Williams, F. A. 1985. *Combustion theory*. 2nd ed., 420. Menlo Park, California: The Benjamin/Cummings.
2. Fristrom, R. M. and A. A. Westenberg. 1965. *Flame structure*. New York: McGraw-Hill.
3. Tsuji, H. 1982. Counterflow diffusion flames. *Progress in Energy and Combustion Science* 8 (2): 93–119. DOI: 10.1016/0360-1285(82)90015-6.
4. Ishizuka, S. 1993. Characteristics of tubular flames. *Progress in Energy and Combustion Science*, 19 (3):187–226. DOI: 10.1016/0360-1285(93)90015-7.
5. Peters, N. 2000. *Turbulent combustion*. New York: Cambridge University Press.
6. Chomiak, J. 1977. *Sixteenth Symposium (International) on Combustion*, 1665. The Combustion Institute.
7. Takeno, T. and S. Ishizuka. 1986. A tubular flame theory. *Combustion and Flame* 64 (1):83–98. DOI: 10.1016/0010-2180(86)90100-8.
8. Nishioka, M. 1987. *M.S. thesis, graduate school of aeronautics*. University of Tokyo (in Japanese).
9. Yuan, S. W. and A. B. Finkelstein. 1956. Laminar pipe flow with injection and suction through a porous wall. *Transactions of the ASME* 78:719–24.
10. Burgers, J. M. 1948. A mathematical model illustrating the theory of turbulence. In *Advances in applied mechanics*, vol. 1, eds. R. von Mises and T. von Karman, 171. New York: Academic Press.
11. Ishizuka, S. 1985. *Twentieth Symposium (International) on Combustion*, 287. The Combustion Institute.
12. Kotani, Y. and T. Takeno. 1983. *Nineteenth Symposium (International) on Combustion*, 1503. The Combustion Institute.
13. Rayleigh, L. 1916. On the dynamics of revolving fluids. *Proceedings of the Royal Society A* 93:148–54. DOI: 10.1098/rspa.1917.0010.
14. Takeno, T., M. Nishioka and S. Ishizuka. 1986. A theoretical study of extinction of a tubular flame. *Combustion and Flame* 66:271.
15. Takeno, T., M. Nishioka, S. Ishizuka, and J. D. Buckmaster. 1987. Extinction behavior of a tubular flame for small lewis numbers. In *Complex chemical reaction systems*, eds. J. Warnatz and W. Jager, 302. Berlin: Springer-Verlag.
16. Takeno, T. 1972. *Tenth Japanese Combustion Symposium*, 17.
17. Wu, C. K. and C. K. Law. 1985. *Twentieth Symposium (International) on Combustion*, 1941. The Combustion Institute.

18. Warnatz, J. and N. Peters. 1984. Dynamics of flames and reactive systems, progress in aeronautics and astronautics. *AIAA* 95:61.
19. Buckmaster, J. 1979. *Seventeenth Symposium (International) on Combustion*, 835. The Combustion Institute.
20. Buckmaster, J. and G. S. S. Ludford. 1983. *Theory of laminar flames*. 179. Cambridge: Cambridge University Press.
21. Klimov, A. M. 1985. In *The mathematical theory of combustion and explosions*, eds. Ya. B. Zeldovich, G. I. Barenbrat, V. B. Librovich, and G. M. Makhviladze, 352. New York: Consultant Bureau.
22. Ishizuka, S. and C. K. Law. 1983. *Nineteenth Symposium (International) on Combustion*, 327. The Combustion Institute.
23. Sivashinsky, G. I. 1976. On a distorted flame front as a hydrodynamic discontinuity. *Acta Astronautica* 3:889.
24. Sato, J. and H. Tsuji. 1983. Extinction of Premixed Flames in a Stagnation Flow Considering General Lewis Number *Combustion Science and Technology* 33:193.
25. Takeno, T. 1977. Ignition criterion by thermal explosion theory. *Combustion and Flame* 29:209.
26. Takeno, T. and K. Sato. 1980. Effect of oxygen diffusion on ignition and extinction of self-heating porous bodies. *Combustion and Flame* 38:75.
27. Dixon-Lewis, G., V. Giovangigli, R. J. Kee, B. Rogg, M. Smooke, G. Stahl, and J. Warnatz. 1990. Numerical modeling of the structure and properties of tubular strained laminar premixed flames. In *Dynamics of deflagrations and reactive systems: flames*, Progress in Astronautics and Aeronautics vol. 131, eds. A. L. Kuhl, J.-C. Leyer, A. A. Borisov, and W. A. Sirignano, 125. Washington D.C.: AIAA.
28. Smoke, M. and V. Giovangigli. 1989. Yale University, Mechanical Engineering Report ME-102-89.
29. Kobayashi, H. and M. Kitano. 1989. Extinction characteristics of a stretched cylindrical premixed flame. *Combustion and Flame* 76:285.
30. Daneshyar, H. and G. Hill. 1987. The structure of small-scale turbulence and its effect on combustion in spark ignition engines. *Progress in Energy and Combustion Science* 13:47.
31. Ishizuka, S. 1989. An experimental study on extinction and stability of tubular flames. *Combustion and Flame* 75:367.
32. Kee, R. J., J. F. Grcar, M. D. Smooke, and J. A. Miller. 1985. Sandia Report, SAND 85-8240.
33. Zhu, D. L., F. N. Egolfopouls, and C. K. Law. 1988. *Twenty-Second Symposium (International) on Combustion*, 1537. The Combustion Institute.
34. Egolfopouls, F. N., P. Cho, and C. K. Law. 1989. *Combustion and Flame* 76:375.
35. Gardiner, W. C. Jr., and J. Troe. 1984. Rate coefficients of thermal dissociation, isomerization, and recombination reactions. In *Combustion chemistry*, ed. W. C. Gardiner, Jr., 173. Berlin: Springer-Verlag.
36. Kee, R. J., J. A. Miller, G. H. Evans, and G. Dixon-Lewis. 1988. *Twenty-Second Symposium (International) on Combustion*, 1479. The Combustion Institute.
37. Kee, R. J., J. A. Miller, and T. H. Jefferson. 1980. Sandia Report, SAND 80-8003.
38. Kee, R. J., G. Dixon-Lewis, J. Warnatz, M. Coltrin, and J. A. Miller. 1986. Sandia Report, SAND86-8246.
39. Kee, R. J., F. M. Rupley, and J. A. Miller. 1987. Sandia Report, SAND87-8215.
40. Garforth, A. M. and C. J. Rallis. 1978. Laminar burning velocity of stoichiometric methane-air: pressure and temperature dependence. *Combustion and Flame* 31:53.
41. Sharma, S. P., D. D. Agrawl, and C. P. Gupta. 1981. *Eighteenth Symposium (International) on Combustion*, 493. The Combustion Institute.
42. Iijima, T. and T. Takeno. 1986. Effects of temperature and pressure on burning velocity. *Combustion and Flame* 65:35.

43. Andrews, G. E. and D. Bradley. 1972. The burning velocity of methane-air mixtures. *Combustion and Flame* 19:275.

44. Nishioka, M., T. Takeno, and S. Ishizuka. 1988. Effects of Variable Density on a Tubular Flame. *Combustion and Flame* 73:287.

45. Libby, P. A., N. Peters, and F. A. Williams. 1989. Cylindrical premixed laminar flames. *Combustion and Flame* 75:265.

46. Ishizuka, S. 1983. *Twenty-Second Japanese Combustion Symposium, Preprint*, 161.

47. Tsuji, H. and I. Yamaoka. 1983. *Nineteenth Symposium (International) on Combustion*, 1533. The Combustion Institute.

48. Sato, J. 1983. *Nineteenth Symposium (International) on Combustion*, 1541. The Combustion Institute.

CHAPTER 3

MATHEMATICAL FORMULATION AND COMPUTATIONAL SIMULATION OF TUBULAR FLAMES

Yuyin Zhang, Huayang Zhu, Robert J. Kee

3.1 INTRODUCTION

Tubular flows represent a class of stagnation flows that satisfy mathematical similarity.[1] Generally speaking, flow is introduced into a tube or tubular annulus radially and the flow exits axially. Tubular flows and flames can be established in several ways, as illustrated in Fig. 3.1. For example, the initial theory was established using purely radial injection through a porous wall as shown in Fig. 3.1(a). Later investigations introduced a circumferential velocity component by rotating the tube wall (e.g., Fig. 3.1(c)). Practical devices introduced the flow streams via tangential slots as illustrated in Fig. 3.1(d). Figure 3.1(b) shows radial outward injection through a centrally placed porous tube. Yet another alternative could be to combine Fig. 3.1(a) and (b), introducing flow inward from the outer shell as well as injecting flow outward through a porous inner tube. The important aspect of these tubular flows is that the temperature and composition profiles are independent of the axial position, depending only on the radial coordinate. Thus, as illustrated in Fig. 3.1, tubular flames appear with a constant radius over long axial tubes. This axial independence is the physical basis for the mathematical similarity, leading to efficient computational solutions, even when complex combustion chemistry is included.

Tubular flows have fundamental theoretical value as well as practical technological value. In both cases, the value stems from the similarity characteristics that cause the temperature and composition to be independent of axial position. In fundamental studies, models and experiments benefit from the ability to interpret two- to three-dimensional flow situations in a de-facto one-dimensional setting. For example, tubular flames provide an interesting configuration to study the combined effects of curvature and strain in flames. Such studies provide valuable insights about the behavior of turbulent eddies in combustion processes.

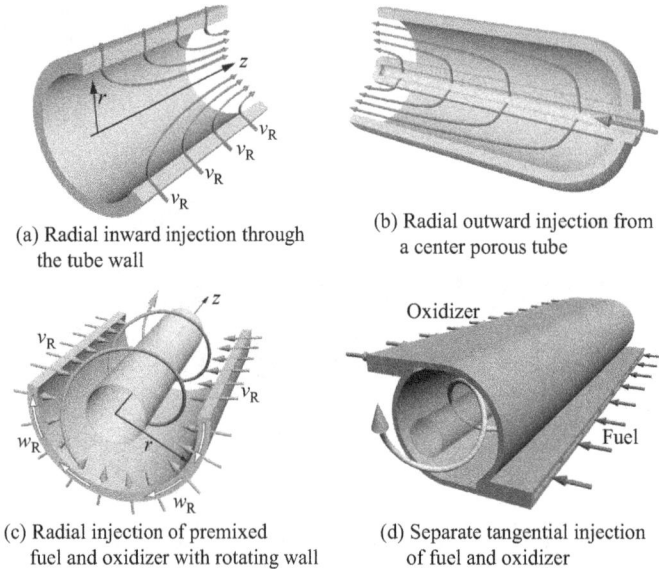

(a) Radial inward injection through the tube wall

(b) Radial outward injection from a center porous tube

(c) Radial injection of premixed fuel and oxidizer with rotating wall

(d) Separate tangential injection of fuel and oxidizer

Figure 3.1. Four possible realizations of tubular flow and flame configurations.

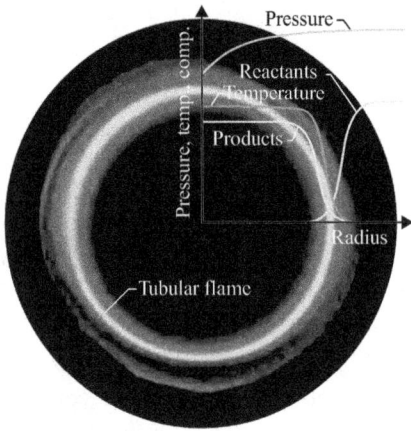

Figure 3.2. An optical photograph along the axis of an optically accessible tubular flame, with superimposed qualitative radial profiles of pressure and composition (comp.).

In technology, the similarity leads to the potential for processes that can produce uniform heat and mass fluxes to the exterior long rods or the interior of cylindrical shells. For example, tubular flame systems have been designed as heaters to deliver uniform heat flux along the length of a long tube.[2]

Figure 3.2 is a photographic image along the axis of a tubular flame. The fact that the flame appears as a thin circle provides direct physical evidence of the similarity. That is, the flame radius is independent of axial position. The superimposed graphs illustrate qualitatively the radial profiles of pressure, temperature, and composition. As is easily anticipated for premixed flames, the reactants are consumed and products are produced within thin flame zone. The heat release as a result of strongly exothermic reactions within the flame zone causes a steep temperature gradient. Of particular interest in the highly swirling tubular flames are the effects of radial pressure profiles, and their effect on flame structure and diffusive mass transport.

Although the similarity characteristics of planar stagnation flows were well known since the early 1900s, the similarity properties of tubular flow were recognized much later by Takeno and Ishizuka.[3] The early theoretical work concentrated on flows in which fluid entered the tube radially through a porous wall (e.g., Fig. 3.1(a)). Although the porous-wall setting is well suited for theoretical purposes, actually constructing such a system is difficult. Ishizuka demonstrated that tubular flames could be established approximately with tangential injection,[4] and Fig. 3.1(c) illustrates such a configuration using tangential flow injectors. Despite the fact that discrete tangential injectors do not strictly satisfy the necessary conditions for tubular-flow similarity, Ishizuka recognized that with sufficiently high injection velocities and swirl rates, the similarity conditions are very well established. The tangential-injection system provided the means to design and build practical tubular burners.

A technologically important attribute of tangentially fed tubular flames is the ability to keep fuel and oxidizer separate prior to entering the tube. In a system such as that illustrated in Fig. 3.1(d), fuel could enter via one row of injectors and air via the other. As a result of rapid mixing induced by the swirling flow within the tube, a de-facto premixed flame can be established. However, the inherent safety characteristics of a diffusion flame (e.g., eliminate flashback hazards) can be preserved. Ishizuka and colleagues have demonstrated several alternative injection strategies that have proven effective for a variety of applications.

3.2   LITERATURE OVERVIEW

The earliest models of tubular flames were based on significantly simplified assumptions, including single-step chemistry, incompressible flow, and high-activation-energy asymptotic methods. Discussion of these models was included in the 1993 review by Ishizuka.[1] The present chapter is concerned primarily with computational simulation, using detailed chemical kinetics. The Ishizuka review also discusses computational models up through 1993.

Takeno and Ishizuka first modeled tubular flames in a rotating flow field, seeking to explain the experimentally observed effects of flow rate and equivalence ratio.[3] This early model developed the similarity framework, but retained assumptions such as incompressible viscous flow field and simplified chemistry with a flame-surface model. The predicted flame structure and stability behavior correlated satisfactorily with the experimental observations. The approach was then extended to analyze flame extinction and compared with computational models.[5] Both the analytical and computational methods were in general agreement and predicted the extinction behavior, with discrepancies being explained in terms of differences in the reaction chemistry models. An important result was to show that because of flame curvature the extinction behavior of tubular flames was different from that of analogous planar flames.

Nishioka et al. extended the models to include the effects of compressible flow.[6] This numerical model revealed strong coupling between flame structure and the flow field that was not as apparent in the earlier incompressible models. The gas flow normal and parallel to the flame are accelerated by the flow expansion, while the circumferential flow is decelerated because of the increased viscosity at high temperature. The flame is shifted outward and the reaction zone is thicker due to enhanced role of molecular transport, resulting in an increase in the unburned fuel remaining on the axis. The extinction Reynolds number is somewhat reduced because of an increase in the stretch due to the flow expansion. However, with the introduction of suitably defined flame stretch and mass burning velocity, the predictions are in reasonable agreement with models using asymptotic analysis and incompressible flow.

By the late 1980s, researchers outside of Japan were beginning to recognize the value of studying tubular flames and several groups were developing computational models with detailed kinetics. In 1990, Prof. Graham Dixon-Lewis hosted a workshop at the University of Leeds in which six groups (University of Leeds, École Polytechnique and CNRS, Sandia National Laboratories, University of Cambridge, Yale University, and Universität Stuttgart) compared different computational methods and solutions with tubular flames.[7] All the groups solved a particular problem that was defined by Prof. Tadao Takeno (Nagoya University), but used different computational methods. All groups used detailed chemical kinetics for methane–air premixed flames, but each group used somewhat different reactions mechanisms.

Adopting detailed C_1 chemical kinetics, Nishioka et al.[8] investigated the effects of pressure (1–8 atm) on structure and extinction of stoichiometric methane–air tubular flames. Results revealed the pressure dependence of burning velocity, flame radius, flame temperature, and critical strain rate at extinction. Smooke and Giovangigli[9] also modeled the extinction of tubular premixed methane–air and propane–air flames with detailed chemistry and complex transport. The conservation equations in similarity form were solved as a two-point boundary value problem, with arc–length continuation methods developed to model the turning-point behavior at extinction. The predicted results agreed well with the experimental observations by Kobayashi and Kitano.[10] As anticipated from the earlier modeling results, the curvature associated with tubular flames caused significant differences from the corresponding planar counterflow flames. Nishioka et al. developed an alternative continuation method to predict ignition–extinction S-curve characteristics for a range of strained flames, including tubular flames.[11]

Modeling premixed tubular methane–air flames, Ju et al.[12] and Chen et al.[13] studied the combined effects of flame curvature and radiation heat loss on flame extinction. These studies found that extinction is related to both factors, with flame curvature being an important factor in extending the radiation extinction limit. Mosbacher et al. investigated the response of the premixed H_2–air tubular flame to the stretch rate and curvature using a tubular flame model with complex chemistry and detailed transport.[14] Comparisons between Raman measurements and numerical simulations showed excellent agreement at low stretch rates (i.e., $\kappa \leq 127$ s^{-1}). However, at a higher flame stretch, hence increased curvature, the computational predictions were less satisfactory. The experimental observations showed extinction at $\kappa \approx 227$ s^{-1}, while the numerical simulations significantly overpredicted the extinction limit ($\kappa \approx 750$ s^{-1}). This discrepancy may be attributed to the very high sensitivity of extinction behavior to transport properties and flame curvature.

Hayashi et al. used tubular flame simulations to study the effects of flame suppressants on flame extinction.[15] The study considers the effects of three suppression agents (water vapor, carbon dioxide, and halon) on premixed methane–air flames. Using tubular flames enabled the study to include the effects of flame curvature.

In a series of papers, Zhang et al. studied the effects of high rotation rates on the structure of swirl-type premixed tubular flames.[16–18] The simulation software was an extension of the Oppdif package.[19] These models incorporated detailed chemical reaction mechanisms for propane–air,[16] methane–air,[18] and hydrogen–methane–air[17] flames. Because high rotation rates cause large radial pressure variations, pressure diffusion (Dufour effect) was included and, unlike all other previous models, the radial momentum equation was retained. Results predict how rotation rate affects pressure variation, flame position, flame temperature, and flame velocity. Pressure diffusion is most significant when the flame position is close to the maximum circumferential velocity. Pressure diffusion strengthens the flame when it assists ordinary diffusion of the rate-controlling species, and vice versa. The direction of pressure diffusion depends on the molecular weight of the rate-controlling species relative to the gas mixture. The swirl rate, and hence the pressure diffusion, can influence methane–air flames differently from comparable propane–air flames. These differences are explained in terms of the pressure fields and the first Damköhler number. Yamamoto also investigated the flame characteristics in a stretched, rotating flow by numerical simulation of tubular laminar flames for lean hydrogen–, methane–, and propane–air mixtures.[20] Twin planar flames in counterflow were also simulated for comparison. However, Yamamoto's predictions are different from those predicted by Zhang et al.[16,18] Yamamoto reported that the temperature of a lean methane–air tubular flame should increase monotonically with an increasing rotation rate. The difference may be related to Yamamoto neglecting the effects of reduced pressure in the flame and using a global reaction mechanism. Subsequent sections in this chapter discuss in detail the role of pressure diffusion in tubular flames.

Hu et al. developed a non-premixed tubular flame model to correspond with measurements using spontaneous Raman spectroscopy.[21] This study considers tubular flames over a range of stretch rates, with the fuel being 15% H_2 diluted by N_2 and the air as the oxidizer. Comparisons between measurements and computational predictions for the opposed-jet flat flames showed that the curvature weakens the effects of preferential diffusion for diluted hydrogen flames where the curvature is concave toward the fuel stream. The effect of curvature on flame behavior depends on Lewis number. For Lewis number close to unity, curvature has a minimal effect on tubular non-premixed flames. Wang et al. compared the tubular diffusion flames with opposed-jet diffusion flames numerically to show the effect of curvature on diffusion flames.[22] Positive curvature strengthens the preferential diffusion and negative curvature weakens the preferential diffusion as in premixed flames. The magnitude of the strengthening or weakening effect depends on the ratio of flame thickness to flame radius. To further study the effects of curvature on extinction and flame instability, Hu et al. compared their numerical results of premixed H_2–air, CH_4–air, and C_3H_8–air tubular flames with measurements based on laser-induced Raman spectroscopy.[23] Discrepancies between model and experiment suggested that the reaction mechanisms may need to be improved for lean premixed near-extinction flames.

Kee et al. developed a model for simulating tubular flames with tangential injection.[24] This extension provided the capability needed to model the kinds of tubular burners that are used in practical applications (e.g., Fig. 3.1(d)). Formulation of the tangential-injection model is discussed in a following section of this chapter.

3.3   MATHEMATICAL FORMULATION

The general derivation of the similarity formulation of tubular flows begins with the axisymmetric Navier–Stokes equations (i.e., r–z coordinates), which is the same starting point as for deriving the axisymmetric stagnation–flow similarity equations for flow impinging onto a flat plate. In the tubular case, the flow is introduced radially and exhausted axially. In the planar case, the flow is introduced axially and exhausted radially. Nevertheless, the derivations are completely analogous.[25]

The governing equations for the tubular flames in similarity form are derived from the three-dimensional steady-state Navier–Stokes equations.[25] Although a three-dimensional flow field is retained, axial variations in temperature T and composition (mass fraction Y_k) vanish. The radial and circumferential velocity components v and w are independent of the axial position z. The axial velocity component u is scaled such that $U = u / z$ is independent of axial position. Additionally, all circumferential variations (i.e., derivatives in θ) are neglected, retaining only z and r as the independent variables.

The following system of conservation equations form the starting point for deriving the tubular-flow similarity equations. Source terms $S(r)$ for mass, momentum, and thermal energy have been included because, as discussed subsequently, they can be used to approximate practical configurations such as tangential injection.[24]

$$\frac{\partial(\rho u)}{\partial z} + \frac{1}{r}\frac{\partial(\rho r v)}{\partial r} = S(r), \tag{3.1}$$

$$\rho u \frac{\partial u}{\partial z} + \rho v \frac{\partial u}{\partial z} + u S(r) = -\frac{\partial p}{\partial z} + \frac{\partial}{\partial z}\left[2\mu\frac{\partial u}{\partial z} - \frac{2}{3}\mu\nabla\cdot\mathbf{V}\right] + \frac{1}{r}\frac{\partial}{\partial r}\left[\mu r\left(\frac{\partial v}{\partial z} + \frac{\partial u}{\partial r}\right)\right], \tag{3.2}$$

$$\rho u \frac{\partial v}{\partial z} + \rho v \frac{\partial v}{\partial r} - \rho \frac{w^2}{r} + v S(r) = -\frac{\partial p}{\partial r} + \frac{\partial}{\partial z}\left[\mu\left(\frac{\partial v}{\partial z} + \frac{\partial u}{\partial r}\right)\right]$$

$$+ \frac{\partial}{\partial r}\left[2\mu\frac{\partial v}{\partial r} - \frac{2}{3}\mu\nabla\cdot\mathbf{V}\right] + 2\mu r\left[\frac{\partial v}{\partial r} - \frac{v}{r}\right]$$

$$+ (1 - \mathcal{F})S_{\mathrm{m}}(r), \tag{3.3}$$

$$\rho u \frac{\partial w}{\partial z} + \rho v \frac{\partial w}{\partial r} + \rho \frac{vw}{r} + w S(r) = \frac{\partial}{\partial z}\left[\mu\left(\frac{\partial w}{\partial z}\right)\right] + \frac{\partial}{\partial r}\left[\mu\left(\frac{\partial w}{\partial r} - \frac{w}{r}\right)\right]$$

$$+ \frac{2\mu}{r}\left[\frac{\partial w}{\partial r} - \frac{w}{r}\right] + \mathcal{F}S_{\mathrm{m}}(r), \tag{3.4}$$

$$\rho v c_{\mathrm{p}} \frac{dT}{dr} = \frac{1}{r}\frac{d}{dr}\left(\lambda r \frac{dT}{dr}\right) - \sum_{k=1}^{K}\rho Y_k V_k c_{\mathrm{p}k} \frac{dT}{dr}$$

$$- \sum_{k=1}^{K} h_k W_k \dot{\omega}_k + \sum_{k=1}^{K} S_k(h_{\mathrm{in},k} - h_k), \tag{3.5}$$

$$\rho v \frac{dY_k}{dr} = -\frac{1}{r}\frac{d}{dr}\left(r\rho Y_k V_k\right) + W_k \dot{\omega}_k + S_k(r) - Y_k S(r), \tag{3.6}$$

Dependent variables include the mass density ρ, pressure p, temperature T, and mass fractions Y_k. Other variables include the species molecular weights W_k, specific enthalpies h_k, and heat capacities $c_{\mathrm{p}k}$. The species diffusion velocities are represented as V_k; $\dot{\omega}_k$ are the species molar production rates due to the homogeneous chemical reactions. Transport properties include the dynamic viscosity μ and thermal conductivity λ. The universal gas constant is represented as R. The source term $S(r)$ represents the radial variation in injected mass, with $S_k(r)$ representing the injection mass rates of the kth species. The specific enthalpies of the injected species are represented as $h_{in,k}$. The total momentum of the injected mass is reprinted as $S_m(r)$. If the injectors are directed purely tangentially, then the injected momentum is only in the circumferential direction. However, the injectors may be designed to introduce a small radial component. The analysis considers the possibility that a fraction of the momentum may be introduced radially. In this case it is assumed that a fraction F of the momentum is introduced circumferentially and a fraction $(1 - F)$ is introduced radially.

The diffusion velocities of the gas-phase species are evaluated as

$$V_k = \sum_{j=1}^{K} \frac{W_j D_{j,k}}{X_k \overline{W}} \left[\frac{dX_j}{dr} + (X_j - Y_j) \frac{1}{p} \frac{dp}{dr} \right] - \frac{D_k^T}{\rho Y_k} \frac{1}{T} \frac{dT}{dr}, \tag{3.7}$$

which includes the effects of pressure diffusion (Dufour) and thermal diffusion (Soret). Thermal diffusion can be especially important in the vicinity of flame fronts. Pressure diffusion is often negligible, except in situations with a very high swirl velocity.[16] In this expression, X_k are mole fractions, $D_{j,k}$ are ordinary multicomponent diffusion coefficients (note that the ordinary multicomponent diffusion coefficients are different from the binary diffusion coefficients, $D_{j,k} \neq D_{j,k}$[25]), $\overline{W}$ is mean molecular weight, and D_k^T are thermal diffusion coefficients.

3.3.1 SIMILARITY FORM

Using a stagnation–flow similarity approach, the continuity and momentum equations can be reduced to ordinary differential equation form.[25] Assume that the axial and radial velocity fields can be represented as

$$\rho u r = -z \frac{dV}{dr}, \quad \rho v r = V(r), \tag{3.8}$$

where $V(r)$ is an as-yet unspecified function of r alone. Assume further that the temperature and composition, and hence the density, have no axial variations (i.e., they are functions of r alone), the axial and radial momentum equations can be rewritten to isolate the pressure gradients as

$$\frac{1}{z} \frac{\partial p}{\partial z} = \frac{\rho}{r} \frac{dV}{dr} S(r) - \frac{1}{\rho r^2} \left(\frac{dV}{dr} \right)^2 + \frac{V}{r} \frac{d}{dr} \left(\frac{1}{\rho r} \frac{dV}{dr} \right) - \frac{1}{r} \frac{d}{dr} \left[\mu r \frac{d}{dr} \left(\frac{1}{\rho r} \frac{dV}{dr} \right) \right], \tag{3.9}$$

$$\frac{\partial p}{\partial r} = -\frac{V}{r} \frac{d}{dr} \left(\frac{1}{\rho} \frac{V}{r} \right) + \rho \frac{w^2}{r} - \frac{V}{\rho r} S(r) + 2 \frac{d}{dr} \left[\mu \frac{d}{dr} \left(\frac{1}{\rho} \frac{V}{r} \right) \right]$$

$$- \frac{2}{3} \frac{d}{dr} \left[\mu \left(-\frac{1}{\rho r} \frac{dV}{dr} + \frac{1}{r} \frac{d}{dr} \left(\frac{V}{\rho} \right) \right) \right] + \frac{2\mu}{r} \left[\frac{d}{dr} \left(\frac{1}{\rho} \frac{V}{r} \right) - \frac{1}{\rho} \frac{V}{r^2} \right]$$

$$- \mu \frac{d}{dr} \left(\frac{1}{\rho r} \frac{dV}{dr} \right) + (1 - F) S_m(r). \tag{3.10}$$

Assuming that w is a function of r alone, it is clear that both momentum equations are functions of r alone. By differentiating the axial-momentum equation with respect to r, it can be concluded that

$$\frac{\partial}{\partial r}\left(\frac{1}{z}\frac{\partial p}{\partial z}\right) = F(r) \tag{3.11}$$

is a function of r alone. However, since the pressure is a continuous differentiable function of the z and r, the order of differentiation can be interchanged as

$$\frac{\partial}{\partial r}\left(\frac{1}{z}\frac{\partial p}{\partial z}\right) = \frac{1}{z}\frac{\partial}{\partial z}\left(\frac{\partial p}{\partial r}\right) = 0. \tag{3.12}$$

Because $\partial p / \partial r$ is known to be a function of r alone, its axial derivative must vanish. Therefore, it must be the case that

$$\frac{1}{z}\frac{\partial p}{\partial z} = \Lambda_z = constant. \tag{3.13}$$

The equations can be written in a somewhat more recognizable form by replacing the functions of V with physical velocities. From Eq. 3.8, where V is defined in terms of the physical velocities,

$$\rho\frac{u}{z} = \rho U = -\frac{1}{r}\frac{dV}{dr}, \qquad \rho v = \frac{V}{r}. \tag{3.14}$$

The scaled axial velocity $U = u / z$ is seen to be a function of r alone.

The system of governing equations in similarity form, which form a boundary-value problem, is summarized as follows:

$$\frac{d\rho v}{dr} + \rho rU = rS(r), \tag{3.15}$$

$$\rho v\frac{dU}{dr} + \rho U^2 + US(r) = -\Lambda_z + \frac{1}{r}\frac{d}{dr}\left(\mu r\frac{dU}{dr}\right), \tag{3.16}$$

$$\rho v\frac{dv}{dr} - \rho\frac{w^2}{r} + vS(r) = -\frac{dp}{dr} + 2\frac{d}{dr}\left(\mu\frac{dv}{dr}\right) - \frac{2}{3}\frac{d}{dr}\left[\mu\left(U + \frac{1}{r}\frac{drv}{dr}\right)\right]$$
$$+ \frac{2\mu}{r}\left(\frac{dv}{dr} - \frac{v}{r}\right) + (1 - \mathcal{F})S_m(r), \tag{3.17}$$

$$\rho v\frac{dw}{dr} + \rho\frac{vw}{r} + wS(r) = \frac{d}{dr}\left[\mu\left(\frac{dw}{dr} - \frac{w}{r}\right)\right]$$
$$+ \frac{2\mu}{r}\left(\frac{dw}{dr} - \frac{w}{r}\right) + \mathcal{F}S_m(r), \tag{3.18}$$

$$\rho v c_{\mathrm{p}} \frac{dT}{dr} = \frac{1}{r} \frac{d}{dr} \left(\lambda r \frac{dT}{dr} \right) - \sum_{k=1}^{K} \rho Y_k V_k c_{\mathrm{p}k} \frac{dT}{dr}$$

$$-\sum_{k=1}^{K} h_k W_k \dot{\omega}_k + \sum_{k=1}^{K} S_k (h_{\mathrm{in},k} - h_k), \tag{3.19}$$

$$\rho v \frac{dY_k}{dr} + Y_k S(r) = -\frac{1}{r} \frac{d}{dr} (r \rho Y_k V_k) + W_k \dot{\omega}_k + S_k(r), \tag{3.20}$$

The radius r is the independent variable. The dependent variables are the radial velocity v, scaled axial velocity $U = u/z$, circumferential velocity w, temperature T, mass fractions Y_k, and pressure p. The scaled axial pressure gradient

$$\Lambda_z = \frac{1}{z} \frac{dp}{dz} \tag{3.21}$$

is an eigenvalue. The density ρ is determined from the equation of state,

$$\rho = \frac{p}{RT \sum_{k=1}^{K} Y_k / W_k}. \tag{3.22}$$

The continuity equation (Eq. 3.15) is a first-order equation, requiring only one boundary condition. However, two boundary conditions must be satisfied for the radial velocity v. That is, at the tube wall the radial inlet velocity is specified and at the centerline symmetry requires $v = 0$. The axial momentum equation (Eq. 3.16) is elliptic and second order, requiring boundary conditions for U at the tube wall and the centerline. Thus, it might appear that there are too many boundary conditions for the first-order continuity equation. This apparent dilemma is resolved by requiring the pressure-gradient eigenvalue Λ_z in the axial momentum equation to determine in such a way that all the boundary conditions are satisfied. This situation is frequently encountered in stagnation-like similarity problems in finite domains.[25,26] As with the momentum equation, the thermal-energy and species-continuity equations require boundary conditions at the tube wall and the centerline.

3.3.2 RADIAL INJECTION

As illustrated in Fig. 3.1(c), fluid enters the tube radially through a rotating porous wall. A swirling flow field is established within the tube. In this case, all the injection source terms within the tube vanish (i.e., $S_k(r) = 0$, $S(r) = 0$, and $S_{\mathrm{m}}(r) = 0$). At the tube wall ($r = R_{\mathrm{w}}$), the velocity, temperature, species composition, and fluxes are specified as

$$U = 0, \quad v = -v_R, \quad w = w_R, \quad T = T_R, \tag{3.23}$$

$$\rho Y_k (v + V_k) = (\rho Y_k v)_R. \tag{3.24}$$

At the tube centerline ($r = 0$), the symmetry boundary condition requires that

$$\frac{dU}{dr} = 0, \quad v = 0, \quad w = 0, \quad \frac{dT}{dr} = 0, \quad \frac{dY_k}{dr} = 0. \tag{3.25}$$

3.3.3 TANGENTIAL INJECTION

Modeling tubular flames with radial inlets and rotating outer walls is a valuable setting for fundamental studies. In practice, however, such a configuration is difficult, if not impossible, to fabricate and operate. In fact, Ishizuka's first experimental realization of tubular flame used a single-slot tangential injection.[4] In other words, the traditional theory of the tubular flame burner has limited ability to represent practical burner designs. Instead, the tangential-injection approaches for the tubular flame (Fig. 3.1(d)) are found in practice. Using the traditional modeling approximation for the tangential-injection tubular flame, the radial flow at the wall is inconsistent with the actual situation of a solid tube wall. The tangential-injection system introduces flow in the circumferential direction, usually without any radial component. However, because of a no-slip condition at the tube wall, the wall boundary layer and mass continuity induce a radial velocity component. Another approach approximates the inlet conditions with interior source terms.[24] The source terms added to the continuity equations, circumferential momentum equation, and energy equation of the tubular-flow conservation equations are used to represent the effects of tangential injection. Therefore, there is no explicit statement of the inlet flow via the boundary conditions. Rather, the inlet conditions are communicated via the interior source terms. At the outer tube wall ($r = R_w$), all velocity components are set to zero, indicating a no-slip solid wall. The wall boundary conditions are specified as

$$U = 0, \quad v = 0, \quad w = 0, \quad T = T_w, \quad \frac{dY_k}{dr} = 0. \tag{3.26}$$

At the centerline ($r = 0$), symmetry conditions remain valid (i.e., Eq. 3.25).

To represent the effects of tangential injection, it is assumed that the injection rates and swirl velocities are high, mixing of the tangential inlet flow streams is rapid in the circumferential direction, and the injected mass can be approximated by a Gaussian function of the radial coordinate that is centered at the midpoint of the injection ports (i.e., near the tube wall as illustrated in Fig. 3.3),

$$G(r) = A_i \exp\left[-\frac{(r - R_i)^2}{W_i^2}\right]. \tag{3.27}$$

The Gaussian is centered at R_i and its width is proportional to W_i. Although the Gaussian function is chosen somewhat arbitrarily, it has the correct qualitative features that the mass, momentum, and energy addition are concentrated near the outer wall, where the injected flow enters the tube. The peak A_i and width W_i of the function can be adjusted to approximate the particular features of the injection system.

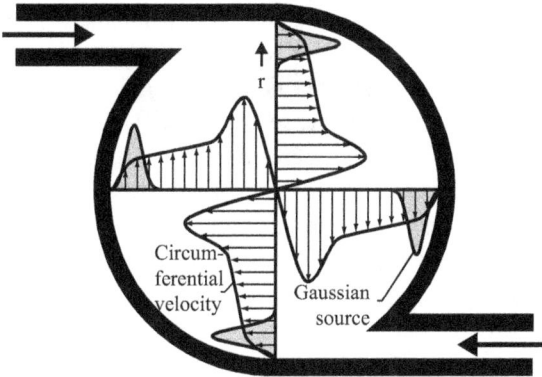

Figure 3.3. Illustration of a tubular flow, showing functional forms of the Gaussian sources and circumferential velocity profiles.

Table 3.1. Source terms of the tangential injection

Source terms	Peak value, A_i	SI dimensions
Species mass, S_k	$\dot{M}'_k/f_g$	$\text{kgs}^{-1}\text{m}^{-3}$
Total mass, S	$\dot{M}'/f_g$	$\text{kgs}^{-1}\text{m}^{-3}$
Momentum, S_m	$\dot{M}'V_i/f_g$	$\text{kgs}^{-2}\text{m}^{-2}$
Energy, S_h	$\dot{M}'h_{in}/f_g$	Wm^{-3}
Mechanical work, S_w	$\dot{M}'V_i^2/f_g$	Wm^{-3}

Assuming that R_i and W_i are specified, the peak values for all the source terms can be calculated at the given injection mass flow rates for species $\dot{M}'_k$, total injection mass rate $\dot{M}'$, velocity V_i, and the mixture-specific enthalpy h_{in}, and are listed in Table 3.1. The factor f_g is determined such that the Gaussian represents the net mass-insertion rates as

$$f_g = 2\pi \int_0^{R_w} \exp\left[-\frac{(r-R_i)^2}{W_i^2}\right] r\,dr \tag{3.28}$$

$$= \pi W_i^2 \left\{ \exp\left[-\frac{R_i^2}{W_i^2}\right] - \exp\left[-\frac{(R_i-R_w)^2}{W_i^2}\right] \right\}$$

$$+ \pi^{3/2} W_i R_i \left\{ \text{erf}\left[\frac{R_i}{W_i}\right] - \text{erf}\left[\frac{(R_i-R_w)}{W_i}\right] \right\},$$

where $\text{erf}(z)$ is the error function (i.e., $\text{erf}(z) = (2/\sqrt{\pi})\int_0^z e^{-t^2}\,dt$).

Although a Gaussian works well, other shapes could easily be used as well. The only requirement is the function is smooth and can be easily integrated. For example, the velocity profile within the entrance slot could be parabolic, which might suggest a parabolic source function. However, there is not a straightforward relationship between the assumed shape of the source and the velocity profiles, either in the entrance slot or in the resulting tubular flow. Rather, a certain level of empiricism is required in establishing the source shape and its specific parameters. Because of high swirl rates and rapid mixing, the source applies at all circumferential locations, not only near the entrance slots. The Gaussian function is nonzero at all radial locations, although, of course it should become vanishingly small away from the peak value. However, depending on parameter values (i.e., R_i and W_i), the source could have a significant value at the tube wall. Initially, this may seem to be physically inappropriate. However, consider rotating the profiles shown in Fig. 3.3 by 45°. In this case, the profiles extend into the slot region, and do not directly intersect the tube wall. Furthermore, as the flow moves away from the slot, the radial behavior of the source profile can distort. It must be recognized that the approach developed here seeks to approximate a complex three-dimensional situation with a relatively simple one-dimensional model.

Additionally, there is no simple relationship between the source profiles and the solution profiles. For example, Fig. 3.3 shows a nominal Gaussian shape as well as a circumferential velocity profile. Note that relative to the peak in the Gaussian, the peak circumferential velocity is shifted significantly toward the center of the tube. Solutions show that the tangential velocity profiles near the centerline qualitatively take the shape of a Burgers vortex.[27,28]

3.3.4 PRACTICAL CONSIDERATIONS

Practical tubular flame burners[4] often use a chimney attached at the burner exit. Figure 3.4 illustrates a possible configuration for a tubular flame with an attached chimney. The chimney serves as a kind of flame holder, helping to stabilize the flame. Strictly speaking, however, the similarity theory is valid only for the flow and combustion processes inside the tubular section without chimney. In the chimney section, where mass is no longer injected, the similarity assumption is not directly applicable. Nevertheless, flames are observed to retain the tubular structure even for relatively long chimneys.

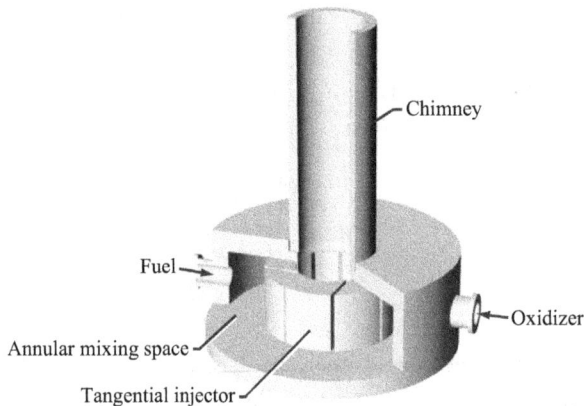

Chimney

Fuel

Oxidizer

Annular mixing space

Tangential injector

Figure 3.4. Illustration of a short tangentially fed tubular burner with an attached chimney.

3.3.5 COMPUTATIONAL PROCEDURE

The governing equations are discretized using a finite-volume approach and the mesh points are placed adaptively to resolve regions of high gradient or curvature in the solution profiles. The resulting system of nonlinear algebraic equations is solved by the hybrid Newton method.[25,29] One approach to software development is to modify either the Oppdif or Spin packages.[19,30,31] Thermodynamic and transport properties as well as reaction rates can be evaluated using the Chemkin package.[25,30,32]

3.4 MODEL VALIDATION

3.4.1 TUBULAR FLAME WITH A RADIAL INLET FLOW

Hu et al.[23] investigated the premixed tubular flames using several fuels (H_2, CH_4, and C_3H_8) and fuel–air stoichiometries. These studies were based upon an experimental configuration that supplied the radial inlet flow and also provided for optical access. Major species and temperature profiles were measured with Raman spectroscopy. Models were developed using a modified version of the Oppdif software.[14,19] For lean H_2–air flames at low strain rate, the results showed very good agreement between model and experiment. For methane and propane flames, the agreement between model and experiment was not as good. For example, the measured CO at near-extinction conditions indicated less-complete reaction than predicted by the model. The authors found significant sensitivities in predicted results using different reaction mechanisms, suggesting that discrepancies between model and experiment may be due to issues with reaction mechanisms. In any case, the results revealed significant differences between tubular flames and equivalent planar strained flames, revealing the major influence caused by flame curvature.

3.4.2 SWIRLING TUBULAR FLAME WITH A SINGLE INLET SLOT

Figure 3.5 schematically illustrates a single-slot tubular-flame configuration as developed by Ishizuka and colleagues.[4,33] The cylindrical tube is 19 mm in diameter and fabricated from

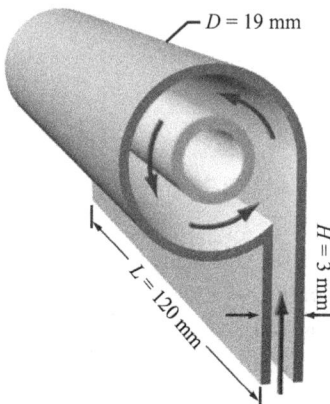

Figure 3.5. A sketch of the single-slot tubular flame developed by Ishizuka.[4,33]

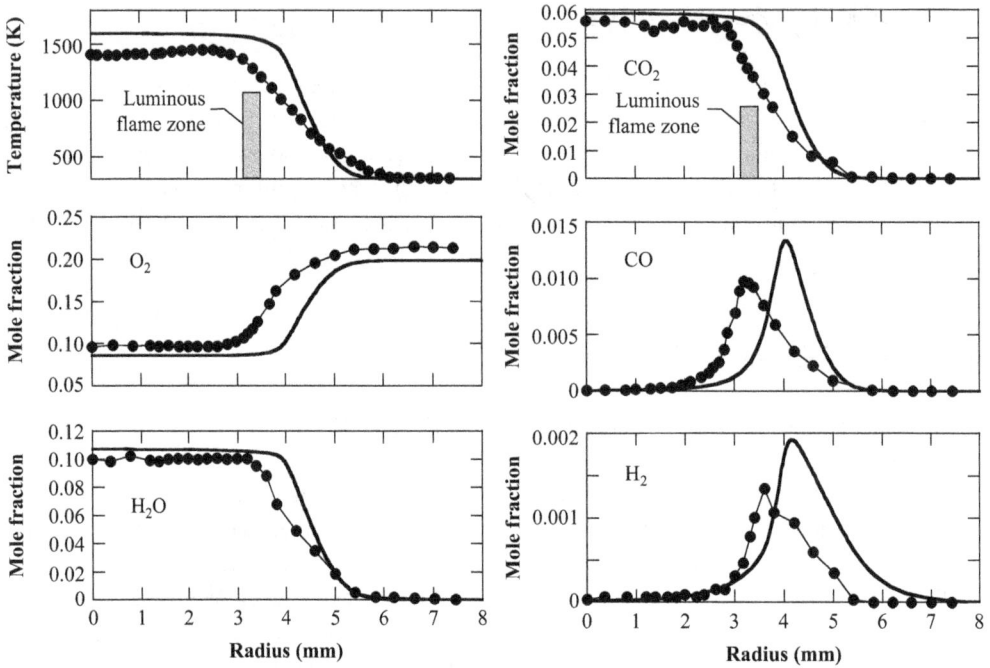

Figure 3.6. Comparison of predicted and measured temperature and species profiles for a lean methane–air tubular flame.[24,33]

glass. The inlet slot is 3 mm wide and the device is 120 mm long. Figure 3.6 shows a comparison between the model predictions with the temperature and species profiles measured by Sakai and Ishizuka[33] for the configuration shown in Fig. 3.5. The inlet mixture is 5.59% CH_4 in air, with an injection velocity of 3.3 m s^{-1}. The geometric parameters for the source function are $R_i = 8$ mm, $W_i = 0.5$ mm (Eq. 3.27), and $F = 1$ (i.e., pure circumferential injection).

Temperature profiles were measured with a coated thermocouple and species profiles were measured using microprobe gas chromatography. The simulation uses GRI-Mech 3.0 to describe the reaction kinetics.[34] Although certainly not perfect, the measurements and predictions are in very good agreement. Not only are the species profiles and flame position predicted accurately, the extinction conditions are also predicted. The simulation predicts flame extinction at 5% methane, which agrees with experimental observation reported by Sakai and Ishizuka.[33] Figure 3.7 shows additional model predictions for the 5.59% methane flame reported by Sakai and Ishizuka.[33]

3.5 FLAME STRUCTURE AND PRESSURE DIFFUSION

The effects of pressure diffusion (i.e., Dufour transport as represented by the pressure-gradient term in Eq. 3.7) on flame structure was first investigated theoretically a half century ago.[35] Nevertheless, because it usually offers only a weak contribution to the species flux, pressure diffusion is usually neglected. Recently, using computational simulations for tubular flames with complex chemistry, Zhang et al. have shown that the pressure-driven molecular transport

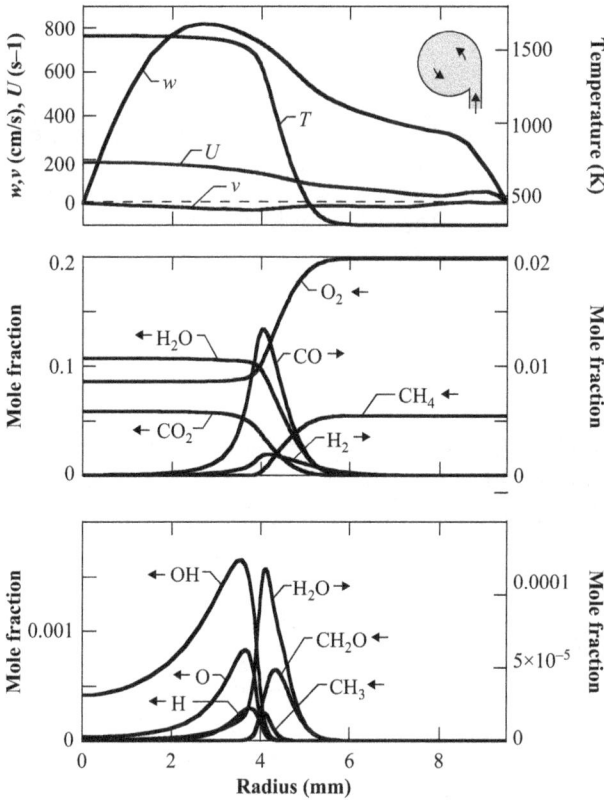

Figure 3.7. Predicted velocity, temperature, and species profiles for a lean methane–air tubular flame.

in tubular flows with very high rotation rates can be significant.[16–18] The following sections discuss the influences of pressure diffusion on the tubular premixed flame structure for propane– and methane–air flames such as seen in Fig. 3.1(c).

3.5.1 PREMIXED PROPANE–AIR FLAMES

Consider a premixed propane–air tubular flame operating at atmospheric pressure, an inlet temperature of 300 K, and a radial inlet velocity of $v_R = 1.91$ m s^{-1}. The tube diameter is $d = 3$ cm. Under these conditions, the resulting axial stretch rate is

$$\varepsilon = \frac{4v_R}{d} = 255 \text{ s}^{-1}. \qquad (3.29)$$

To illustrate the effects of certain operating conditions, the wall rotation rates are varied as $0 \le w_R / v_R \le 15$ and the fuel–air equivalence ratio is varied as $0.9 \le \phi \le 1.6$.

The model incorporates an elementary propane–air reaction mechanism adopted from Sung et al.[36] The mechanism has 92 species (including aromatic C$_6$) and 621 elementary reactions. To validate the mechanism itself, Fig. 3.8 shows the predicted and measured burning velocities as a function of propane–air equivalence ratio.[37,38] These predictions were made using the Premix software.[29,39]

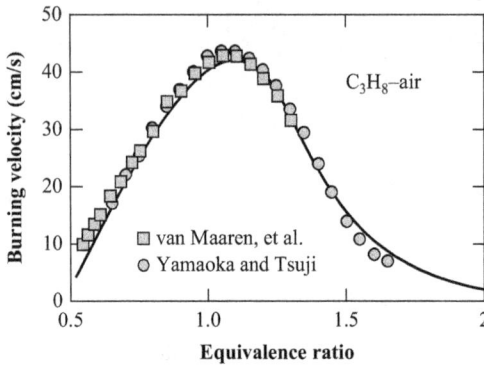

Figure 3.8. Predicted and measured laminar burning velocities. The measurements are from van Mareen et al.[38] and Yamaoka and Tsuji.[37] The predictions use the propane–air mechanism from Sung et al.[36]

Figure 3.9 illustrates the radial profiles of pressure, temperature, flow velocities, and heat-release rate for the lean ($\phi = 0.9$) and rich ($\phi = 1.6$) flames with a rotation ratio $w_R / v_R = 11.8$. The results indicate that fuel–air stoichiometry can significantly affect flame position and flame structure.

Figure 3.10 shows the radial variations of factors that are found to affect pressure diffusion for lean ($\phi = 0.9$) and rich ($\phi = 1.6$) flames. For all cases in Fig. 3.10 the inlet velocity ratio is $w_R / v_R = 11.8$. Figure 3.10(a) shows that the molecular weight of the rate-controlling species (C_3H_8) is greater than the mean molecular weight of the mixture. Thus, because $Y_k = X_k W_k / \overline{W}$, the mass fraction of the controlling species is greater than its mole fraction (Fig. 3.10(b)). Consequently, the $(X_k - Y_k)$ factor in Eq. 3.7 for the rate-controlling species is negative. As illustrated in Fig. 3.10(c), the pressure gradient is always positive. Thus the pressure-diffusion contribution to the net diffusive mass flux of the controlling species opposes the ordinary-diffusion contribution associated with the concentration gradients. Figure 3.11 compares the diffusion mass flux of the rate-controlling species (C_3H_8 for the lean mixture and O_2 for the rich mixture) with and without the pressure-diffusion term included in Eq. 3.7. When pressure diffusion is included, the peak diffusive mass flux of the rate-controlling species decreases only slightly for the lean mixture. However, the effect is more significant for the rich mixture. In both the lean and rich cases, the maximum diffusive mass flux shifts slightly toward the centerline when pressure diffusion is included.

Fig. 3.10(c) shows that the radial pressure-gradient profiles have similar trends for the lean and rich cases. The pressure gradients near the outer wall are relatively small, but increase toward local maxima just upstream of the flame. These local peaks are caused by the centrifugal forces due to the high rotation rates. The pressure gradients then fall toward local minima near the flame front, which are caused by the lower density in high-temperature flame zone. Downstream of the flame, the pressure gradients increase sharply because of the significant increase in the circumferential velocity. Finally, the pressure gradients fall toward zero at the centerline due to axisymmetry. As illustrated in Fig. 3.10(d), $\Lambda_z = (1/p)dp/dr$ in Eq. 3.7 at the points of maximum heat release is nearly a factor of 10 greater in the rich case than that in the lean case. This greater pressure gradients in the rich case lead to the larger differences in the diffusion mass fluxes (Fig. 3.11). However, this effect is counterbalanced by the smaller difference between the controlling species molecular weight and the mixture mean molecular weight in the rich case. The result suggests that pressure diffusion may be more significant when the flame position is close to the position of maximum circumferential velocity.

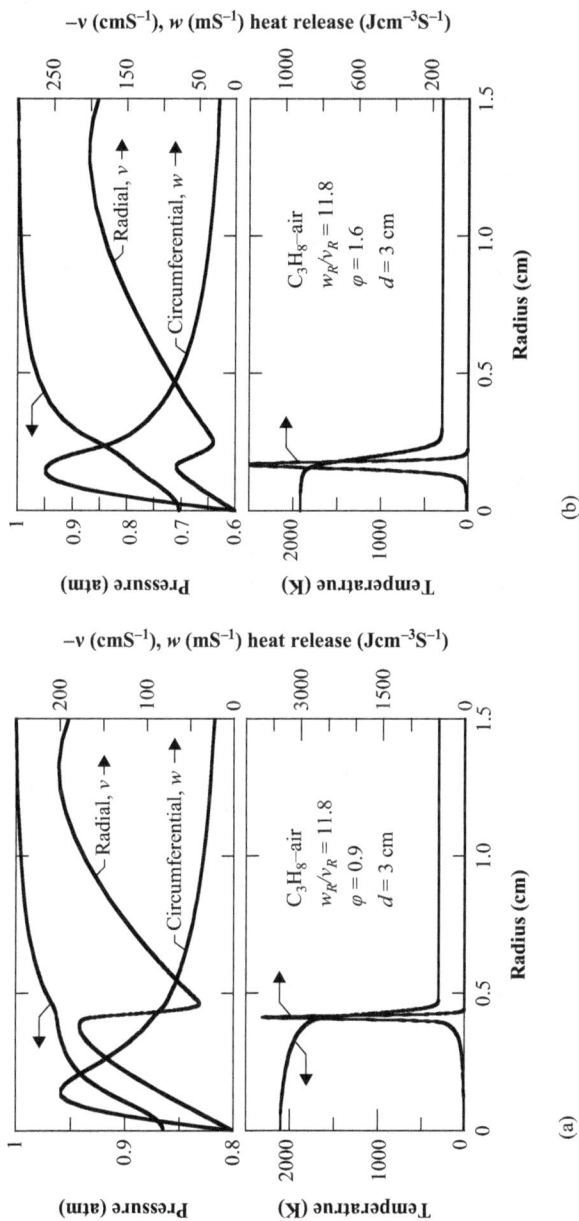

Figure 3.9. Solution profiles for (a) lean flame ($\phi = 0.9$) and (b) rich flame ($\phi = 1.6$) at a rotation ratio $w_R / v_R = 11.8$.

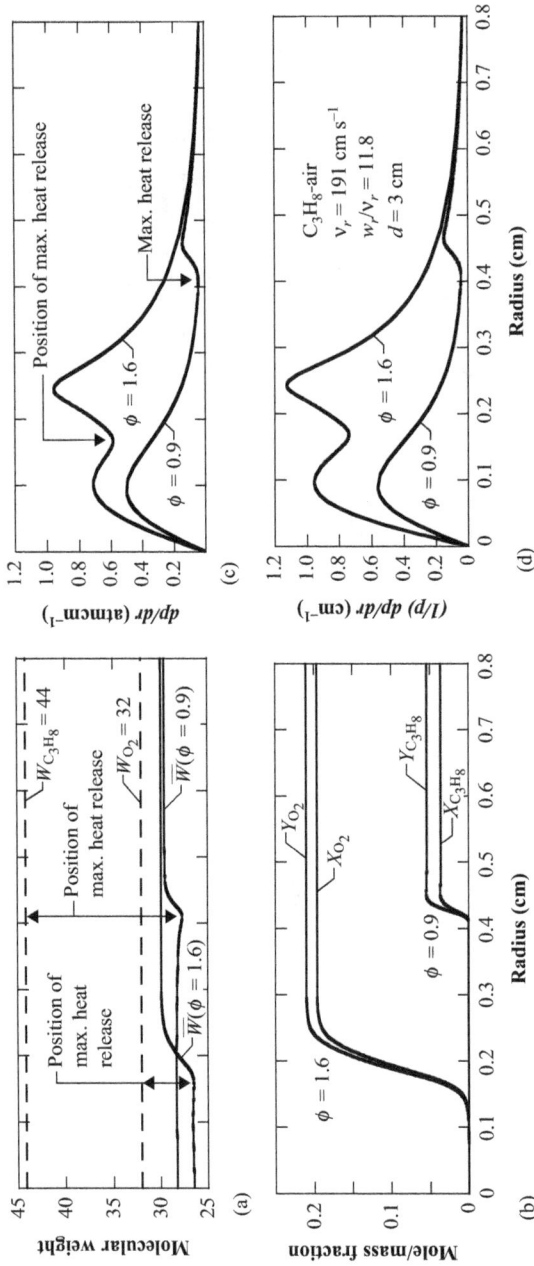

Figure 3.10. Radial profiles of factors affecting the pressure diffusion for lean ($\phi = 0.9$) and rich ($\phi = 1.6$) flames. (a) Molecular weights of the mixture W and the controlling species (C_3H_8 for the rich case and O_2 for the lean case). (b) Mass and mole fractions for the rate-controlling species. (c) Pressure gradient dp/dr, and (d) $\Lambda_z = (1/p)dp/dr$.

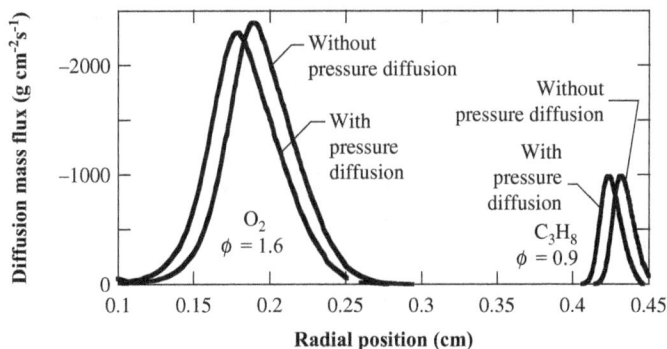

Figure 3.11. Radial profiles of diffusive mass fluxes of the rate-controlling species at two equivalence ratios of 0.9 and 1.6 with and without pressure diffusion. The inlet velocity ratio $w_R / v_R = 11.8$.

High swirl rates can cause large pressure gradients, thus significantly affecting tubular-flame characteristics. Figure 3.12 shows that an increasing rotation rate causes the pressure gradient to increase, the flame diameter and temperature to decrease, and the apparent burning velocity to increase. The apparent burning velocity S_T is defined here as the minimum radial velocity just upstream of the flame front (Fig. 3.9). These trends are more significant in rich mixtures than they are in the comparable lean mixtures. Figure 3.13(a) shows the peak value of the heat-release rate, and Fig. 13(b) shows the total heat-release (i.e., integrated over the entire flame zone from the centerline to the tube wall). As the rotation rate increases, the heat-release rate decreases. The decreases are relatively greater for the rich case. As the rotation rate increases, the pressure gradient increases and thus suppressing the supply of the controlling species to the flame front, causing the decreased heat-release rate. As a result, the flame temperature decreases and the flame location shifts toward the centerline (i.e., downstream). Because the heat-release rate decreases, the corrected flame speed decreases with increasing rotation rate as illustrated in Fig. 3.13(c). It is worth noting that a correction for the burning velocity is needed because the temperature and pressure at the minimum velocity position (i.e., where S_T is evaluated) differs significantly from that of the unburned mixture at the inlet tube wall. The corrected burning velocity is defined to be the product of S_T and the ratio of the mixture densities at the minimum-velocity location and the tube wall.

3.5.2 PREMIXED METHANE–AIR FLAMES

Fig. 3.14 illustrates important trends for a lean premixed methane–air tubular flame at the equivalence ratio $\phi = 0.7$ and radial inlet velocity $v_R = 191$ cm s^{-1} (axial stretch rate, $\varepsilon = 4v_R / d = 255$ s^{-1}). Wall rotation velocity w_R varies as $0 \leq w_R / v_R \leq 16$. By increasing the rotation velocity, the pressure at the flame front (defined as the position of the maximum heat release rate, R_{HRR}) decreases, the maximum flame temperature decreases, and the flame diameter decreases slightly. These trends are similar to those found in the propane–air flames. However, unlike in the propane flame, the molecular weight of the rate-controlling species (i.e., methane) is smaller than the mixture mean molecular weight. Consequently the pressure-diffusion contribution to the diffusion flux is in the same direction as ordinary diffusion (Fig. 3.15). Therefore, it might be anticipated that the pressure diffusion would cause the flame to shift toward the wall side (i.e., upstream) and the flame temperature to increase due to the enhanced chemical enthalpy supply to the reaction zone. Understanding this apparently conflicting behavior requires an analysis of the effects of the pressure, stretch, and curvature.

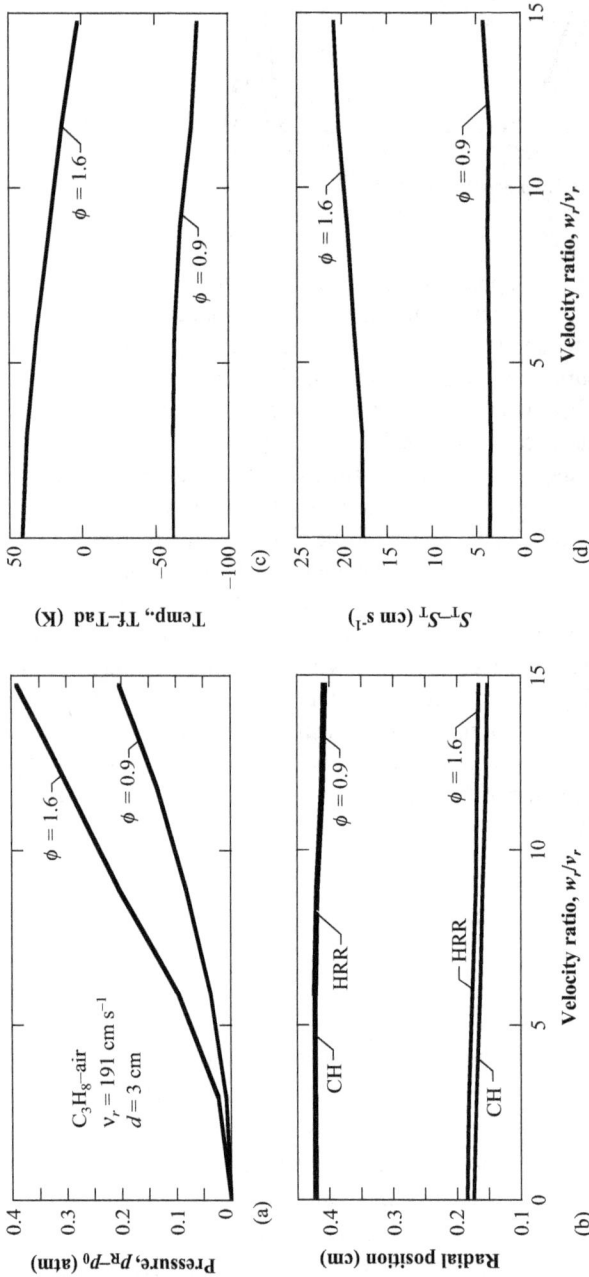

Figure 3.12. Flame characteristics as a function of wall velocity ratio, w_R / v_r, and two equivalence ratios. (a) Pressure difference between the wall, p_R and the centerline, p_0. (b) Flame position is measured in two ways. The position labeled "HRR" is measured by the position of the maximum heat release rate. The position labeled "CH" is measured by the peak in the CH-radical profile. (c) Difference between the peak flame temperature, T_f and the adiabatic flame temperature, T_{ad}. (d) Difference between the apparent tubular-flame burning velocity, S_T and the planar laminar adiabatic burning velocity, S_L.

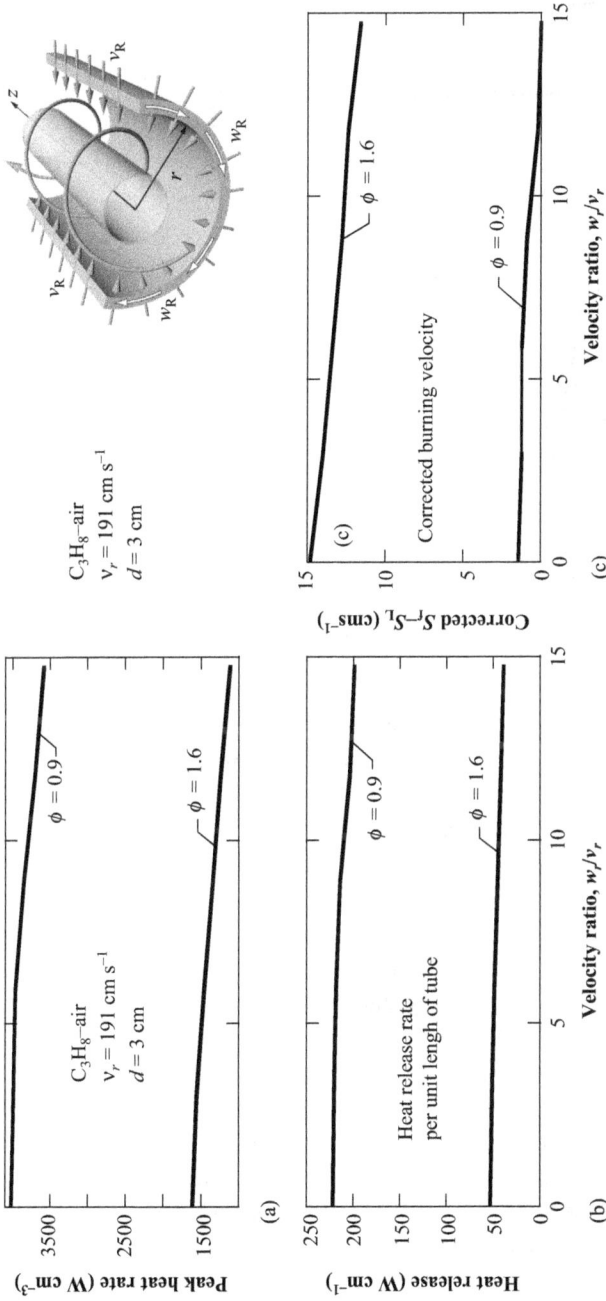

Figure 3.13. Heat release rates and corrected burning velocity as functions of rotation rate for two equivalence ratios. (a) Maximum heat release rate (HRR) within the flame zone (Fig. 9(a)). (b) Total heat release (per unit length of the tube). (c) Corrected burning velocity.

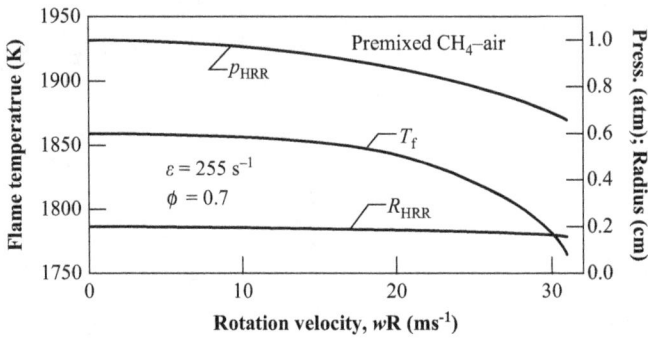

Figure 3.14. Maximum flame temperature and pressure at the flame front as functions of rotation velocity.

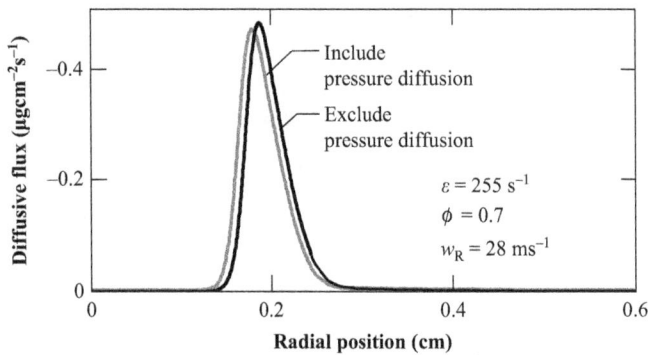

Figure 3.15. Radial dependence of diffusive mass fluxes for the rate-controlling species. Profiles are shown for situations that include and exclude pressure diffusion.

Figure 3.16 compares the flame temperature T_f, flame position R_{HRR}, and local pressure at the flame front p_{HRR} as functions of rotation rate at low stretch rate ($\varepsilon = 64$ s^{-1}) for two lean equivalence ratios ($\phi = 0.7$ and $\phi = -0.55$). At $\phi = 0.7$, the flame is closer to the tube wall, but the flame temperature and position are affected only slightly by increasing the wall rotation rate. However, at $\phi = 0.55$ the flame temperature first increases and then decreases as the rotation rate increases. The decrease at high rotation is caused by higher curvature and lower pressure. Figure 3.16(b) includes comparative results to evaluate the effect of neglecting pressure diffusion. Neglecting pressure diffusion causes the flame to move closer to the centerline and the flame temperature to decrease significantly, particularly at a high rotation rate ($w_R > 50$ m s^{-1}) and a low stretch rate ($\varepsilon = 64$ s^{-1}).

Figure 3.17 compares the radial profiles of methane (the rate-controlling species) diffusive mass flux at the rotation rate $w_R = 70$ m s^{-1} and stretch rate $\varepsilon = 64$ s^{-1}, with and without the pressure-diffusion contribution. Neglecting the pressure-diffusion contribution causes the diffusive mass flux to decrease and the peak to shift toward the centerline, thus lowering the supply of chemical enthalpy to the flame. The pressure-diffusion contribution is also found to extend significantly the extinction flame limit at a high rotation rate. In contrast to the propane–air flames, the pressure-diffusion contribution for the methane–air flames enhances mass diffusion of the rate-controlling species to the flame and thus increases the flame temperature at a high rotation rate and a small stretch rate. However, the high rotation rates lead to reduced pressure in the flame zone, lowering the reaction rate and flame temperature. Thus, the flame temperature depends on a balance between the pressure diffusion and pressure itself. As illustrated in

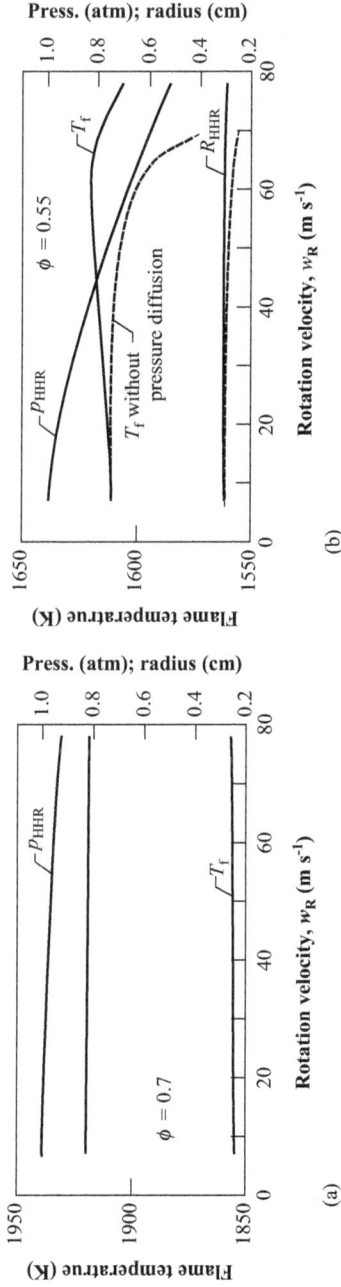

Figure 3.16. Effect of wall rotation on flame temperature, position of maximum heat release rate (R_{HRR}), and flame-front pressure p_{HRR} in swirling tubular flames with a relatively low stretch rate for two equivalence ratios. The dashed lines show results that exclude the pressure-diffusion contribution in Eq. 3.7.

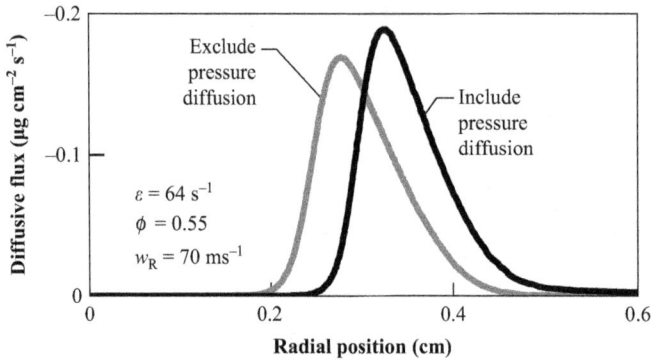

Figure 3.17. Diffusive mass fluxes for the rate-controlling species (CH_4) as a function of radius. Profiles are shown for models that include and exclude pressure diffusion.

Fig. 3.16(b) for $\phi = 0.55$ flame, the slight increase of the flame temperature upon increasing the rotation rate up to about $w_R = 60$ m s^{-1} indicates that the pressure-diffusion effect slightly dominates the effects of pressure itself. As the rotation rate increases further, pressure diffusion (a positive contribution to flame temperature) cannot overcome the effect of reduced pressure (a negative effect). As a consequence, the flame temperature begins to decrease as rotation rate increases.

At a high stretch rate $\varepsilon = 255$ s^{-1}, Fig. 3.14 shows that the flame position is close to the centerline, even at an equivalence ratio $\phi = 0.7$. As the rotation rate increases, pressure decreases in the flame zone, which causes the Damköhler number (i.e., the ratio of a characteristic flow time and a characteristic chemical reaction time) to decrease significantly. Consequently, despite the "positive" pressure-diffusion contribution, the flame temperature decreases.

3.5.3 SUMMARY OF PRESSURE DIFFUSION

Based on simulating premixed propane–air and methane–air tubular flames under a range of operating conditions, some broad observations can be made. High rotation rates cause significant radial pressure variations and when the pressure gradients are sufficiently high, the pressure diffusion of species is significant. In premixed propane–air tubular flames, increasing the rotation rate shifts the flame position toward the centerline and the flame temperature decreases. The reason for this behavior is that the supply of the rate-controlling species (propane for lean mixtures and oxygen for rich mixtures) into the flame front is suppressed as the rotation rate increases. The consequence is a decreasing supply of chemical enthalpy to the flame, decreasing the heat release rate. For lean premixed methane–air tubular flames, however, the situation is opposite. In this case, the rate-controlling species is methane. Because the molecular weight of methane is lower than the mean molecular weight, pressure diffusion manifests itself as a positive contribution to the total mass flux of the rate-controlling species into the flame reaction zone. Table 3.2 provides a compact summary of these results.

3.6 POTENTIAL TECHNOLOGY APPLICATIONS

It is at least interesting to note that the application planar axisymmetric stagnation flows have enormous technological impact. Configurations such as that illustrated in Fig. 3.18 are used in

Table 3.2. Summary of pressure diffusion effects in premixed tubular flames

	Premixed Propane–Air	Premixed Methane–Air
Rate-controlling species	$X_{C_3H_8} - Y_{C_3H_8} < 0$ (lean) $X_{O_2} - Y_{O_2} < 0$ (rich)	$X_{CH_4} - Y_{CH_4} > 0$ (lean)
Direction of pressure diffusion for rate-controlling species	Opposes ordinary diffusion	Assists ordinary diffusion
Contribution of pressure diffusion to diffusive mass flux of rate-controlling species toward the flame	Inhibits mass flux	Assists mass flux
Contribution of pressure diffusion to the transport of chemical enthalpy to flame	Suppresses flux	Enhances flux
Contribution of pressure diffusion to flame characteristics	Decreases heat release rate Decreases flame temperature	Increases heat release rate Increases flame temperature (if contributions of stretch and pressure variations are excluded)

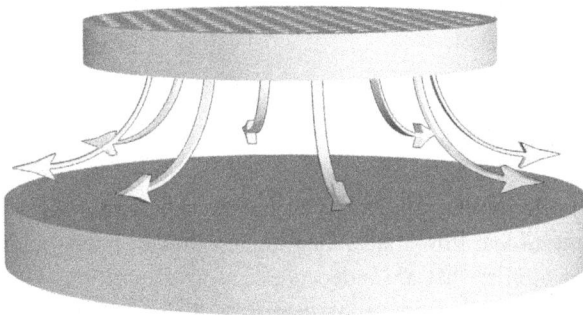

Figure 3.18. Illustration of a planar axisymmetric stagnation flow.

every modern semiconductor fabrication facility worldwide. In this setting, the flow is introduced axially and exhausted radially. It may also be noted that the lower stagnation surface may be rotating, which does not interrupt the similarity properties. The stagnation–flow similarity causes the temperature and composition profiles to be independent of radius (i.e., flat and parallel to the upper inlet manifold and lower stagnation surface. Thus, in processes such as chemical vapor deposition, the deposition rate onto a semiconductor wafer is nearly perfectly uniform over the entire surface of the wafer, which can be more than 300 mm in diameter. This is an extraordinarily valuable attribute. As noted earlier in this chapter, the similarity attributes of the planar stagnation flows and the tubular flows are completely analogous.[25] Thus, one must be at least intrigued about technological opportunities that can beneficially exploit the special characteristics of the tubular flows.

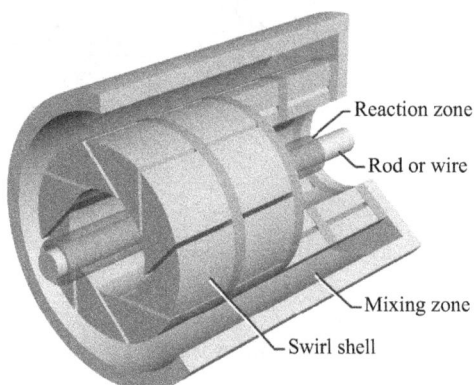

Figure 3.19. A possible reactor configuration for the uniform surface modification or the application of uniform coatings onto rods or wires.

Figure 3.19 illustrates a configuration that might be valuable for the uniform surface modification of rods or wires. In this configuration, reactive gases are introduced tangentially through slots. Ishizuka and colleagues have developed and demonstrated various implementations of such tangential injectors. Because of the tubular similarity, the composition and temperature profiles around the rod are uniform along the length of the rod. The "reaction zone" could be a flame, primarily producing heat for a heat-treatment process. Alternatively, the reaction zone could be responsible for producing species that lead to chemical vapor deposition or other thin-film coating processes. The rod could be stationary or it could be drawn through the reactor in a continuous process. One can imagine numerous potential technological applications for tubular flows; yet unlike planar stagnation flow, very few have actually been developed and deployed.

3.7 SUMMARY AND CONCLUSIONS

Tubular flames have a mathematical similarity that has important fundamental and practical implications. Despite two- or three-dimensional flow fields, ideal tubular flames have radial temperature, pressure, and composition profiles that are independent of axial position. This behavior facilitates scientific investigation of flames, concentrating especially on flame curvature and strain. At high swirl rates the radial pressure gradients can be sufficiently large as to affect the molecular mass transport via pressure diffusion (i.e., Dufour transport). The effects of pressure diffusion can be understood by comparing swirling and nonswirling tubular flames. The effects of curvature can be understood by comparing tubular flames with analogous planar opposed-flow flames. Because the mathematical similarity reduces the simulation problem to a one-dimensional boundary-value problem, computation is very efficient even when large reaction mechanisms are considered.

In addition to fundamental combustion research, tubular flames may also have practical value in technology. Here the similarity behavior (i.e., axial independence of temperature and composition profiles) may enable industrial processes such as the uniform surface modification of cylindrical rods or wires. However, the technological exploitation of tubular flows is not as mature as the practical use of mathematically and physically analogous planar stagnation flows.

ACKNOWLEDGMENTS

Profs. Tadao Takeno and Satoru Ishizuka have made important scientific and engineering contributions by elucidating and demonstrating the theoretical and technical attributes and advantages of tubular flows and flames. We gratefully acknowledge numerous insightful discussions over many years with Profs. Takeno and Ishizuka.

REFERENCES

1. Ishizuka, S. 1993. Characteristics of tubular flames. *Progress in Energy and Combustion Science* 19:187–226. DOI: 10.1016/0360-1285(93)90015-7.
2. Tseng, L. K., K. Abhishek, and J. P. Gore. 1955. An experimental realization of premixed methane/air cylindrical flames. *Combustion and Flame* 102:519–22. DOI: 10.1016/0010-2180(95)00122-M.
3. Takeno, T. and S. Ishizuka. 1986. A tubular flame theory. *Combustion and Flame* 64:83–98. DOI: 10.1016/0010-2180(86)90100-8.
4. Ishizuka, S. 1984. On the behaviour of premixed flames in a rotating flow field: Establishment of tubular flames. *Proceedings of the Combustion Institute* 20:287–94.
5. Takeno, T., M. Nishioka, and S. Ishizuka. 1986. A theoretical study of extinction of a tubular flame. *Combustion and Flame* 66:271–83. DOI: 10.1016/0010-2180(86)90140-9.
6. Nishioka, M., T. Takeno, and S. Ishizuka. 1988. Effects of variable density on a tubular flame. *Combustion and Flame* 73:287–301. DOI: 10.1016/0010-2180(88)90024-7.
7. Dixon-Lewis, G., V. Giovangigli, R. J. Kee, J. A. Miller, B. Rogg, M. D. Smooke, G. Stahl, and J. Warnatz. 1991. Numerical modeling of the structure and properties of tubular strained laminar premixed flames. *Program Astronautics and Aeronautics* 131:125–44.
8. Nishioka, M., K. Inagaki, S. Ishizuka, and T. Takeno. 1991. Effects of pressure on structure and extinction of tubular flame. *Combustion and Flame* 86:90–100. DOI: 10.1016/0010-2180(91)90058-J.
9. Smooke, M. D. and V. Giovangigli. 1991. Extinction of tubular premixed laminar flames with complex chemistry. *Proceedings of the Combustion Institute* 23:447–54. DOI: 10.1016/S0082-0784(06)80290-0.
10. Kobayashi, H. and M. Kitano. 1989. Extinction characteristics of a stretched cylindrical premixed flame. *Combustion and Flame* 76:285–95. DOI: 10.1016/0010-2180(89)90111-9.
11. Nishioka, M., C. K. Law, and T. Takeno. 1996. A flame-controlling continuation method for generating S-curve responses with detailed chemistry. *Combustion and Flame* 104:328–42. DOI: 10.1016/0010-2180(95)00132-8.
12. Ju, Y., H. Matsumi, K. Takita, and G. Masuya. 1999. Combined effects of radiation, flame curvature, and stretch on the extinction and bifurcations of cylindrical CH4/air premixed flame. *Combustion and Flame* 116:580–92. DOI: 10.1016/S0010-2180(98)00051-0.
13. Chen, Z. and Y. Ju. 2008. Combined effects of curvature, radiation, and stretch on the extinction of premixed tubular flames. *International Journal of Heat and Mass Transfer* 51:6118–25. DOI: 10.1016/j.ijheatmasstransfer.2008.05.002.
14. Mosbacher, D. M., J. A. Wehrmayer, R. W. Pitz, C. J. Sung, and J. L. Byrd. 2002. Experimental and numerical investigation of premixed tubular flames. *Proceedings of the Combustion Institute* 29:1479–86. DOI: 10.1016/S1540-7489(02)80181-X.
15. Hayashi, A. K., S. Nakano, N. Saito, C. Liao, and T. Tsuruda. 2002. Water vapor suppression to tubular flame. *Journal of Physics. IV France,* 12:291–7. DOI: 10 lOSl/Ip4.20020296.
16. Zhang, Y., S. Ishizuka, H. Zhu, and R. J. Kee. 2007. The effects of rotation rate on the characteristics of premixed propane/air swirling tubular flames. *Proceedings of the Combustion Institute* 31:1101–7. DOI: 10.1016/j.proci.2006.07.030.

17. Zhang, Y., J. Wu, and S. Ishizuka. 2009. Hydrogen addition effect on laminar burning velocity, flame temperature and flame stability of a planar and a curved CH4–H2-air premixed flame. *International Journal of Hydrogen Energy* 34:519–27. DOI: 10.1016/j.ijhydene.2008.10.065.

18. Zhang, Y., S. Ishizuka, H. Zhu, and R. J. Kee. 2009. Effects of stretch and pressure on the characteristics of premixed swirling tubular methane-air flames. *Proceedings of the Combustion Institute* 32:1149–56.

19. Lutz, A. E., R. J. Kee, J. F. Grcar, and F. M. Rupley. 1996. OPPDIF: A Fortran program for computing opposed-flow diffusion flame. Technical Report SAND96-8243, Sandia National Laboratories.

20. Yamamoto, K. 1999. Pressure change and transport process on flames formed in a stretched, rotating flow. *Combustion and Flame* 118:431–44. DOI: 10.1016/S0010-2180(99)00013-9.

21. Hu, S., P. Wang, R. W. Pitz, and M. D. Smooke. 2007. Experimental and numerical investigation of non-premixed tubular flames. *Proceedings of the Combustion Institute* 31:1093–9. DOI: 10.1016/j.proci.2006.08.058.

22. Wang, P., S. Hu, and R. W. Pitz. 2007. Numerical investigation of the curvature effects on diffusion flames. *Proceedings of the Combustion Institute* 31:989–96. DOI: 10.1016/j.proci.2006.07.223.

23. Hu, S., P. Wang, and R. W. Pitz. 2009. A structural study of premixed tubular flames. *Proceedings of the Combustion Institute* 32:1133–40. DOI: 10.1016/j.combustflame.2008.07.017.

24. Kee, R. J., A. M. Colclasure, H. Zhu, and Y. Zhang. 2008. Modeling tangential injection into ideal tubular flames. *Combustion and Flame* 152:114–24. DOI: 10.1016/j.combustflame.2007.07.019.

25. Kee, R. J., M. E. Coltrin, and P. Glarborg. 2003. *Chemically reacting flow: Theory and practice.* Hoboken, NJ: Wiley.

26. Kee, R. J., J. A. Miller, G. H. Evans, and G. Dixon-Lewis. 1989. A computational model of the structure and extinction of strained, opposed-flow, premixed, methane-air flames. *Proceedings of the Combustion Institute* 22:1479–94. DOI: 10.1016/S0082-0784(89)80158-4.

27. Ishizuka, S. Flame propagation along a vortex axis. 2002. *Progress in Energy and Combustion Science* 28:477–542.

28. Burgers, J. M. 1948. A mathematical model illustrating the theory of turbulence. In *Advances in applied mathematics,* vol. 1, 171–99. New York: Academic Press.

29. Grcar, J. F., R. J. Kee, M. D. Smooke, and J. A. Miller. 1988. A hybrid Newton/time-integration procedure for the solution of steady, laminar, one-dimensional, premixed flames. *Proceedings of the Combustion Institute* 21:1773–82. DOI: 10.1016/S0082-0784(88)80411-9.

30. Kee, R. J., F. M. Rupley, E. Meeks, and J. A. Miller. 1996. Chemkin-III: A Fortran chemical kinetics package for the analysis of gas-phase chemical and plasma kinetics. Technical Report SAND96-8216, Sandia National Laboratories.

31. Coltrin, M. E., R. J. Kee, G. H. Evans, E. Meeks, F. M. Rupley, and J. F. Grcar. 1991. SPIN: A Fortranpprogram for modeling one-dimensional rotating-disk/stagnation-flow chemical vapor deposition reactors. Technical Report SAND91-8003, Sandia National Laboratories.

32. Kee, R. J., G. Dixon-Lewis, J. Warnatz, M. E. Coltrin, and J. A. Miller. 1986. A Fortran computer code package for the evaluation of gas-phase multicomponent transport properties. Technical Report SAND86-8246, Sandia National Laboratories.

33. Sakai, Y. and S. Ishizuka. 1991. Structures of the tubular flames of lean methane/air mixtures in a rotating stretched-flow field. *JSME International Journal* 34:234–41.

34. Smith, G. P., D. M. Golden, M. Frenklach, N. W. Moriarty, B. Eiteneer, M. Goldenberg, C. T. Bowman, R. K. Hanson, S. Song, W. C. Gardiner, V. Lissianski, and Z. Qin. 1999. Gri-mech—an optimized detailed chemical reaction mechanism for methane combustion. Technical Report http://www.me.berkeley.edu/gri mech, Gas Research Institute.

35. Laranjeira, M. F. 1960. An elementary theory of thermal and pressure diffusion in gaseous binary and complex mixtures: I. General theory. *Physica* 26:409–16. DOI: 10.1016/0031-8914(60)90030-6.

36. Sung, C. J., B. Li, H. Wang, and C. K. Law. 1998. Structure and sooting limits in counterflow methane/air and propane/air diffusion flames from 1 to 5 atmospheres. *Proceedings of the Combustion Institute* 27:1523–30.

37. Yamaoka, I. and H. Tsuji. 1984. Determination of burning velocity using counterflow flames. *Proceedings of the Combustion Institute* 20:1883–92. DOI: 10.1016/S0082-0784(85)80687-1.

38. Van Maaren, A. and L. P. H. de Goey. 1994. Stretch and the adiabatic burning velocity of methane- and propane-air flames. *Combustion Science and Technology* 102:309–14. DOI: 10.1080/00102209408935483.

39. Kee, R. J., J. F. Grcar, J. A. Miller, and E. Meeks. 1998. Premix: A Fortran program for modeling steady laminar one-dimensional premixed flames. Technical report, Sandia National Laboratories.

CHAPTER 4

RAMAN SPECTROSCOPIC MEASUREMENTS OF TUBULAR FLAMES

Robert W. Pitz

4.1   INTRODUCTION

Although the Raman scattering effect was first discovered by C. V. Raman and K. S. Krishnan in 1928,[1] it was not applied to flame measurement until the advent of lasers in the 1970s. The first spontaneous Raman scattering measurements in flames were made using a CW argon-ion laser to measure the vibrational Raman scattering spectra of H_2O, N_2, and O_2 in flames where the flame temperature was interpreted from spectral fits.[2] Since that time, Raman scattering has been widely used to measure major species concentrations and temperatures in both laminar and turbulent flames. There are excellent reviews of Raman scattering theory and measurement in flames.[3-8]

Raman scattering was applied to measure temperature and major species concentration profiles in planar opposed-jet flames using pulsed UV lasers[9-11] and visible lasers.[12-14] Subsequently, Raman scattering was applied to measure species concentrations and temperature in premixed tubular flames[15,16] and in nonpremixed opposed-flow tubular flames.[17,18] In this chapter, the Raman scattering technique and its calibration for concentration and temperature measurements in flames are described. The design of a new tubular burner with optical access is presented. Raman scattering measurements of temperature and composition are compared with computer simulations with detailed molecular transport and complex chemistry in hydrogen–air and hydrocarbon–air lean premixed tubular flames. Finally, Raman scattering is applied to measure cellular tubular flames.

4.2   RAMAN SCATTERING TECHNIQUE

As a laser beam passes through the flame, light is scattered from molecules both elastically (Rayleigh scattering) and inelastically (Raman scattering). Rayleigh scattered light is shifted slightly by the gas velocity (Doppler shift) and its spectrum can be used to measure temperature, density, and velocity.[19,20] Both Rayleigh and Raman scattering processes are shown in Fig. 4.1 (a) and the corresponding spectra in Fig. 4.1 (b). Raman scattering is about 1000 times weaker than Rayleigh scattering. In the Raman process, light scattered from rotating and vibrating molecules is shifted by a change in the rotational and/or vibrational energy level of the molecule. In pure rotational Raman scattering for linear molecules, the rotational quantum number J can change as $\Delta J = \pm 2$. Rotational energy changes are small and pure rotational Raman scattering lies close to the laser wavelength. For vibrational Raman scattering, the Q branch ($\Delta J = 0$) gives the strongest spectral line with selection rules for the vibrational quantum number v of $\Delta v = \pm 1$ where $\Delta v = +1$ is termed "Stokes" and $\Delta v = -1$ is termed "anti-Stokes." Vibrational energy changes are much larger leading to vibrational Raman lines that are far from the exciting laser line. In vibrational Stokes Raman scattering ($\Delta v = +1$), the molecule absorbs some of the incident photon energy leading to a red-shifted photon. In anti-Stokes Raman scattering ($\Delta v = -1$), the scattered photon is blue shifted. Because vibrational Stokes scattering is much stronger than anti-Stokes, Stokes scattering is used in all laminar flame measurements.

An example of Stokes vibrational Raman scattering from a lean premixed H_2–air tubular flame is shown in Fig. 4.2.[15] Raman scattering from a 532-nm YAG laser produces strong Stokes lines from O_2, N_2, H_2O, and H_2 at the red-shifted wavelengths (vibrational frequency shifts) of 580 nm (1556 cm^{-1}), 607 nm (2330.7 cm^{-1}), 660 nm (3657 cm^{-1}), and 683 nm (4160 cm^{-1}).

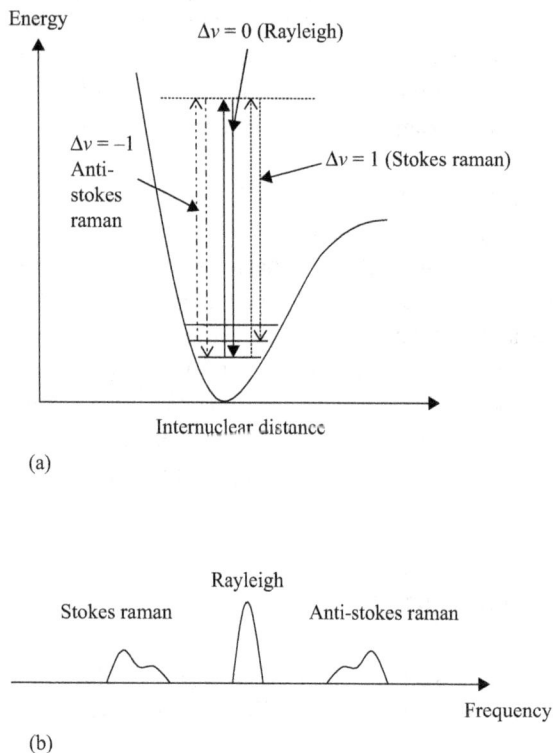

Figure 4.1. Vibrational Raman scattering and Rayleigh scattering processes: (a) energy diagram of a diatomic molecule; (b) light spectrum. (*Source*: Cheng, Z. 2001. *Visible laser Raman diagnostics application in partially premixed flames of premixed fuel versus hot products*. Nashville, Tennessee USA: MS Thesis, Mechanical Engineering Department, Vanderbilt University.)

Vibrational frequency shifts are proportional to the inverse of mass of the molecular oscillator and H_2 has the largest vibrational frequency Raman shift. Vibrational Raman shifts and their corresponding Stokes wavelength for a 532-nm laser are given in Table 4.1 for combustion species commonly measured in tubular flames.[21] The Raman shifts of other molecules are published elsewhere.[6,22] Pure rotational Raman scattering from H_2 is seen in Fig. 4.2 at two lines near 560 nm because H_2 has the largest rotational Raman shifts. The rotational Raman scattering from the other molecules (N_2, O_2) is too close to the laser line to be seen. Vibrational Raman

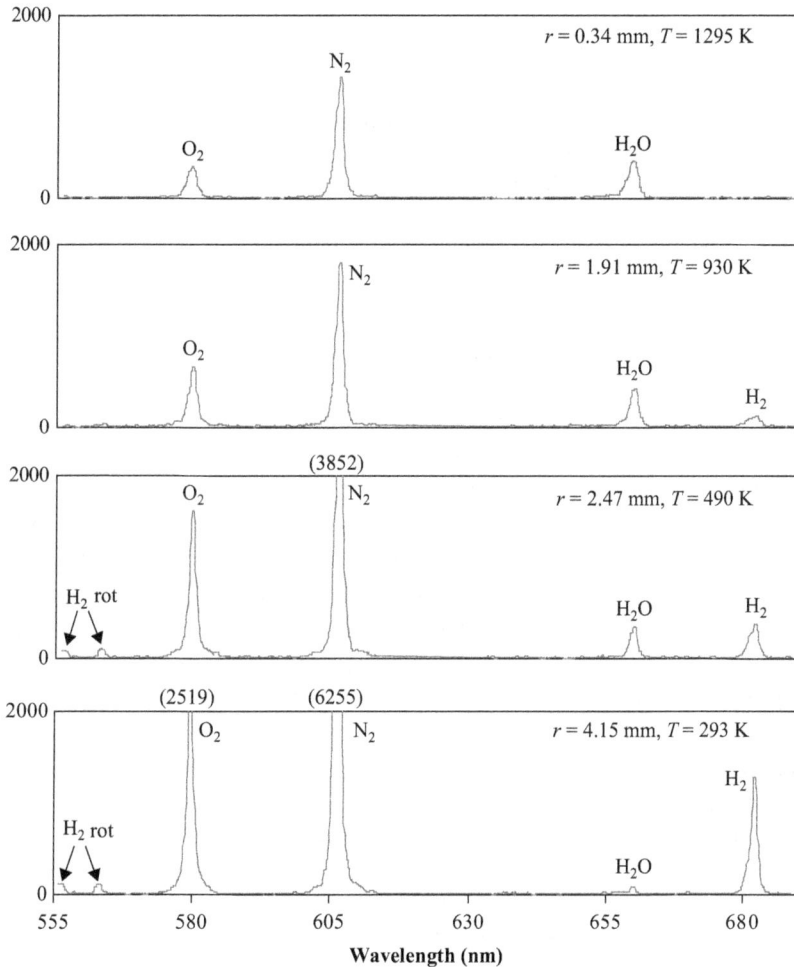

Figure 4.2. Time-averaged, Raman spectra induced by a 532 nm laser in a laminar, lean premixed H_2–air tubular flame ($\phi = 0.175$, $k = 163$ s^{-1}). Spectra correspond to four radial positions from the stagnation centerline of the flame: burned gas, initial reaction, preheat, and unburned gas regions. Some peaks are cut off for clarity and their relative peak values are given in parentheses. (*Source*: Mosbacher, D. M. et al. 2002. Experimental and numerical investigation of premixed tubular flames. *Proceedings of the Combustion Institute* 29:1479–86.)

Table 4.1. Frequency-shifts and related emission wavelengths with the excitation of 532-nm Nd:YAG laser for various species formed in hydrocarbon flames

Laser Wavelength	Species	Frequency Shift (cm^{-1})	Detected Wavelength (nm)
532-nm	C_3H_8	872, 2886	557.8, 629
YAG laser	CO_2	1285, 1388	571.0, 574.4
	O_2	1556	580.0
	CO	2145	600.5
	N_2	2330.7	607.3
	CH_4	2915, 3017	629.6, 633.7
	H_2O	3657	660.5
	H_2	4160	683.2

Source: Cheng, Z. 2001. *Visible laser Raman diagnostics application in partially premixed flames of premixed fuel versus hot products*. Nashville, Tennessee USA: MS Thesis, Mechanical engineering department, Vanderbilt University.[21]

scattering spectra from a lean premixed CH_4–air tubular flame is shown in Fig. 4.3, where Stokes Raman lines from CO_2 and CH_4 appear.[16]

A schematic of a Raman scattering system for measuring concentration and temperature in tubular flames is shown in Fig. 4.4.[18] The YAG laser is either a Continuum Powerlite 9010 (~1000 mJ/pulse max. at 523 nm, 10 Hz) or a Continuum Surelite II-10 (~400 mJ/pulse max. at 532 nm). To avoid laser-induced breakdown, the laser energy is reduced from its maximum setting and temporally stretched. The laser pulse is stretched temporally from a ~10-ns pulse length to a ~150-ns pulse length using a three-stage cavity[23] to avoid laser-induced breakdown in the flame. The laser pulse energy is reduced at the exit of the laser by a rotatable zero-order wave plate followed by a thin film plate polarizer mounted at its Brewster angle to enable continuous adjustment of the laser power. Typically, ~140 mJ/pulse can be used in the flame measurements without breakdown for a laser focal diameter of 150 µm and 150-ns pulse length.[16] A laser energy of ~225 mJ/pulse was used for a larger focal diameter of 250 µm and a beam pulse length of ~35 ns.[15]

As seen in Fig. 4.4, vertically polarized laser light is scattered in the flame and collected at 90 degrees by an f/2 achromatic lens (76 mm dia.) that collimates the light which is subsequently focused onto the entrance slit of a 0.65-m spectrometer by an f/7.5 achromat lens (76 mm dia.). A Displaytech ferroelectric liquid crystal shutter (heated to 40–45 °C to give a 40 µs response, 28% transmission, rotates polarization by 90°) and a mechanical shutter (4 ms) reduce background light. Rayleigh-scattered light and laser light scattered from surfaces are rejected by a Schott OG-550 filter. In addition, an IR filter (dielectric short pass filter, 750 nm) reduces infrared radiation from the flame. The spectrometer is mounted on its side so that the entrance slit is horizontal. This allows about 4.6 mm of the laser line to be imaged onto the 20-mm slit. The sample volume is divided into 30–40 sections along the laser line. The spatial resolution of the optical system is 98 µm along the laser line as verified by backlighting a 0.169-mm/line Ronchi grating placed in the measurement zone.[16]

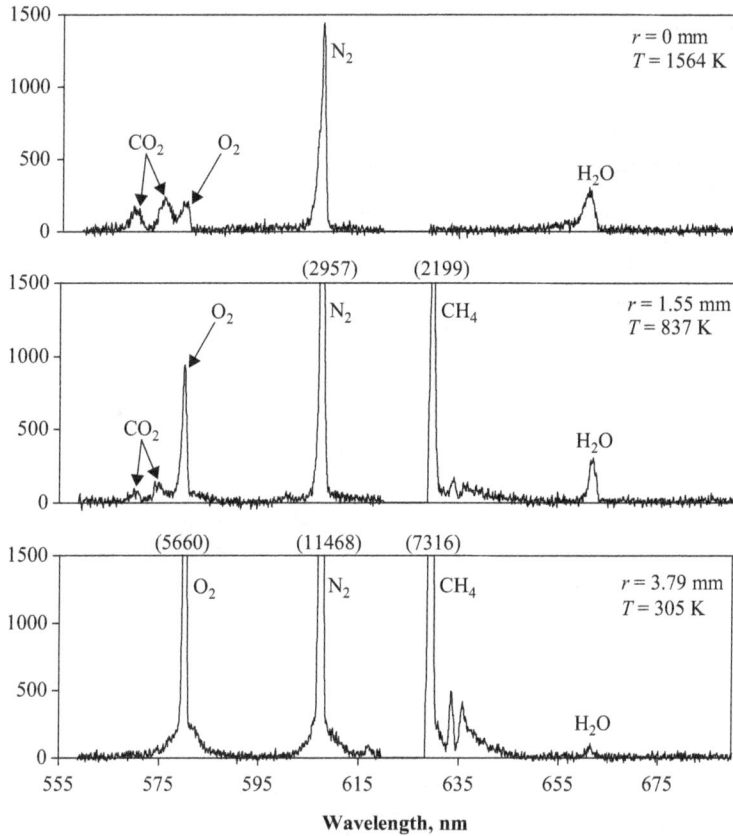

Figure 4.3. Representative Raman spectra of a CH_4/air premixed tubular flame at four radial locations ($\phi = 0.58$, k = 257 s^{-1}). For clarity, some strong peaks are cutoff and the peak values given in parentheses. (*Source*: Hu, S. T., P. Y. Wang, and R. W. Pitz. 2009. A structural study of premixed tubular flames. *Proceedings of the Combustion Institute* 32:1133–40. Reprinted with permission from Elsevier.)

The Czerny Turner spectrometer (0.75 m SPEX 1800) that has 0.75 m entrance collection mirror, ruled grating (110 mm × 110 mm, 600 grooves/mm, 500 nm blaze, 80% reflective at 600 nm) and has been modified with a 0.65-m focusing mirror.[24] The final focusing mirror is very large to avoid vignetting of the Raman spectrum. The spectrometer has a final dispersion of 0.39 mm/nm. The Raman spectra is recorded on a liquid-nitrogen cooled back-illuminated CCD camera (Princeton Instruments, 1024 × 1024 array, 24 μm × 24 μm pixel, 70% quantum efficiency). Using a back-illuminated LN/CCD, we integrate the Raman spectra resulting from about 1000 laser shots to reduce the shot noise.

The number of detected photons (or photoelectrons) for a Q-branch line for ith chemical species can be estimated by,

$$S_i = E_L\ \sigma_{zz}\ X_i\ N\ L\ \Omega\ \eta\ Q_e\ \Gamma(T)\ \lambda/(h\ c) \qquad (4.1)$$

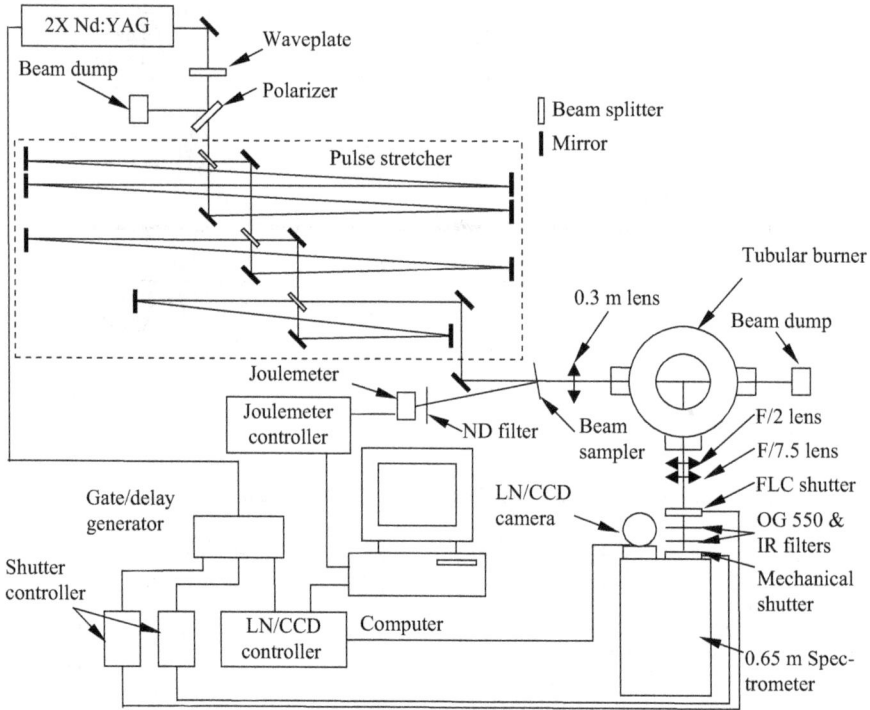

Figure 4.4. Schematic of the visible 532 nm spontaneous Raman scattering system. (*Source*: Hu, S. T. and R. W. Pitz. 2009. Structural study of non-premixed tubular hydrocarbon flames. *Combustion and Flame* 156 (1):51–61. Reprinted with permission from Elsevier.)

where E_L is the laser energy (130 mJ), σ_{zz} is the vibrational Raman cross section (0.46 × 10^{-34} m^2/sr for N$_2$), X$_i$ is the species mole fraction, N is the total number density (#/m^3), L is the spatial resolution along the laser beam (0.2 mm), Ω is the collection solid angle (0.2 sr), η is the efficiency of the detection system (0.1), Q_e is the quantum efficiency of the back-illuminated CCD camera (0.7), $\Gamma(T)$ is the temperature-dependent factor that accounts for the distribution of the molecules among the vibrational states (~1), λ is the Stokes Raman wavelength (607.3 nm for N$_2$), h is Planck's constant (6.63 × 10^{-34} J-s), and c is the speed of light (3.0 × 10^8 m/s). For flames burning in air, the strongest Raman scattering comes from N$_2$ as seen in Figs 4.2 and 4.3. Estimates of the Raman signals (photoelectrons) from room air (1 atm, 300 K) and a flame (1 atm, 1500 K) from a single-laser pulse (or "shot") are given in Table 4.2.

One can estimate the signal-to-noise ratio of the vibrational Raman signals. For a CCD camera detection of Raman scattering, the major source of noise comes from the photoelectric effect. The noise is Poisson distributed and the signal-to-noise ratio SNR is given by:

$$SNR = (S_i)^{1/2} \tag{4.2}$$

Using this relation, the signal-to-noise ratios are given in Table 4.2. For a single-laser pulse, the noise level is high in room air and a flame (3% and 7% respectively). However, by integrating 1000 laser pulses on the back-illuminated camera, the Poisson noise from the photoelectric effect is greatly reduced to less than 1%. The remaining uncertainty comes from

Table 4.2. N_2-vibrational Stokes Raman signals and signal-to-noise ratio (SNR) for 532 nm laser for room air and a flame (0.2 mm length of beam, 140 mJ pulse, f/2 collection optics)

	N_2 Vibrational Raman Line	
	Room Air 79% N_2 (300 K, 1 atm)	Flame 75% N_2 (1500 K, 1 atm)
Single-shot signal (photoelectrons)	1000	200
Single-shot SNR	32 (3%)	14 (7%)
1000 shot average signal (photoelectrons)	1,000,000	200,000
1000 Shot average SNR	1000 (0.1%)	450 (0.2%)

systematic errors in the calibration, flame emission background, and laser-induced fluorescence interference.

To calibrate the Raman signals, a single-calibration factor is determined for each species:

$$C(T)_i = hc/[\sigma_{zz} L \Omega \eta Q_e \Gamma(T) \lambda] \qquad (4.3)$$

Once this calibration factor and the temperature are known, the number density of any species can be determined from the vibrational Raman signal:

$$N_i = C(T)_i S_i / E_L \qquad (4.4)$$

The temperature is given by,

$$T = P/[k\Sigma N_i] \qquad (4.5)$$

where k is Boltzmann's constant (1.38×10^{-23} J/K), P is the pressure, and N_i is the number density of the ith species. The calibration factors are determined by making Raman measurements in the post-flame region of a Hencken burner (Research Technologies, Inc.) which is a multielement diffusion flame that has very little heat loss at a high flow rate. Calibration measurements are made about 2 cm downstream where the flame radicals are recombined and the temperature and species are very close to adiabatic equilibrium conditions. Thus an adiabatic equilibrium program can be used to calculate the composition and temperature in the post-flame zone. Measuring many different H_2–air and CH_4–air flames, all the temperature-dependent calibrations factors are determined from 300 K to 2200 K.

An example calibration factor for N_2 is shown in Fig. 4.5 compiled from data in lean, stoichiometric and rich H_2–air Hencken burner flames.[25] An additional complicating factor is that some vibrational Raman lines overlap as seen in Fig. 4.6, where the O_2 Raman signal contributes to the CO_2 Raman line and vice versa. These Raman line overlaps must be taken into

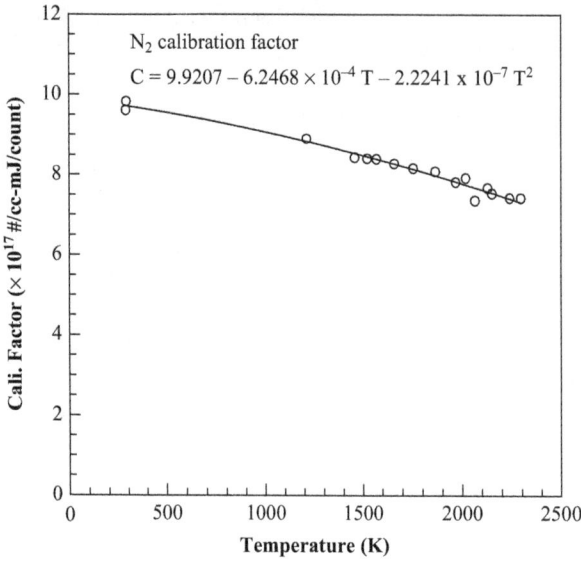

Figure 4.5. Temperature-dependent calibration factor for N_2: C(T), N_2. (*Source*: Hu, S. T. 2007. *Measurements and modeling of non-premixed tubular flames: Structure, extinction and instability.* Nashville, Tennessee USA: Ph.D. Thesis, Mechanical Engineering Department, Vanderbilt University.)

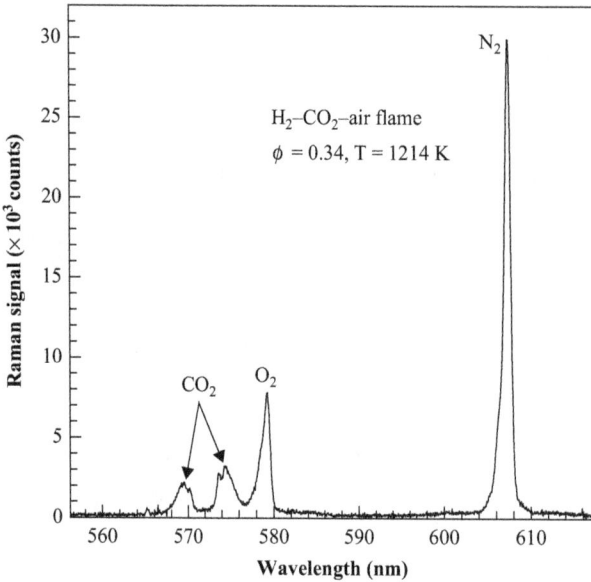

Figure 4.6. Raman spectra of H_2–CO_2–air flame at $\phi = 0.34$ showing the overlap between O_2 and CO_2. (*Source*: Hu, S. T. 2007. *Measurements and modeling of non-premixed tubular flames: Structure, extinction and instability.* Nashville, Tennessee USA: Ph.D. Thesis, Mechanical Engineering Department, Vanderbilt University.)

account. For example, the CO_2 number density is determined from the O_2 and CO_2 signals and the temperature as follows:

$$N_{CO_2} = [C(T)_{CO_2} S_{CO_2} - K(T)_{O_2\text{-}CO_2} S_{O_2}] / E_L \tag{4.6}$$

Example curves for the CO_2 calibration factor [$C(T)_{CO_2}$] and the interference of O_2 on CO_2 [$K(T)_{O_2\text{-}CO_2}$] are shown in Figs 4.7 and 4.8.[25] Similar equations are written for all the major species.

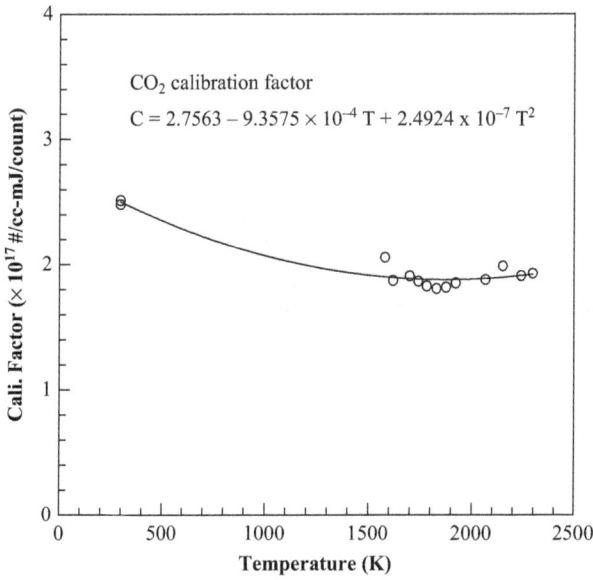

Figure 4.7. Temperature-dependent calibration factor of CO_2: $C(T)$, CO_2. (*Source*: Hu, S. T. 2007. *Measurements and modeling of non-premixed tubular flames: Structure, extinction and instability.* Nashville, Tennessee USA: Ph.D. Thesis, Mechanical Engineering Department, Vanderbilt University.)

CO_2 calibration factor

$C = 2.7563 - 9.3575 \times 10^{-4}\, T + 2.4924 \times 10^{-7}\, T^2$

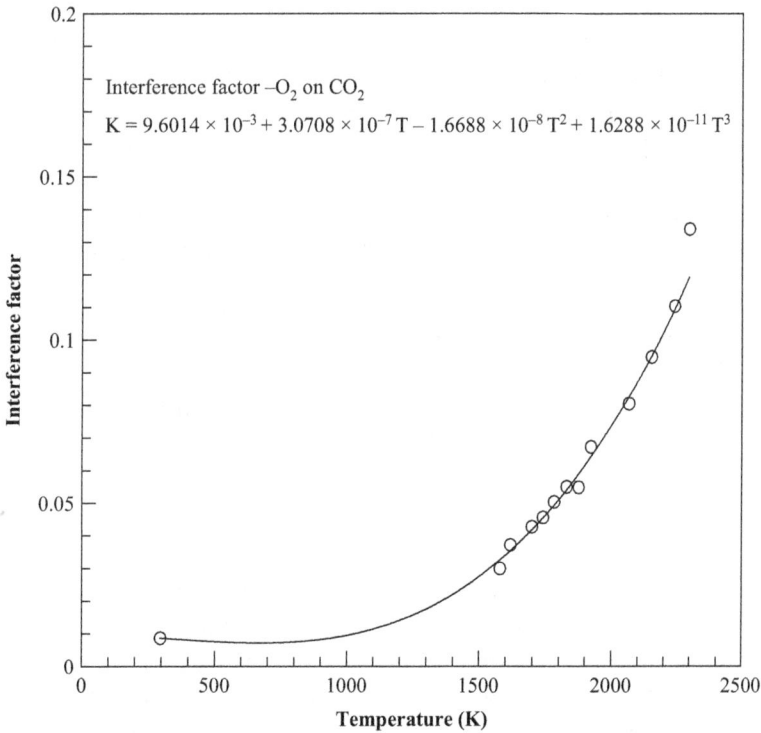

Interference factor $-O_2$ on CO_2

$K = 9.6014 \times 10^{-3} + 3.0708 \times 10^{-7}\, T - 1.6688 \times 10^{-8}\, T^2 + 1.6288 \times 10^{-11}\, T^3$

Figure 4.8. Temperature-dependent interference factor of O_2 on CO_2 $K(T)$ O_2-CO_2. (*Source*: Hu, S. T. 2007. *Measurements and modeling of non-premixed tubular flames: Structure, extinction and instability.* Nashville, Tennessee USA: Ph.D. Thesis, Mechanical Engineering Department, Vanderbilt University.)

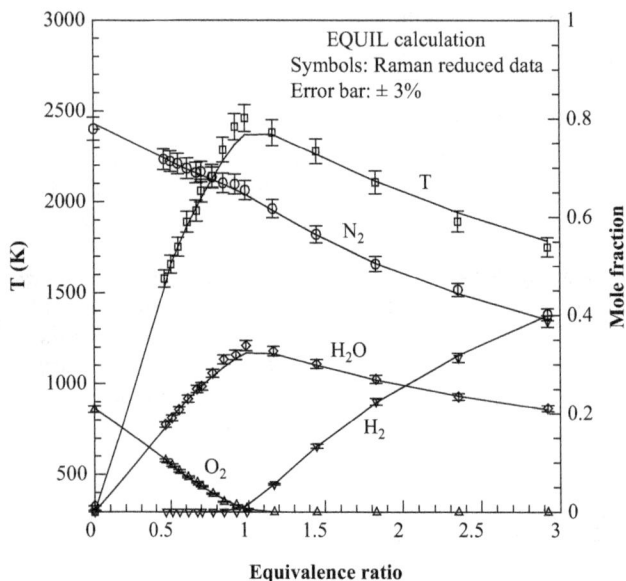

Figure 4.9. Comparison between measured temperature and species mole fractions using Raman scattering vs adiabatic equilibrium calculations in the post flame zone of a H_2–air flame in a Hencken burner. (*Source*: Hu, S. T. 2007. *Measurements and modeling of non-premixed tubular flames: Structure, extinction and instability.* Nashville, Tennessee USA: Ph.D. Thesis, Mechanical Engineering Department, Vanderbilt University.)

Due to the temperature dependence of the calibration constant, the calibration factors have to be determined by iteration using equations that include all the major species found in the flame (e.g., CO, CH_4, CO_2, H_2, H_2O, N_2, and O_2). The results of a calibration in the post-flame zone of a H_2–air Hencken burner flame are shown in Fig. 4.9. In this calibration, the photoelectron shot noise given by Eqn. 4.2 is less than 1%. The reported error of ±3% shown in Fig. 4.9 is due to other error sources such as gas flow meter uncertainty, flame background radiation at the Raman spectral wavelengths, laser energy measurement error, etc. For line Raman measurements, calibration factors are determined for each ~0.2 mm segment along a ~5 mm line; this accounts for variations in the spectral throughput of the light collection–spectrometer system along the ~5 mm line.[25]

4.3   TUBULAR FLAME BURNER

A tubular flame burner was designed and built for good optical access[15] and is shown in Fig. 4.10. The tubular burner is an annular nozzle design based on an earlier radial flow nozzle burner.[26] To provide optical access, two windows (25 mm dia.) are mounted at 180° to allow the laser beam to enter and exit the burner. The windows are mounted at an angle to reduce laser scattering inside the burner. A third window (50 mm dia.) is mounted at 90° to the axis of the laser optical ports to measure the Raman scattered light. The premixed reactants enter the stagnation chamber through 16 circumferentially spaced inlet ports. The stagnation chamber is packed with fine gauge (00) stainless-steel wool to distribute the 16 jets and give a uniform radial pressure drop. Cylindrical screen cavities are placed in front of each window to keep the stainless steel from interfering with the optical access.

The annular nozzle creates a uniform velocity of premixed reactants flowing radially inward. Above and below the central nozzle, two co-flow nozzles produce radial inward N_2 gas flows to provide shielding of the tubular flame from laboratory disturbances and cool the burner walls. The tubular burner also has internal water cooling. The burner is mounted on a xyz

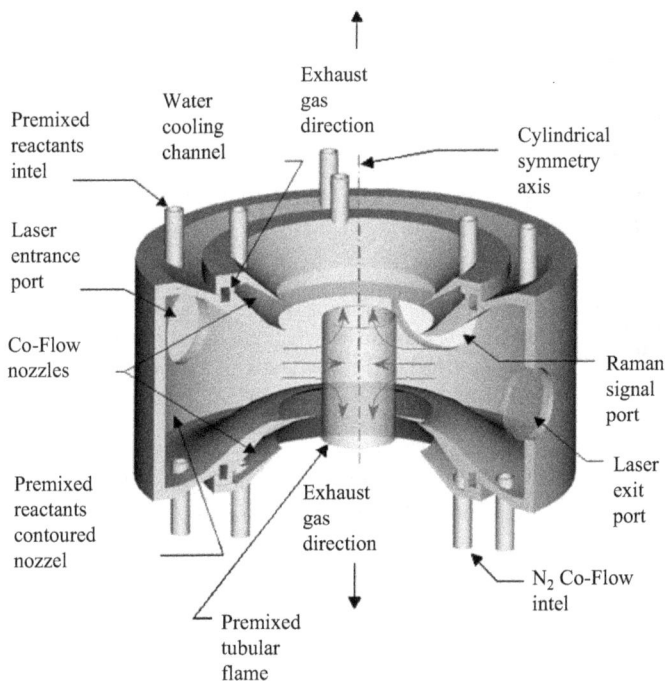

Figure 4.10. Schematic of premixed tubular burner design. (*Source*: Mosbacher, D. M. et al. 2002. Experimental and numerical investigation of premixed tubular flames. *Proceedings of the Combustion Institute* 29:1479–86. Reprinted with permission from Elsevier.)

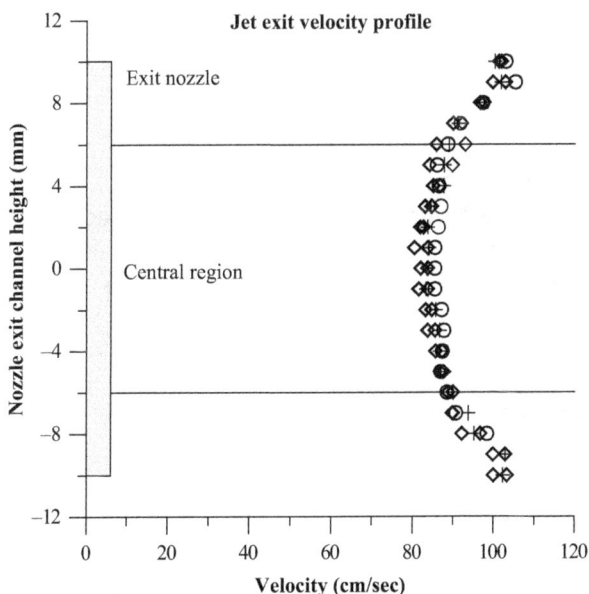

Figure 4.11. Circumferential profiles (measurements taken at 90° increments) of the jet exit velocity for an air flow rate of 85.9 slpm. (*Source*: Mosbacher, D. M. 2002. *An application of non-intrusive laser-based diagnostics in tubular flames.* Nashville, Tennesssee, USA: MS Thesis, Mechanical Engineering Department, Vanderbilt University.)

translation stage. Hot wire measurements in Fig. 4.11 show the uniformity of the radial velocity at different 90° increments in the central region of the annular nozzle.

The annular nozzle radius, R, is 15 mm and the annual nozzle height, H, is 20 mm. Based on an analytical solution of the two-dimensional flow field, the stretch rate at the flame surface is given by,[28]

$$k = \pi V/R \qquad\qquad (4.7)$$

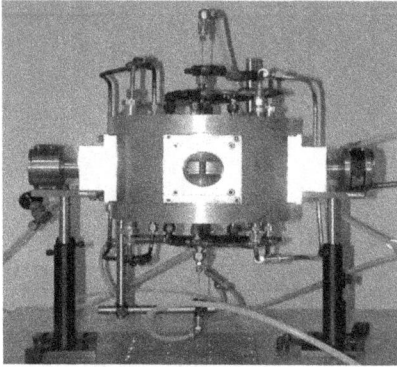

(a)

→| |← 7 mm

(b)

Figure 4.12. (a) Picture of the tubular burner with a central fuel nozzle installed (not used for premixed tubular flames). (b) Side image of a tubular premixed CH_4–air flame: $\phi = 0.5$, $k = 157$ s^{-1}. (*Source*: Mosbacher, D. M. 2002. *An application of non-intrusive laser-based diagnostics in tubular flames*. Nashville, Tennesssee, USA: MS Thesis, Mechanical Engineering Department, Vanderbilt University.)

where V is the radial velocity of inflowing premixed reactants at the nozzle exit.

As seen in Fig. 4.12, the burner and the tubular flame are vertically oriented. The picture of the burner in Fig. 4.12(a) includes a central fuel nozzle that is removed for premixed tubular flame studies. Very uniform tubular flames are produced by the burner as seen by the image of a premixed CH_4–air tubular flame ($\phi = 0.5$, $k = 157$ s^{-1}) in Fig. 4.12(b).

4.4 RAMAN SCATTERING MEASUREMENTS IN TUBULAR FLAMES

4.4.1 HYDROGEN–AIR TUBULAR FLAMES

The Raman scattering system was applied to measure temperature and major species composition in a H_2–air tubular flame[a] ($\phi = 0.175$) and the results are shown in Figs 4.13 and 4.14 for a stretch rate of $k = 141$ s^{-1}.[15] The laser beam passes though the center of the tubular flame. The burner is translated to measure two 4.6-mm line segments on either side of the tubular flame

---

[a] In the original study (Mosbacher et al. 2002), the stretch rate was given as $k = 2V/R$. The values given here are for $k = \pi V/R$.

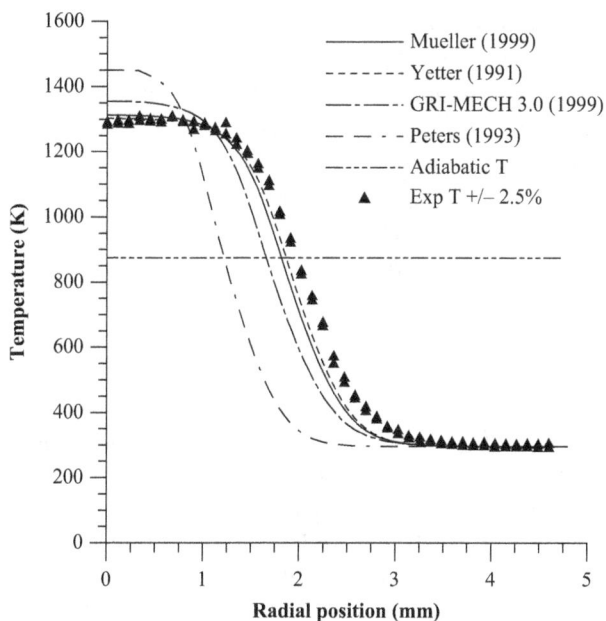

Figure 4.13. Comparison of predicted temperature profiles from four reaction mechanisms in a premixed H_2-air tubular flame ($\phi = 0.175$, $k = 141$ s^{-1}). (*Source*: Mosbacher, D. M. et al. 2002. Experimental and numerical investigation of premixed tubular flames. *Proceedings of the Combustion Institute* 29:1479–86. Reprinted with permission from Elsevier.)

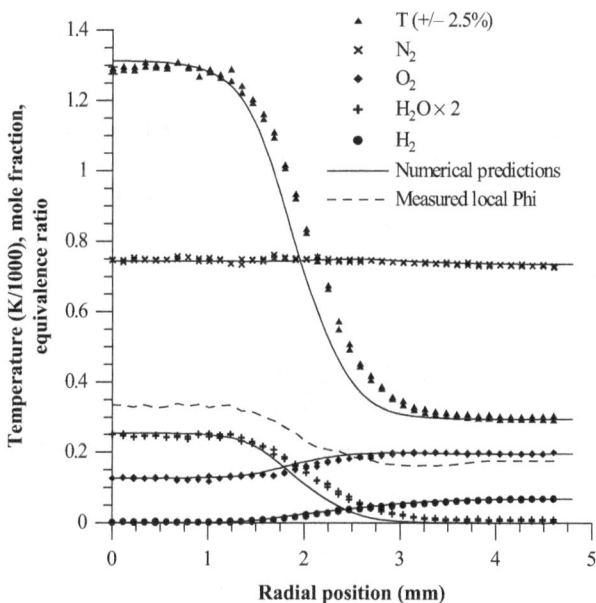

Figure 4.14. Raman-derived and numerically predicted product measurements using Mueller et al. chemistry with thermal diffusion in $\phi = 0.175$, $k = 141$ s^{-1} H_2–air tubular flame. (*Source*: Mosbacher, D. M. et al. 2002. Experimental and numerical investigation of premixed tubular flames. *Proceedings of the Combustion Institute* 29:1479–86. Reprinted with permission from Elsevier.)

cylindrical axis. The data are plotted versus radius giving two data points at each radial point as seen in Fig. 4.13. The points nearly overlap demonstrating the symmetry of the tubular flame structure.

The measured temperature and composition profiles in the H_2–air flame are compared to a detailed molecular transport and chemistry numerical simulation of the two-dimensional tubular flame. The standard Oppdif program[29] was modified for the tubular flame geometry[15] following the tubular flame equations derived previously.[30] The molecular transport model included thermal diffusion of light species (molecular weights <4). Mixture-averaged transport formulas were used to decrease the computational time as the multicomponent transport gave only marginally different results for the flame shown in Figs 4.13 and 4.14.[15] Finally, gas phase radiation in the thin optical limit was included in the numerical simulation.

In Fig. 4.13, the Raman temperature profile is compared with numerical simulations using four different chemical kinetic mechanisms.[31–34] The premixed tubular flame structure is very sensitive to the specific chemical kinetic mechanism. GRI-Mech and Peters mechanisms both give a flame radius that is too small and a peak temperature that is too high. The Yetter and Mueller mechanisms gave the best fit to the H_2–air tubular flame and the Mueller mechanism is chosen for the rest of the H_2–air flame comparisons. The computer simulations all show a smaller flame radius.

In Fig. 4.14, the temperature and major species composition (H_2, H_2O, O_2, N_2) data show a good fit to the computer simulations, using the Mueller mechanism. The tubular flame ($k = 141$ s^{-1}) is highly curved with a flame radius of 2.0 mm (curvature = 0.5/mm) based on the temperature profile. The combined effect of curvature and stretch leads to strong preferential diffusion effects. The local ϕ value is calculated from the Raman data and is plotted in Fig. 4.14. Preferential diffusion effects in Le = 0.33 lean positively curved flame cause ϕ to decrease ahead of the flame (e.g., $\phi = 0.16$ at r = 3 mm) and ϕ to increase behind the flame (e.g., $\phi = 0.33$ at r = 0.2 mm). The light H_2 molecules diffuse more rapidly into the flame than the heavy O_2 molecules. The measured peak flame temperature of 1295 K is 420 K above the adiabatic flame temperature of the incoming mixture (875 K for $\phi = 0.175$). However, the peak temperature of 1295 K is equal to the adiabatic flame temperature of the local enriched mixture ($\phi = 0.33$). This demonstrates the strengthening effects of stretch and positive flame curvature in this low Lewis number (Le = 0.33) lean H_2–air flame. For a premixed flame, the Lewis number is defined by Le = α/D_i where α is the diffusivity of the incoming fuel–air mixture and D_i is the molecular diffusivity of the ith deficient reactant (either oxidizer or fuel). Here we define positive flame curvature as a convex flame to the incoming reactants and negative flame curvature as a concave flame to the incoming reactants. The tubular flame always has positive flame curvature.

The effect of H_2 diffusion on the flame temperature can be seen clearly in Fig. 4.15, where the H_2 diffusion coefficient is varied by +32% or –29% and the thermal diffusion of light species is turned on and off. The predicted temperature profile is very sensitive to the value of the H_2 diffusion coefficient and to thermal diffusion of light species (Soret effect). The flame radius is decreased by ~25% when the Soret effect is omitted or when the H_2 diffusion coefficient is reduced by 29%.

The structure of a lean H_2–air tubular flame near extinction ($\phi = 0.175$, $k = 363$ s^{-1}) is shown in Fig. 4.16.[16] This flame ($k = 363$ s^{-1}) has a flame radius of 0.9 mm (curvature = 1.1/mm) that is half of the previous flame ($k = 141$ s^{-1}). The flame was numerically simulated with the modified Oppdif code used previously including thermal diffusion of light species, optically thin radiation, and the Mueller chemical kinetic mechanism. The only change is that the molecular transport is computed with full multicomponent molecular diffusion. Mixture-averaged formulas for molecular transport gave distinctly different results and were not used here. At this high stretch rate and flame curvature, preferential diffusion effects are increased and the maximum flame temperature of 1350 K is 475 K above the adiabatic flame temperature of the incoming mixture (875 K) and 200 K above the peak flame temperature of a corresponding planar opposed-jet flame (1150 K).

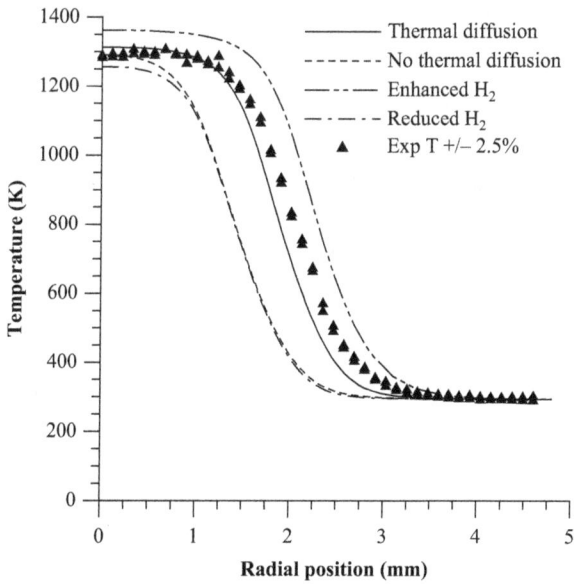

Figure 4.15. Simulation using Mueller et al. chemistry showing effects of thermal diffusion coefficient and increasing/decreasing the binary diffusion coefficient of H_2 into N_2 by +32%/−29% in a H_2–air tubular flame ($\phi = 0.175$, $k = 141$ s^{-1}). (*Source*: Mosbacher, D. M. et al. 2002. Experimental and numerical investigation of premixed tubular flames. *Proceedings of the Combustion Institute* 29:1479–86. Reprinted with permission from Elsevier.)

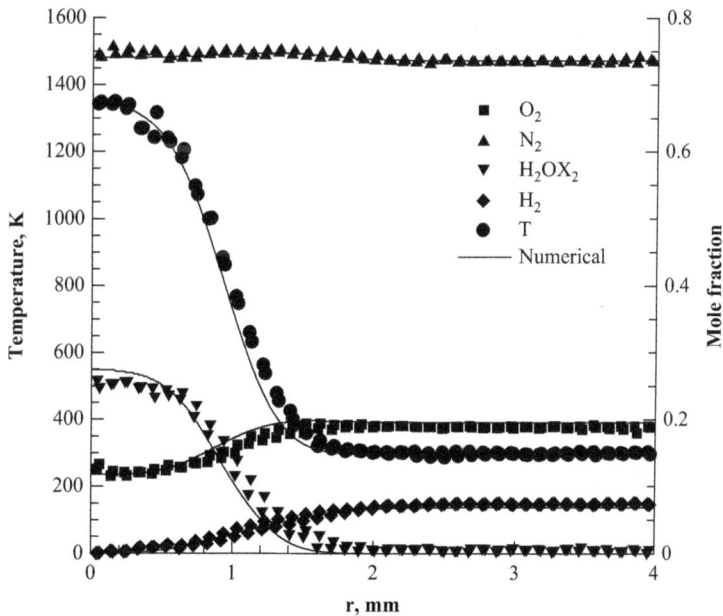

Figure 4.16. Measured and calculated temperature and species profiles for a $\phi = 0.175$, $k = 363$ s^{-1}, Le = 0.33, H_2–air premixed tubular flame. (*Source*: Hu, S. T., P. Y. Wang, and R. W. Pitz. 2009. A structural study of premixed tubular flames. *Proceedings of the Combustion Institute* 32:1133–40. Reprinted with permission from Elsevier.)

The important parameter that determines the strengthening or weakening of preferential diffusion by flame curvature for tubular flames is the corrected Karlovitz number:

$$Ka_c = (1 + \delta/R) \, Ka \tag{4.8}$$

where δ is the flame thickness and R is the flame radius.[35] The Karlovitz number is given by $Ka = k\delta/S_b$ where S_b is the flame speed in the burned mixture and k is the flame stretch. If the stretch rate strengthens a planar flame through preferential diffusion, then a tubular flame is further enhanced by curvature according to the corrected Karlovitz number. The corrected Karlovitz number can be generalized to other curved flames,

$$Ka_c = (1 + \delta \nabla \cdot \mathbf{n}) \, Ka \tag{4.9}$$

where $\mathbf{n}$ is the normal vector to the flame surface. Here the flame curvature ($1/R$) is substituted for a generalized curvature ($\nabla \cdot \mathbf{n}$) that can be positive or negative. For lean H_2–air mixtures ($\phi = 0.4$), analytical simulations of peak flame temperatures of planar opposed-jet flames and tubular flames collapse onto a single curve when plotted against $(1 + \delta \nabla \cdot \mathbf{n}) \, Ka$ as shown in Figs 4.17 and 4.18.[35] Analytical solutions of the peak flame temperature versus stretch are shown in Fig. 4.17 for various curved stretched flames with positive and negative curvature. The peak temperatures differ by as much as 100 K due to curvature effects. When the peak temperatures are plotted against the corrected Karlovitz number, Ka_c, the peak temperatures nearly collapse onto a single curve as shown in Fig. 4.18.

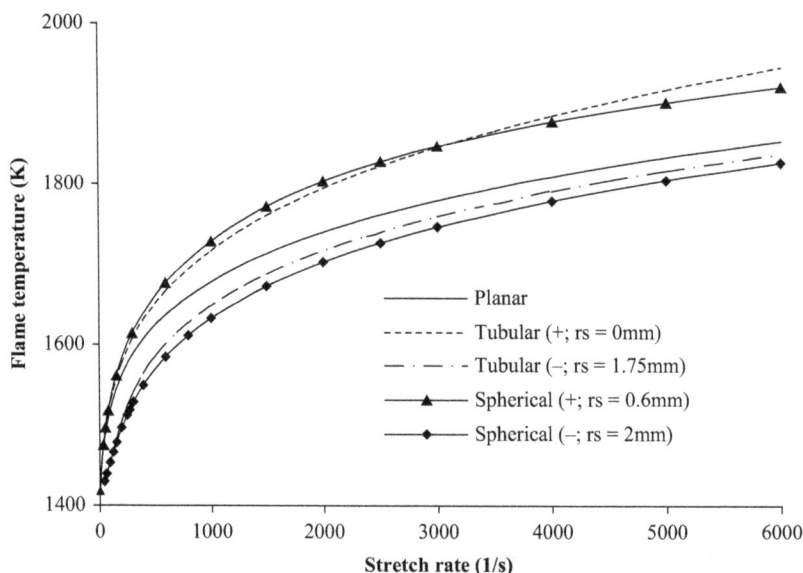

Figure 4.17. Peak flame temperature variation with a stretch rate for the planar opposed-jet and curved flames (H_2–air, $\phi = 0.4$). The tubular flame has positive curvature (+) and a stagnation radius of $r_s = 0$. (*Source*: Wang, P. Y. and R. W. Pitz. 2005. The premixed flame parameters for the stretched and curved flames. In *4th Joint Meeting of the U.S. Sections of the Combustion Institute*. Philadelphia, PA.)

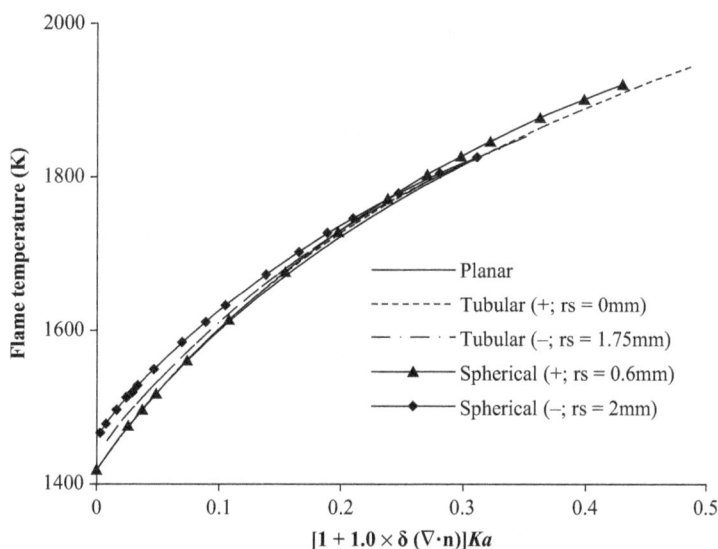

Figure 4.18. Peak flame temperature variation with corrected Karlovitz number for planar opposed-jet and curved stretched flames (H_2–air, $\phi =$ 0.4). The tubular flame has positive curvature (+) and a stagnation radius of $r_s = 0$. (*Source*: Wang, P. Y. and R. W. Pitz. 2005. The premixed flame parameters for the stretched and curved flames. In *4th Joint Meeting of the U.S. Sections of the Combustion Institute*. Philadelphia, PA.)

The strong strengthening effect of positive curvature for lean H_2–air flames has a great effect on the extinction strain rate. In Fig. 4.19, the measured and numerically predicted extinction strain rates are compared between tubular flames and planar opposed-jet flames using the Mueller mechanism. The measured extinction strain rates of the tubular flame differ from the numerical predictions for $\phi \geq 0.16$ due to the onset of turbulence in the nozzle inlet. At $\phi = 0.16$, the measured extinction strain rate is $k = 450$ s^{-1} which corresponds to a nozzle exit Reynolds number of Re = HV/$\nu \cong 2600$ where H is the height of the annular nozzle (20 mm), V is the inlet velocity, and ν is the kinematic viscosity.

Tubular flames are able to exist at strain rates well beyond the extinction strain rates of planar opposed-jet flames. For example, for a H_2–air mixture at $\phi = 0.16$, the tubular flame has an extinction strain rate of nearly eight times larger than the planar opposed-jet flame. The increase in extinction strain rate is due to the strengthening effect of positive flame curvature on the flame. For subunity Lewis number lean H_2–air flames, preferential diffusion strengthens flames with positive curvature (flame surface convex toward the unburned reactants) and weakens flames with negative curvature (flame surface concave toward the unburned reactants).

4.4.2 METHANE–AIR TUBULAR FLAMES

Raman measurements were made in lean CH_4–air tubular flames and the results are shown in Figs 4.20 and 4.21 for a flame near extinction ($\phi = 0.58$, $k = 257$ s^{-1}).[16] The measured temperature profile in the lean CH_4–air tubular flame is compared with numerical simulations using five

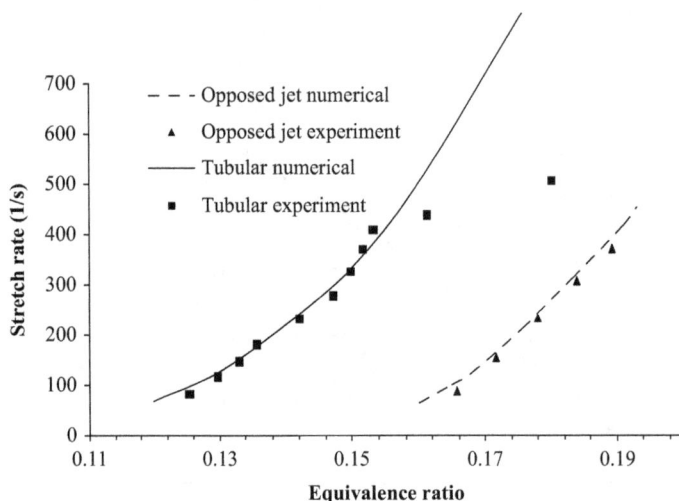

Figure 4.19. Measured and calculated extinction stretch rates for lean H_2–air premixed tubular and flat opposed-jet flames. (*Source*: Hu, S. T., P. Y. Wang, and R. W. Pitz. 2009. A structural study of premixed tubular flames. *Proceedings of the Combustion Institute* 32:1133–40. Reprinted with permission from Elsevier.)

different chemical kinetic mechanisms: C1 and C2,[34] San Diego,[36] Kee,[37] and GRI-3.0.[33] For this near-extinction flame, the predictions were very sensitive to the particular chemical kinetic mechanism and the C1 and GRI-3.0 mechanisms caused extinction at this condition. Predictions using the C2, Kee, and San Diego mechanisms enabled stable flames and the temperature profiles are shown in Fig. 4.20. The San Diego mechanism predicted a flame radius greater than measured. Both the San Diego and Kee mechanisms gave similar predictions and the Kee mechanism was used for comparison to chemical species measurements shown in Fig. 4.21.

Unlike lean H_2–air flames, preferential diffusion effects are minimal in lean CH_4–air tubular flames as the Lewis number of the mixture is close to one (Le = 0.98). Stretch rate and curvature do not have a strong effect on the flame temperature. For the CH_4–air tubular flame shown in Fig. 4.20, the adiabatic flame temperature is T_{ad} = 1626 K which is close to the peak measured and calculated flame temperature. The lack of preferential diffusion effects for methane tubular flames can be seen in the numerically predicted variation of peak flame temperature versus stretch as shown in Fig. 4.22. Peak tubular flame temperatures are at most ~20 K above corresponding planar opposed-jet flames and ~40 K above their corresponding adiabatic flame temperatures (shown at zero stretch rate in Fig. 4.22).

4.4.3 PROPANE–AIR TUBULAR FLAMES

In lean propane–air flames, the Lewis number of the reactants is greater than one and preferential diffusion causes the flame temperature to decrease from its adiabatic equilibrium value. Raman measurements of temperature and species concentrations (O_2, N_2, H_2O, CO_2, C_3H_8) in a C_3H_8–air premixed tubular flame (ϕ = 0.64, k = 168 s^{-1}) are shown in Fig. 4.23. For this

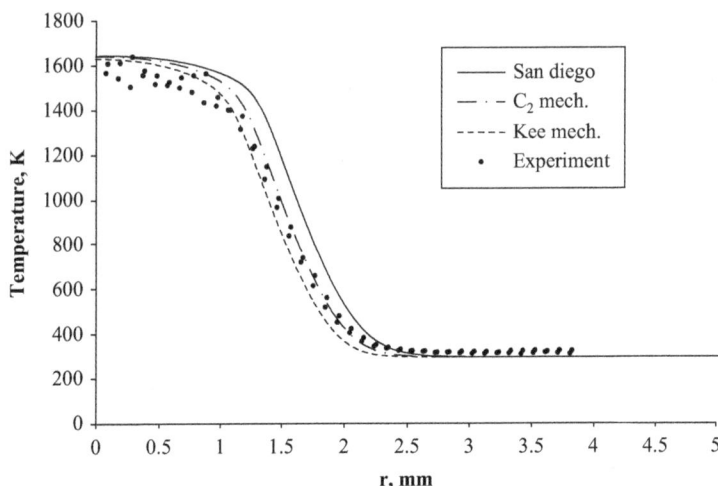

Figure 4.20. Measured temperature profiles for a $\phi = 0.58$, $k = 257$ s^{-1} CH$_4$–air premixed tubular flame compared with calculation with different chemical kinetic mechanisms. (*Source*: Hu, S. T., P. Y. Wang, and R. W. Pitz. 2009. A structural study of premixed tubular flames. *Proceedings of the Combustion Institute* 32:1133–40. Reprinted with permission from Elsevier.)

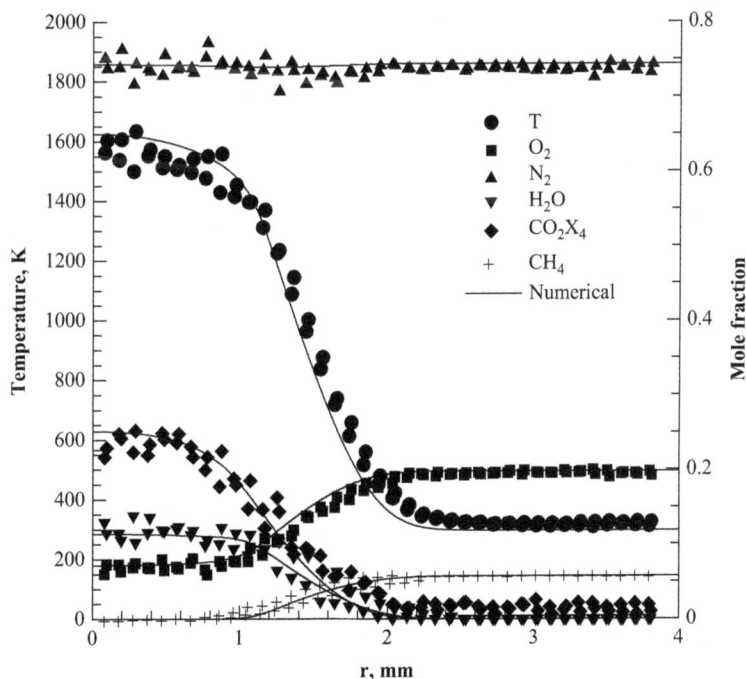

Figure 4.21. Measured and calculated temperature and species profiles for a $\phi = 0.58$, $k = 257$ s^{-1} CH$_4$–air premixed tubular flame. (*Source*: Hu, S. T., P. Y. Wang, and R. W. Pitz. 2009. A structural study of premixed tubular flames. *Proceedings of the Combustion Institute* 32:1133–40. Reprinted with permission from Elsevier.)

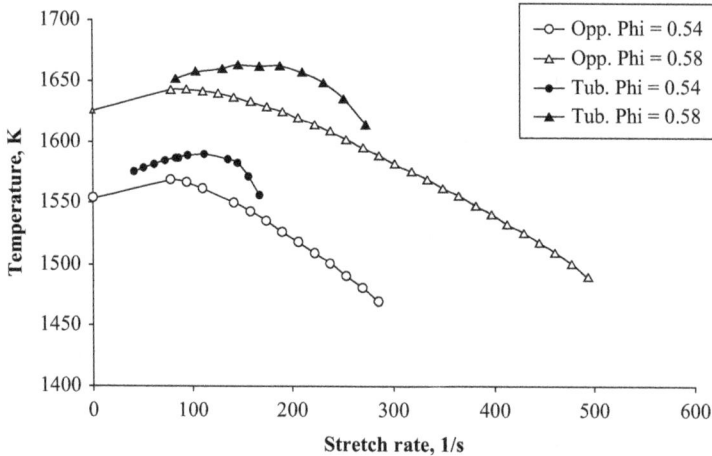

Figure 4.22. Peak temperatures in tubular vs flat CH_4/air premixed flames at various stretch rates (adiabatic flame temperature shown at zero stretch rate and extinction occurs beyond the right end of the curves). (*Source*: Hu, S. T., P. Y. Wang, and R. W. Pitz. 2009. A structural study of premixed tubular flames. *Proceedings of the Combustion Institute* 32:1133–40. Reprinted with permission from Elsevier.)

mixture, the Lewis number is ~1.9. The measured peak flame temperature is ~1500 K which is less than the adiabatic flame temperature of 1771 K. In Fig. 4.23, the Raman measurements are compared with numerical simulations using the San Diego mechanism[36] with and without radiation in the optically thin limit. Due to difficulty in achieving numerical convergence for the propane–air flames, mixture-averaged transport properties are used instead of the more computationally expensive multicomponent formulas. The numerical simulation predicts a peak flame temperature of 1700 K that is below the adiabatic flame temperature of 1771 K showing the deleterious effects of preferential diffusion. Radiation has a minor effect at this stretch rate and only lowers flame temperature by ~20 K.

There is a large ~200 K difference observed between the measured flame temperature and the predicted flame temperature using the San Diego mechanism. These differences could be due to limitations in the numerical simulation or in the Raman scattering measurement. As seen earlier for H_2–air and CH_4–air tubular flames, model-data comparisons are very sensitive to the specific chemical kinetic mechanism. Other propane chemical kinetic mechanisms as well as the more accurate multicomponent molecular transport formula may give better agreement with the data. Propane–air flames also produce more laser-induced background that interferes with the Raman scattering measurement which can lead to increased measurement error.

4.5 CELLULAR TUBULAR FLAMES

4.5.1 INSTABILITIES IN TUBULAR FLAMES

In premixed flames, cellular instabilities arise from (1) gas thermal expansion across the flame surface (hydrodynamic or Darrieus–Landau instability) and from (2) thermal diffusive or preferential

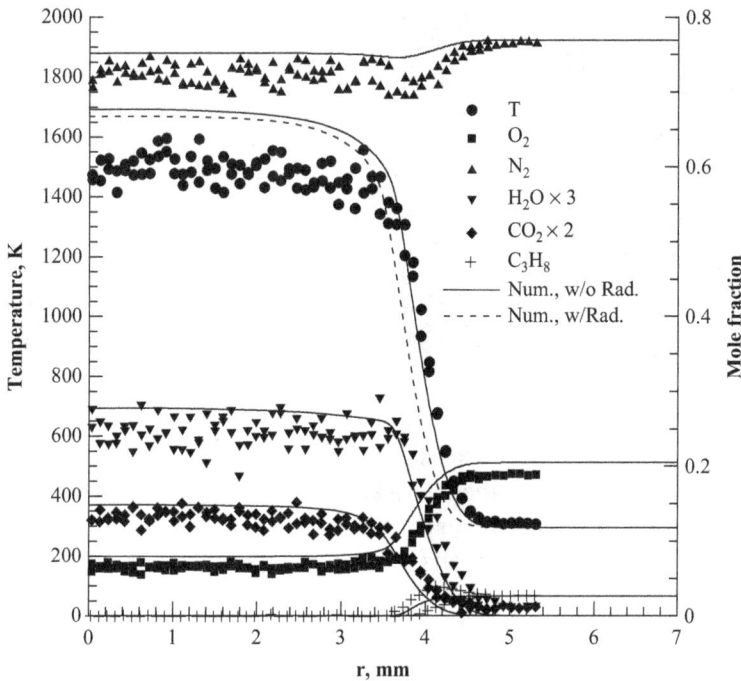

Figure 4.23. Measured and calculated temperature and species profiles for a $\phi = 0.64$, $k = 168$ s^{-1} C$_3$H$_8$/air premixed tubular flame. (*Source*: Hu, S. T., P. Y. Wang, and R. W. Pitz. 2009. A structural study of premixed tubular flames. *Proceedings of the Combustion Institute* 32:1133–40. Reprinted with permission from Elsevier.)

diffusion effects.[38,39] In strongly stretched flames such as opposed-jet or tubular flames, the hydrodynamic instability is suppressed and instabilities are due to thermal-diffusive effects.

In stretched flames (Le <1), the cellular instabilities occur periodically along the unstretched coordinate.[b] In opposed jet flames, the flame surface is radially stretched and instabilities are periodic in the azimuthal angle creating "star shaped flames."[40] In tubular flames, the flame is axially stretched along the cylindrical flame surface and axial cellular ribbons appear that are periodic in the azimuthal angle. When viewed from the top, the flame instabilities appear as a "multipetal" flame.[41–44]

Examples of multipetal H$_2$–air tubular flames are shown in Fig. 4.24.[44] The flame emission is imaged from the bottom of the tubular burner using a 45° mirror and an ICCD video camera (Xybion ISG-250, 60 Hz). This is a lean hydrogen–air flame which has a low Lewis number (Le = 0.3). The multipetal cellular flames are formed far from extinction. Starting from a non-cellular tubular flame as seen in Fig. 4.24(a), ϕ is gradually increased and the flame diameter becomes larger in order to balance the increased flame speed with an increased velocity near the nozzle. Eventually, instabilities occur in the cylindrical flame surface. Instabilities that are positively curved (flame surface convex toward the reactants) are strengthened by preferential

---

[b] Consider a curtain hanging from a rod. The curtain is stretched vertically and its folds occur periodically along the horizontal coordinate.

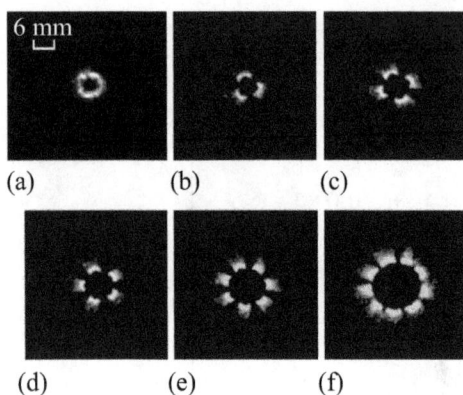

Figure 4.24. Cellular structure in premixed tubular flames of H_2/air: (a) $\phi = 0.2$, $k = 406$ s^{-1} (b) $\phi = 0.25$, $k = 422$ s^{-1} (c) $\phi = 0.27$, $k = 434$ s^{-1} (d) $\phi = 0.274$, $k = 444$ s^{-1} (e) $\phi = 0.30$, $k = 457$ s^{-1} (f) $\phi = 0.336$, $k = 472$ s^{-1}. (*Source*: Wang, Y., S. T. Hu, and R. W. Pitz. 2009. Extinction and cellular instability of premixed tubular flames. *Proceedings of the Combustion Institute* 32:1141–7. Reprinted with permission from Elsevier.)

diffusion; their flame speed is increased and they move to higher radial locations. Instabilities that are negatively curved (flame surface concave toward the reactants) are weakened and the flame speed is decreased resulting in extinction. As seen in Fig. 4.24, as ϕ is increased, the overall diameter of the multipetal flame increases and the number of petals increase. Stable cellular lean H_2–air tubular flames (Le < 1) can be observed with 3–9 petals.[44] Cellular tubular flames also have been observed in rich propane–air flames (Le < 1) with as many as 10 petals.[43] For methane–air tubular flames, the Lewis number is near unity (Le = 0.98) and the tubular flame only shows mild wrinkling with no local extinction.[44]

4.5.2 RAMAN SCATTERING MEASUREMENTS IN CELLULAR TUBULAR FLAMES

Curvature and stretch are found in turbulent premixed flames where preferential diffusion can cause flame strengthening or flame weakening leading to extinction. However, the study of preferential diffusion in turbulent premixed flames is difficult due to their transient nature. Laminar cellular tubular flames consist of strongly curved reaction cells adjacent to extinction zones. These flames are stable and can be studied with chemiluminescence imaging and Raman scattering. Cellular tubular flames are an important canonical flame for the study of the effect of curvature and stretch on preferential diffusion that leads to the formation of stable flame cells adjacent to extinction zones.

To study cellular tubular flames, a high-stretch-rate tubular burner was developed with a more uniform reactant inflow into the stagnation chamber.[25] A schematic of the burner is shown in Fig. 4.25. The reactants flow into the stagnation chamber through a porous annular plate rather than 16 individual jets. With the uniform inflow, the stainless steel wool in the stagnation chamber is no longer required. The nozzle height and radius are reduced (H = 8 mm, R = 12 mm). Compared with the tubular burner in Fig. 4.12, for the same inlet velocity (V), the inlet Reynolds number (Re = HV/v) is lowered delaying the onset of turbulence and the stretch rate is increased ($k = \pi V/R$). The cellular flame images shown in Fig. 4.24 were taken in the new high-stretch-rate burner with stretch rates as high as 472 s^{-1}. In the previous burner, high stretch rates ($k > 400$ s^{-1}) were accompanied by unwanted turbulence in the inlet nozzle leading to premature extinction (Fig. 4.19). More details on the high-stretch tubular burner can be found elsewhere.[25]

The Raman scattering system described in Section 4.2, was applied to the cellular tubular flame.[45] For these measurements, a Quanta Ray Lab-150 YAG laser (532-nm 9-ns pulse at

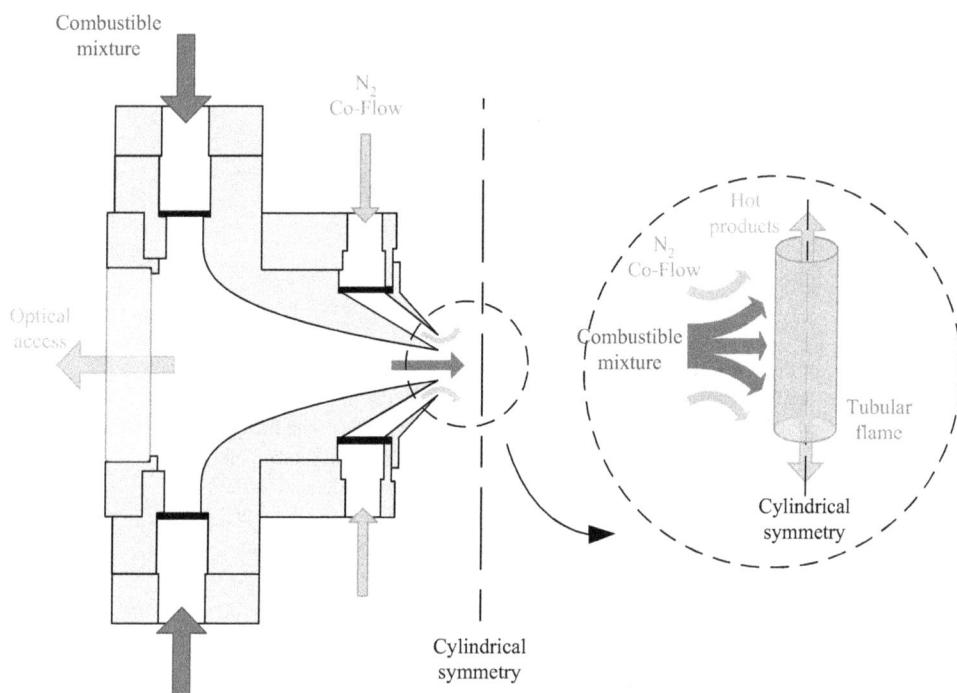

Figure 4.25. Schematic of the cross section of the high-stretch-rate tubular burner with annular porous plates for uniform inflow of the combustible mixture. (*Source*: Hall, C. A. and R. W. Pitz. 2013. A structural study of premixed hydrogen-air cellular tubular flames. *Proceedings of the Combustion Institute* 34:973–80. Reprinted with permission from Elsevier.)

10 Hz) stimulated the Raman scattering in the flame. A three-stage pulse stretcher lengthened the pulse and the laser energy was reduced to 70 mJ to avoid laser-induced breakdown. The laser beam was focused by a 0.3-m lens to 180 μm diameter in the flame as established by burn patterns. The spatial resolution along the laser beam was 86 μm as confirmed by imaging a back-lighted Ronchi grating in the measurement zone. The burner was translated to scan the cellular flame with an overall resolution of 86 × 200 μm in the 2D plane of the cellular flame.

An ICCD camera (576 × 384, 5-ms gate) coupled to a UV-Nikkor lens (F/4.5, 105 mm) axially imaged chemiluminescence from the tubular flame using a 45° mirror mounted at the bottom of the burner. A filter (Schott UG11 filter, 280–380 nm transmission) was mounted in front of the camera to record OH chemiluminescence (280–310 nm) from the flame. Examples of OH chemiluminescence images taken in the lean hydrogen–air tubular flame ($\phi = 0.25$, Le = 0.32) are shown in Fig. 4.26 for a four-cell, low stretch flame ($k = 200$ s^{-1}) and for a three-cell, high stretch flame (400 s^{-1}). The cells of both flames have approximately the same local curvature (~1 mm^{-1}).

Although the cellular flames in Fig. 4.26 are fairly stable, the cells do occasionally fluctuate rotationally. Raman scattering measurements require the cellular flame to be stable over periods of up to one hour. To keep the flames stable, a small wire is introduced in the inflow upstream of one of the petals. The wire produces minimal distortion to the cellular image but keeps the multipetal flame from rotating.[45]

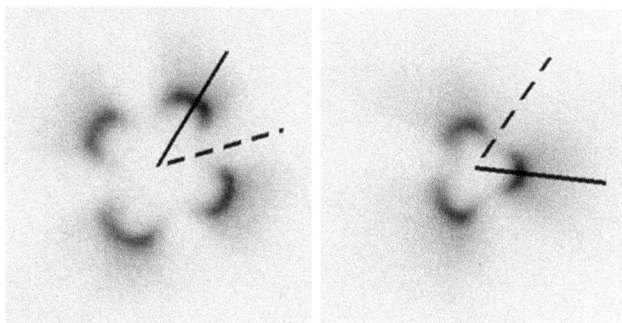

Figure 4.26. OH chemiluminescence images of hydrogen-air cellular tubular flames ($\phi = 0.25$, Le = 0.32) with stretch rates of 200 s^{-1} (left, 4 cells) and 400 s^{-1} (right, 3 cells). The image dimensions are 14 mm × 14 mm. One-dimensional interpolated profiles are generated along the solid (reaction cell) and dashed (extinction cell) lines. (*Source*: Hall, C. A. and R. W. Pitz. 2013. A structural study of premixed hydrogen-air cellular tubular flames. *Proceedings of the Combustion Institute* 34:973–80. Reprinted from Elsevier.)

Raman scattering measurements of temperature and chemical species mole fractions (O_2, N_2, H_2O, and H_2) in a four-petal lean hydrogen–air flame ($\phi = 0.25$, $k = 200$ s^{-1}, Le = 0.32) are shown in Fig. 4.27.[45] The Raman measurements were made inside the black box region shown in the OH chemiluminescence image in the upper left hand corner of Fig. 4.27. Temperature and H_2O images show strongly curved surfaces while the H_2 reactant image has the least curvature.

To quantitatively compare the reaction zone and extinction zone, the 2D Raman measurements are interpolated along two radii: one through the center of the reaction cell and one through the center of the extinction cell as shown in Fig. 4.26. The interpolated radial profiles of temperature and mole fraction (O_2, N_2, H_2O, and H_2) are shown in Fig. 4.28. Preferential diffusion causes the temperatures to be much higher in the reaction cell. In the reaction cell, the peak temperature of ~1600 K occurs at r = 2.8 mm and is ~550 K above the adiabatic flame temperature (T_{ad} = 1055 K). This ~550 K increase in peak flame temperature above T_{ad} is higher than the noncellular flames discussed earlier (~475 K above T_{ad} in Fig. 4.16). The local equivalence ratio ϕ also increases in the reaction cell to 180% of its initial value in the incoming reactants.[45] Positive flame curvature in the reaction cell has increased the equivalence ratio and flame temperature.

In the extinction cell, the temperature at r = 2.8 mm (radius of peak reaction cell temperature) is only 900 K (~700 K below the corresponding peak temperature in the reaction cell and 155 K below the adiabatic flame temperature) and very little H_2O product is found at this radial location indicating extinction. The local equivalence ratio ϕ at r = 2.8 mm is reduced to 80% of its initial value.[45] Negative curvature in the extinction cell has decreased the equivalence ratio and caused extinction.

Raman measurements were also made in a lean hydrogen–air cellular tubular flame with twice the stretch rate ($\phi = 0.250$, $k = 400$ s^{-1}, Le = 0.324). The chemiluminescence image of this three-cell flame is shown in Fig. 4.26. Radial profiles of temperature and mole fraction were also interpolated along the radial lines. The Raman measurements are not shown here but the

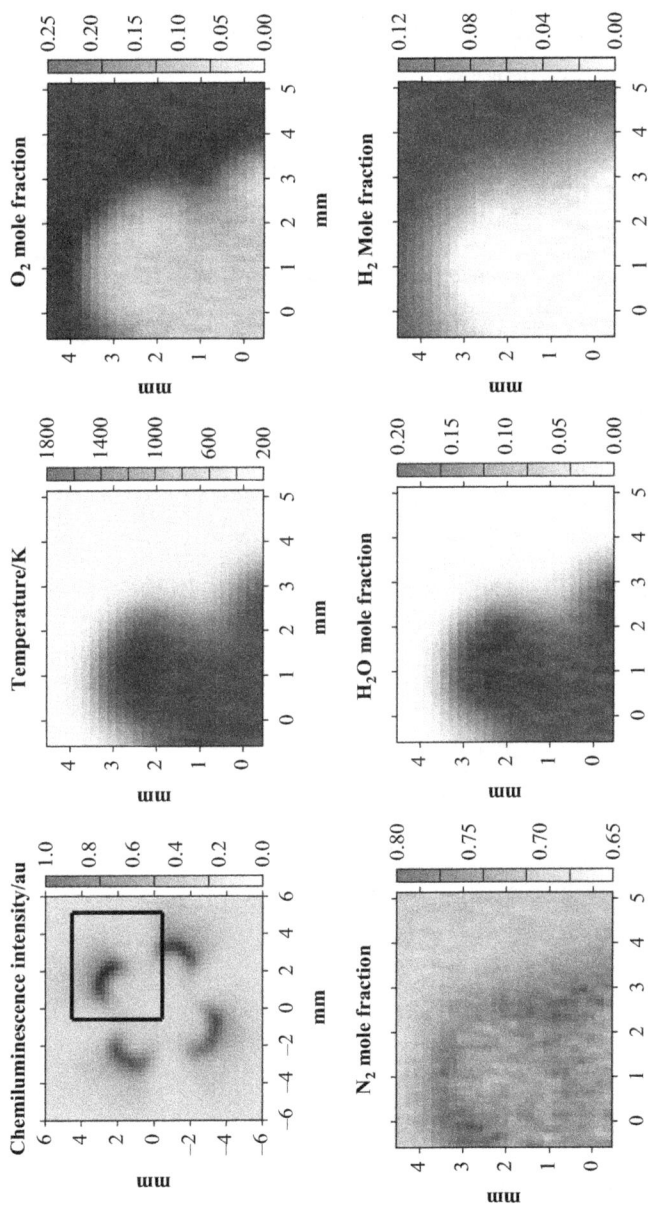

Figure 4.27. Spatially resolved Raman scattering measurements of species concentrations and temperature in a H_2–air cellular tubular flame ($\phi = 0.25$, $k = 200$ s^{-1}). The chemiluminescence image of the cellular flame is shown in the upper left corner and the Raman measurements were made in the region indicated by the black box. (*Source:* Hall, C. A. and R. W. Pitz. 2013. A structural study of premixed hydrogen-air cellular tubular flames. *Proceedings of the Combustion Institute* 34:973–80. Reprinted with permission from Elsevier.)

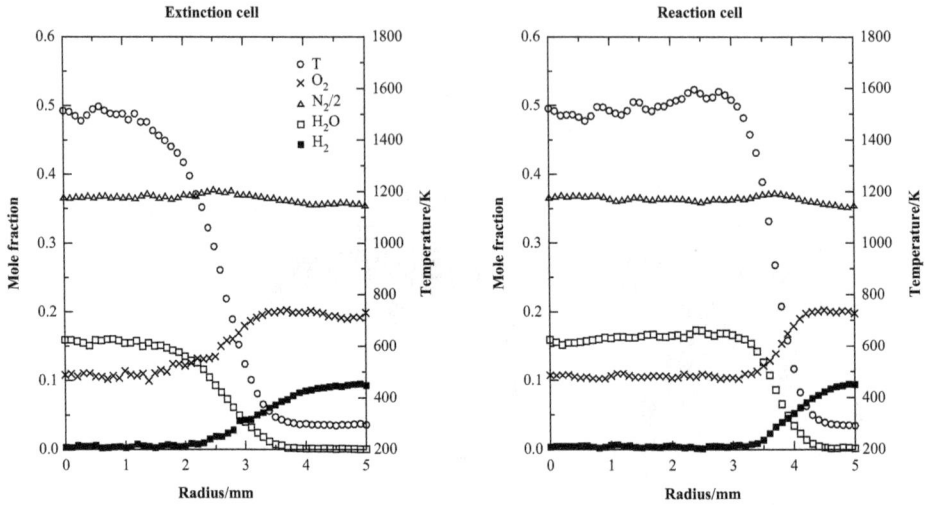

Figure 4.28. Radial interpolated profiles of temperature and species mole fractions in a H_2–air cellular tubular flame ($\phi = 0.25$, $k = 200$ s^{-1}) in the center of the extinction cell (left) and the reaction cell (right). (*Source*: Hall, C. A. and R. W. Pitz. 2013. A structural study of premixed hydrogen-air cellular tubular flames. *Proceedings of the Combustion Institute* 34:973–80. Reprinted with permission from Elsevier.)

reaction cells in the high stretch flame showed about the same peak temperature and enhanced local equivalence ratio as the low stretch flame.[45] The flame cells in both flames have the same curvature ($\sim$1 mm^{-1}) leading to the conclusion that the flame curvature is controlling the extent of preferential diffusion rather than the global stretch rate.

ACKNOWLEDGMENTS

The financial support of the National Science Foundation under Grants CBET-1134268 & CTS-0314704) and the NASA Microgravity Program under Grant NNC04AA14A is gratefully acknowledged by the author.

REFERENCES

1. Raman, C. V. 1928. A change of wave-length in light scattering. *Nature* 121:619.
2. Lapp, M., L. M. Goldman, and C. M. Penney. 1972. Raman scattering from flames. *Science* 175: 1112–15.
3. Lapp, M. and C. M. Penney, eds. 1974. *Laser Raman gas diagnostics*. New York: Plenum Press.
4. Lederman, S. 1977. Use of laser Raman diagnostics in flow fields and combustion. *Progress in Energy and Combustion Science* 3 (1):1–34. DOI: 10.1016/0360-1285(77)90007-7.
5. Long, D. A. 1977. *Raman spectroscopy*. London: McGraw-Hill.
6. Eckbreth, A. C. 1996. *Laser diagnostics for combustion temperature and species*. 2nd ed. United Kingdom: Gordon and Breach.

7. Hassel, E. P. and S. Linow. 2000. Laser diagnostics for studies of turbulent combustion. *Measurement Science & Technology* 11 (2):R37–57. DOI: 10.1088/0957-0233/11/2/201.

8. Kohse-Höinghaus, K. and J. Jeffries, eds. 2002. *Applied combustion diagnostics*. New York: Taylor and Francis.

9. Trees, D., T. M. Brown, K. Seshadri, M. D. Smooke, G. Balakrishnan, R. W. Pitz, V. Giovangigli, and S. P. Nandula. 1995. The structure of nonpremixed hydrogen-air flames. *Combustion Science and Technology* 104 (4–6):427–39.

10. Tanoff, M. A., M. D. Smooke, R. J. Osborne, T. M. Brown, and R. W. Pitz. 1996. The sensitive structure of partially premixed methane-air vs. air counterflow flames. *Proceedings of the Combustion Institute* 26 (1):1121–8.

11. Brown, T. M., M. A. Tanoff, R. J. Osborne, R. W. Pitz, and M. D. Smooke. 1997. Experimental and numerical investigation of laminar hydrogen-air counterflow diffusion flames. *Combustion Science and Technology* 129 (1–6):71–88.

12. Sung, C. J., J. B. Liu, and C. K. Law. 1995. Structural response of counterflow diffusion flames to strain-rate variations. *Combustion and Flame* 102 (4):481–92. DOI: 10.1016/0010-2180(95)00041-4.

13. Wehrmeyer, J. A., S. Yeralan, and K. S. Tecu. 1996. Influence of strain rate and fuel dilution on laminar nonpremixed hydrogen-air flame structure: An experimental investigation. *Combustion and Flame* 107:125–40.

14. Wehrmeyer, J. A., Z. X. Cheng, D. M. Mosbacher, R. W. Pitz, and R. Osborne. 2002. Opposed jet flames of lean or rich premixed propane-air reactants versus hot products. *Combustion and Flame* 128 (3):232–41.

15. Mosbacher, D. M., J. A. Wehrmeyer, R. W. Pitz, C. J. Sung, and J. L. Byrd. 2002. Experimental and numerical investigation of premixed tubular flames. *Proceedings of the Combustion Institute* 29:1479–86.

16. Hu, S. T., P. Y. Wang, and R. W. Pitz. 2009. A structural study of premixed tubular flames. *Proceedings of the Combustion Institute* 32:1133–40. DOI: 10.1016/j.proci.2008.06.183.

17. Hu, S. T., P. Y. Wang, R. W. Pitz, and M. D. Smooke. 2007. Experimental and numerical investigation of non-premixed tubular flames. *Proceedings of the Combustion Institute* 31:1093–9. DOI: 10.1016/j.proci.2006.08.058.

18. Hu, S. T. and R. W. Pitz. 2009. Structural study of non-premixed tubular hydrocarbon flames. *Combustion and Flame* 156 (1):51–61. DOI: 10.1016/j.combustflame.2008.07.017.

19. Pitz, R. W., R. Cattolica, F. Robben, and L. Talbot. 1976. Temperature and density in a hydrogen-air flame from Rayleigh-scattering. *Combustion and Flame* 27 (3):313–20.

20. Miles, R. B., W. R. Lempert, and J. N. Forkey. 2001. Laser Rayleigh scattering. *Measurement Science & Technology* 12 (5):R33–51. DOI: 10.1088/0957-0233/12/5/201.

21. Cheng, Z. 2001. *Visible laser Raman diagnostics application in partially premixed flames of premixed fuel versus hot products*. Nashville, Tennessee, USA: MS Thesis, Mechanical Engineering Department, Vanderbilt University.

22. Stephenson, D. A. 1974. Raman cross-sections of selected hydrocarbons and freons. *Journal of Quantitative Spectroscopy & Radiative Transfer* 14 (12):1291–301. DOI: 10.1016/0022-4073(74)90098-3.

23. Kojima, J. and Q. V. Nguyen. 2002. Laser pulse-stretching with multiple optical ring cavities. *Applied Optics* 41 (30):6360–70.

24. Osborne, R. J., P. A. Skaggs, and R. W. Pitz. 1996. Multi-camera/spectrometer design for instantaneous line Rayleigh/Raman/LIPF measurements in methane/air flames. In *34th AIAA Aerospace Sciences Meeting, Paper No. AIAA-1996-0175*. Reno, Nevada USA.

25. Hu, S. T. 2007. *Measurements and modeling of non-premixed tubular flames: structure, extinction and instability*. Nashville, Tennessee, USA: Ph.D. Thesis, Mechanical Engineering Department, Vanderbilt University.

26. Kobayashi, H. and M. Kitano. 1989. Extinction characteristics of a stretched cylindrical premixed flame. *Combustion and Flame* 76 (3–4):285–95. DOI: 10.1016/0010-2180(89)90111-9.

27. Mosbacher, D. M. 2002. *An application of non-intrusive laser-based diagnostics in tubular flames.* Nashville, Tennesssee, USA: MS Thesis, Mechanical Engineering Department, Vanderbilt University.

28. Wang, P., J. A. Wehrmeyer, and R. W. Pitz. 2006. Stretch rate of tubular premixed flames. *Combustion and Flame* 145 (1–2):401–14. DOI: 10.1016/j.combustflame.2005.09.015.

29. Kee, R. J., F. Rupley, J. Miller, M. Coltrin, J. Grcar, E. Meeks, H. Moffat, A. Lutz, G. Dixon-Lewis, M. Smooke, J. Warnatz, G. Evans, R. Larson, R. Mitchell, L. Petzold, L. Reynolds, M. Caracotsios, W. Stewart, and P. Glarborg. 1999. Oppdif: A Fortran Program for Computing Opposed-flow Diffusion Flames, *User Manual*, The CHEMKIN Collection III.

30. Dixon-Lewis, G., V. Giovangigli, R. J. Kee, J. A. Miller, B. Rogg, M. D. Smooke, G. Stahl, and J. Warnatz. 1991. Numerical modeling of the structure and properties of tubular strained laminar premixed flames. In *Progress in astronautics and aeronautics*. Washington, DC: American Institute of Aeronautics and Astronautics.

31. Mueller, M. A., T. J. Kim, R. A. Yetter, and F. L. Dryer. 1999. Flow reactor studies and kinetic modeling of the H_2/O_2 reaction. *International Journal of Chemical Kinetics* 31 (2):113–25.

32. Yetter, R. A., F. L. Dryer, and H. Rabitz. 1991. A comprehensive reaction-mechanism for carbon monoxide/hydrogen/oxygen kinetics. *Combustion Science and Technology* 79 (1–3):97–128.

33. Smith, G. P., D. M. Golden, M. Frenklach, N. W. Moriarty, B. Eiteneer, M. Goldenberg, C. T. Bowman, R. K. Hanson, S. Song, W. C. Gardiner, Jr., V. V. Lissianski, and Z. Qin. 1999. http://www.me.berkeley.edu/gri_mech/. Chicago, Illinois: Gas Research Institute.

34. Peters, N. and B. Rogg. 1993. *Reduced kinetic mechanisms for applications in combustion systems.* Chapters 1 and 5 vols, *Lecture notes in physics*. Berlin: Springer-Verlag.

35. Wang, P. Y. and R. W. Pitz. 2005. The premixed flame parameters for the stretched and curved flames. In *4th Joint Meeting of the U.S. Sections of the Combustion Institute*. Philadelphia, PA.

36. Williams, F. A. 2005. *San Diego Mechanism,* http://maeweb.ucsd.edu/~combustion/cermech/ (15 June 2005).

37. Kee, R. J., J. F. Grcar, M. D. Smooke, and J. A. Miller. 1985. *Premix: a Fortran program for modeling steady laminar one-dimensional premixed flames.* Livermore, California, USA: Sandia National Laboratories.

38. Law, C. K. 2006. *Combustion physics*. New York: Cambridge University Press.

39. Matalon, M. 2007. Intrinsic flame instabilities in premixed and nonpremixed combustion. *Annual Review of Fluid Mechanics* 39:163–91. DOI: 10.1146/annurev.fluid.38.050304.092153.

40. Ishizuka, S. and C. K. Law. 1982. An experimental study on extinction and stability of stretched premixed flames. *Proceedings of the Combustion Institute* 19:327–35.

41. Ishizuka, S. 1988. An experimental study of tubular flames in rotating and non-rotating stretched flow fields. In *Mathematical modeling in combustion science*, ed. J. D. Buckmaster and T. Takeno, 93–102. Berlin: Springer-Verlag.

42. Ishizuka, S. 1991. Determination of flammability limits using a tubular flame geometry. *Journal of Loss Prevention in the Process Industries* 4 (3):185–93. DOI: 10.1016/0950-4230(91)80035-s.

43. Ishizuka, S. 1993. Characteristics of tubular flames. *Progress in Energy and Combustion Science* 19 (3):187–226. DOI: 10.1016/0360-1285(93)90015-7.

44. Wang, Y., S. T. Hu, and R. W. Pitz. 2009. Extinction and cellular instability of premixed tubular flames. *Proceedings of the Combustion Institute* 32:1141–7. DOI: 10.1016/j.proci.2008.07.012.

45. Hall, C. A. and R. W. Pitz. 2013. A structural study of premixed hydrogen-air cellular tubular flames. *Proceedings of the Combustion Institute* 34:973–80. DOI: 10.1016/j.proci.2012.06.023.

CHAPTER 5

NON-PREMIXED TUBULAR FLAMES

Robert W. Pitz

5.1   INTRODUCTION

Both stretch rate and curvature in the presence of preferential diffusion (nonunity Lewis number) have important effects on the flame structure of non-premixed flames. The effect of stretch rate on preferential diffusion in non-premixed flames has been studied in planar opposed-jet flames[1,2] and in Tsuji burner flames.[3] Preferential diffusion can occur in the oxidizer stream as well as in the fuel stream. It is controlled by the Lewis number in each stream: Le_f and Le_o.[4] The Lewis number is defined as $Le_i = \alpha/D$ where α is the thermal diffusivity of the bulk mixture in either the fuel ($i = f$) or oxidizer ($i = o$) and D is the molecular diffusivity of the reactant in either the fuel or oxidizer. If both fuel and oxidizer Lewis numbers are unity, there is no preferential diffusion. Since air has a near-unity Lewis number, preferential diffusion depends on the Lewis number of the fuel for non-premixed combustion using air. For example, in steady stretched planar opposed-flow flame with a $Le_f < 1$ fuel stream (i.e., H_2 fuel diluted with N_2) versus an air stream ($Le_o = 1$), the flame temperature will be superadiabatic, whereas for a $Le_f > 1$ fuel (i.e., C_3H_8 diluted in N_2), the flame temperature will be subadiabatic.[5]

The effect of flame surface curvature on non-premixed flames has been studied in planar opposed-jet flames perturbed by microjets.[6–8] These studies show that for an air jet versus a fuel jet with $Le_f < 1$ (H_2 diluted in N_2) that convex flame curvature toward the fuel mixture increases the flame temperature and concave flame curvature toward the fuel mixture reduces the flame temperature. There have been many studies of the curvature of the flame tip of laminar jet diffusion flames. When the flame tip has concave curvature toward the $Le_f < 1$ fuel (such as H_2 diluted in N_2), the flame tip can extinguish and "open."[9,10] However, the effect of curvature at the flame tip is complicated by preferential diffusion along the length of the laminar jet diffusion flame[11,12] and negative stretch rate at the flame tip.[13] Also in these previous studies of non-premixed flames, the curvature and stretch rate are not uniform, making the systematic study of flame curvature effects difficult.

Recently, flame curvature has been experimentally and numerically studied in non-pre-mixed opposed-flow tubular flames where flame curvature and stretch rate are uniform.[14–16] The effect of stretch rate and curvature can be varied independently in these flames by changing the boundary conditions. In an optically accessible burner, temperature and species concentration profiles can be measured with Raman scattering and compared with detailed chemistry–molecular transport numerical simulations using a one-dimensional similarity solution that is computationally efficient. In addition, the effect of curvature on flame extinction and cellular instability can be been studied in non-premixed opposed-flow tubular flames.[17–19] These topics are discussed in detail in the next sections.

5.2 NUMERICAL STUDY OF THE NON-PREMIXED TUBULAR FLAMES

To determine the effect of curvature and stretch rate on non-premixed flames, the opposed flow tubular flame was studied numerically in the non-premixed configuration. The schematic of the opposed-flow tubular flame burner is shown in Fig. 5.1. In the non-premixed configuration, the fuel and oxidizer streams are separated and can flow from either side 1 or 2.

The relevant global parameters for the opposed-flow tubular flame have been determined.[20] Assuming the flame is thin, the two gas flows meet at the stagnation surface with a radius,

$$R_s = R_2[1 - (R_2/R_1 - R_1/R_2)/(R_2/R_1 - \sqrt{\rho_1/\rho_2}V_1/V_2)]^{0.5} \qquad (5.1)$$

where R_i, ρ_i, and V_i are the nozzle radius, density, and velocity of the ith side. On sides 1 and 2 of the stagnation surface, the stretch rates are given by,

$$k_1 = -V_2 Q^{0.5}/R_2 \qquad (5.2)$$

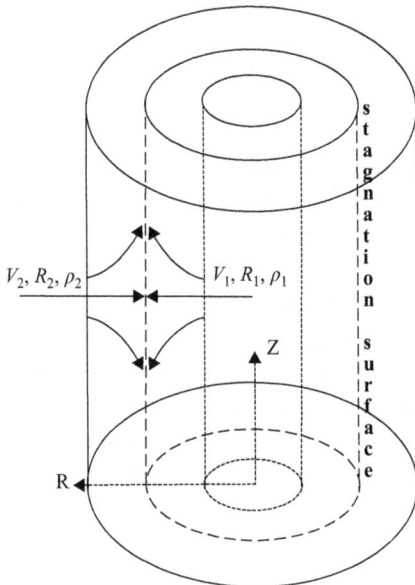

Figure 5.1. Schematic of an opposed-flow tubular burner. (*Source*: Wang, P. Y., S. T. Hu, and R. W. Pitz. 2007. Numerical investigation of the curvature effects on diffusion flames. *Proceedings of the Combustion Institute* 31:989–96. Reprinted with permission from Elsevier.)

$$k_2 = -V_2(\rho_2 Q/\rho_1)^{0.5}/R_2 \qquad (5.3)$$

where Q is determined by:

$$\sqrt{Q} = [(R_2/R_1 - \sqrt{\rho_1/\rho_2} V_1/V_2)/(R_2/R_1 - R_1/R_2)]\pi \qquad (5.4)$$

Since air is the oxidizer in this study (Le$_o$ = 1), the preferential diffusion effects are defined by the stretch rate, Lewis number, and curvature on the fuel side of the flame. Flame surface curvature convex to the fuel is defined as "positive curvature" and a flame surface curvature concave to the fuel is defined as "negative curvature." In non-premixed flames, the flame resides very near the stagnation surface and the flame curvature is given by the stagnation radius R_s (Eq. 5.1).

The effect of curvature on the non-premixed tubular flame has been analyzed numerically for fuel (20% H$_2$/80% N$_2$) flowing against air.[15] The tubular flame was predicted by a self-similarity solution using a modified Oppdif program with detailed molecular transport (including thermal diffusion) and Mueller mechanism chemical kinetics.[21] The conservation equations are given elsewhere.[20]

The predicted flame temperatures for the non-premixed tubular flame (positive and negative) curvature are shown in Fig. 5.2 as a function of stretch rate. The peak temperatures of the curved flames and the corresponding planar opposed-jet flame are compared for 20% H$_2$/80% N$_2$ versus air flames. The stretch rate for side 1 and side 2 of the planar opposed-jet flame is given by,[22]

$$k_1 = 2[|V_2|(\rho_2/\rho_1)^{0.5} + |V_1|]/L \qquad (5.5)$$

$$k_2 = 2[|V_1|(\rho_1/\rho_2)^{0.5} + |V_2|]/L \qquad (5.6)$$

where L is the separation distance between the opposed nozzles. The density and velocity at the jet exits are given by V_i and ρ_i for the ith nozzle. The stretch rate on the fuel side is used because preferential diffusion in air is negligible.

In Fig 5.2, the effect of curvature on the peak flame temperature is seen where the "positively curved" tubular flame (flame surface convex to the fuel) has the highest temperature followed by the planar and negatively curved flames. All three peak flame temperatures decrease with flame stretch due to incompleteness of chemical reaction. It is interesting that the peak temperature of the "negatively curved" flame first increases with stretch rate and then decreases; this will be explained later. The positively curved tubular flame has the highest extinction strain rate of 1590 s^{-1} followed by 1482 s^{-1} and 1192 s^{-1} for the planar and negatively curved flames, respectively. For the Le$_f$ <1 fuel (20% H$_2$ diluted in N$_2$, Le$_f$ = 0.4, Le$_o$ = 1), positive flame curvature intensifies the flame and negative flame curvature weakens the flame.

To clarify the effect of curvature, predictions were also carried out for infinite reaction rate chemistry and are shown in Fig. 5.3. The peak temperature of the planar opposed jet flame (Le$_f$ = 0.4, Le$_o$ = 1) is independent of stretch rate but its peak flame temperature (1771 K) greatly exceeds the adiabatic equilibrium (Le$_f$ = Le$_o$ = 1) temperature (1368 K); this result is consistent with earlier analytical studies under the infinite reaction rate assumption.[4,5] At low stretch rate ($\sim$30 s^{-1}), the peak temperature of the positively curved flame is approximately 2000 K (600 K above the adiabatic equilibrium temperature) and the peak temperature of the

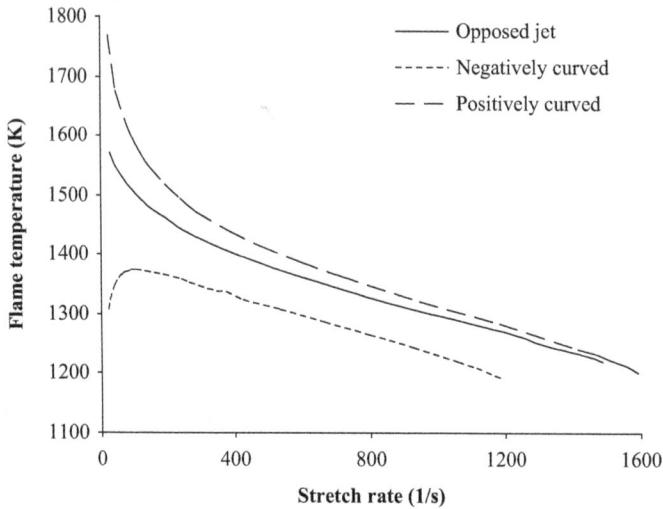

Figure 5.2. Predicted peak flame temperatures for opposed tubular flames with a constant flame radius (R_s = 5 mm) and planar opposed-jet flames with finite-rate chemistry (20% H_2/80% N_2 vs air, Le_f = 0.4, Le_o = 1). The adiabatic equilibrium flame temperature at stoichiometric ($Le_o = Le_f$ = 1) is 1368 K. The extinction occurs at the highest stretch rate shown. (*Source*: Wang, P. Y., S. T. Hu, and R. W. Pitz. 2007. Numerical investigation of the curvature effects on diffusion flames. *Proceedings of the Combustion Institute* 31:989–96. Reprinted with permission from Elsevier.)

negatively curved flame is approximately 1500 K. At low stretch rates, the flame curvature has a very large effect on the flame temperature.

With an increasing stretch rate, the peak temperature of the curved flames approaches the temperature of the stretched planar flame. As the stretch rate increases the flame gets thinner and thinner. The flame thickness of a stretched non-premixed flame decreases as $\delta_f \sim k^{-1/2}$.[1,2] The effect of curvature on the flame temperature depends on the ratio of the flame thickness to the flame radius (δ_f/R) and this parameter decreases with increasing stretch rate.[15] At a very high stretch rate and infinitely fast chemistry, the flame sheet is infinitely thin and flame curvature has little effect on the flame temperature.

Now, the increasing and decreasing trends of the peak temperature of the "negatively curved" flame seen in Fig. 5.2 can be explained. The peak temperature first increases with stretch rate as the weakening effects of the negative curvature are diminished by a thinner reaction zone. Then the flame temperature decreases with stretch rate due to the incompleteness of reaction and eventually the flame extinguishes.

Including finite-rate chemistry effects again, the effect of flame curvature on peak temperature can be seen in Fig. 5.4 for a constant stretch rate of k = 200 s^{-1}. For a constant stretch rate, the flame thickness is approximately constant ($\delta_f \sim k^{-1/2}$). As the flame radius, R $\cong$ R_s, is increased, the effect of curvature on the flame temperature diminishes. This indicates that the effect of curvature on non-premixed flames depends on the ratio of the flame thickness to the flame radius (δ_f/R).

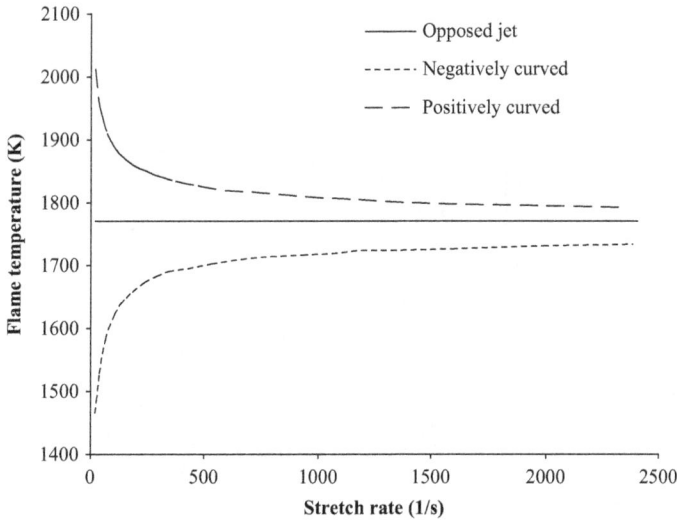

Figure 5.3. Under the assumption of infinitely fast chemistry, predicted peak flame temperature variation with stretch rate for opposed tubular flames with constant flame radius (R_s = 5 mm) and planar opposed-jet flames (20% H_2/80% N_2 vs air; Le_f = 0.4, Le_o = 1). The adiabatic equilibrium flame temperature at stoichiometric (Le_o = Le_f = 1) is 1368 K. (*Source*: Wang, P. Y., S. T. Hu, and R. W. Pitz. 2007. Numerical investigation of the curvature effects on diffusion flames. *Proceedings of the Combustion Institute* 31:989–96. Reprinted with permission from Elsevier.)

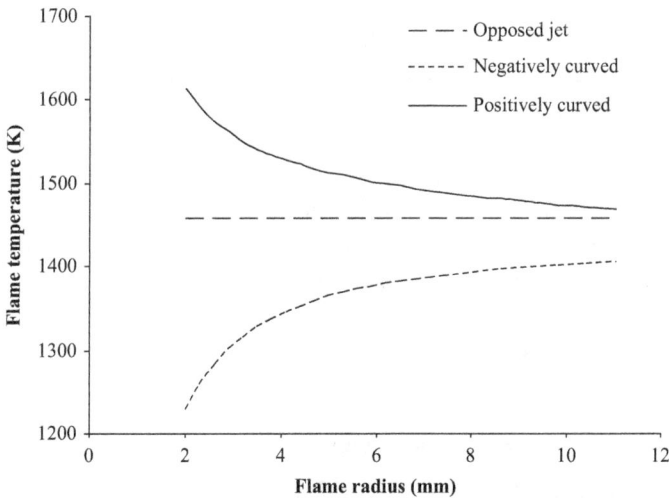

Figure 5.4. For constant stretch rate (200 s^{-1}), predicted peak flame temperature variation with flame radius for opposed tubular flames and planar opposed-jet flames with finite-rate chemistry (20% H_2/80% N_2 vs air, Le_f = 0.4, Le_o = 1). The adiabatic equilibrium flame temperature at stoichiometric (Le_o = Le_f = 1) is 1368 K. (*Source*: Wang, P. Y., S. T. Hu, and R. W. Pitz. 2007. Numerical investigation of the curvature effects on diffusion flames. *Proceedings of the Combustion Institute* 31:989–96. Reprinted with permission from Elsevier.)

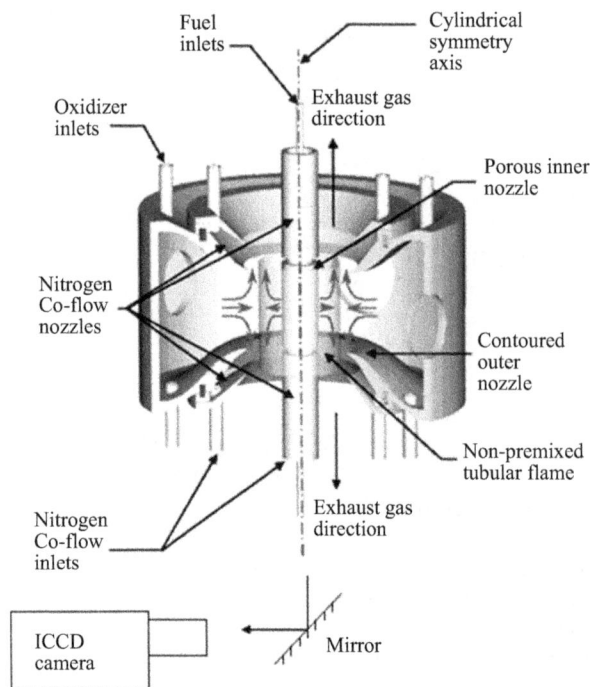

Figure 5.5. Schematic of the non-premixed opposed-flow tubular burner consisting of an outer air nozzle (20 mm height, 15 mm radius) and an inner porous fuel nozzle (20 mm height, 3.2 mm radius). Additional co-flows of nitrogen gas shield the flame from disturbances. (*Source*: Hu, S. T. et al. 2007. Experimental and numerical investigation of non-premixed tubular flames. *Proceedings of the Combustion Institute* 31:1093–9. Reprinted with permission from Elsevier.)

5.3 NON-PREMIXED OPPOSED-FLOW TUBULAR BURNER

The non-premixed opposed-flow tubular flames are studied experimentally in the burner sketched in Fig. 5.5. This is the same tubular burner described in Chap. 4, where it has an additional central fuel nozzle. The exits of the air and fuel nozzles are 20 mm in height. Air flows radially inward from the exit of the outside annular nozzle (R_2 = 15 mm) and fuel flows radially outward from a porous fuel nozzle (R_1 = 3.2 mm). Both the outside and inside nozzles have annular co-flows of nitrogen to cool the burner and isolate tubular flame from laboratory disturbances. The inner porous nozzle is made of a sintered metal (Mott Corporation, 5 μm pore size). A picture of the opposed-flow tubular burner with its central fuel nozzle is shown in Fig. 4.12(a). A complete description of the outer nozzle of the tubular burner and its optical access is given in Chap. 4 (Section 4.3).

5.4 RAMAN SCATTERING MEASUREMENTS IN NON-PREMIXED TUBULAR FLAMES

5.4.1 HYDROGEN / AIR NON-PREMIXED TUBULAR FLAMES

The Raman scattering system described in Chap. 4 (Section 4.2) was applied to measure temperature and major species concentrations in 15% H_2/85% N_2 versus air flames in the non-premixed opposed-flow tubular burner. The Raman system schematic is shown in Fig. 4.4. The burner is translated a number of times to cover the non-premixed opposed tubular flame

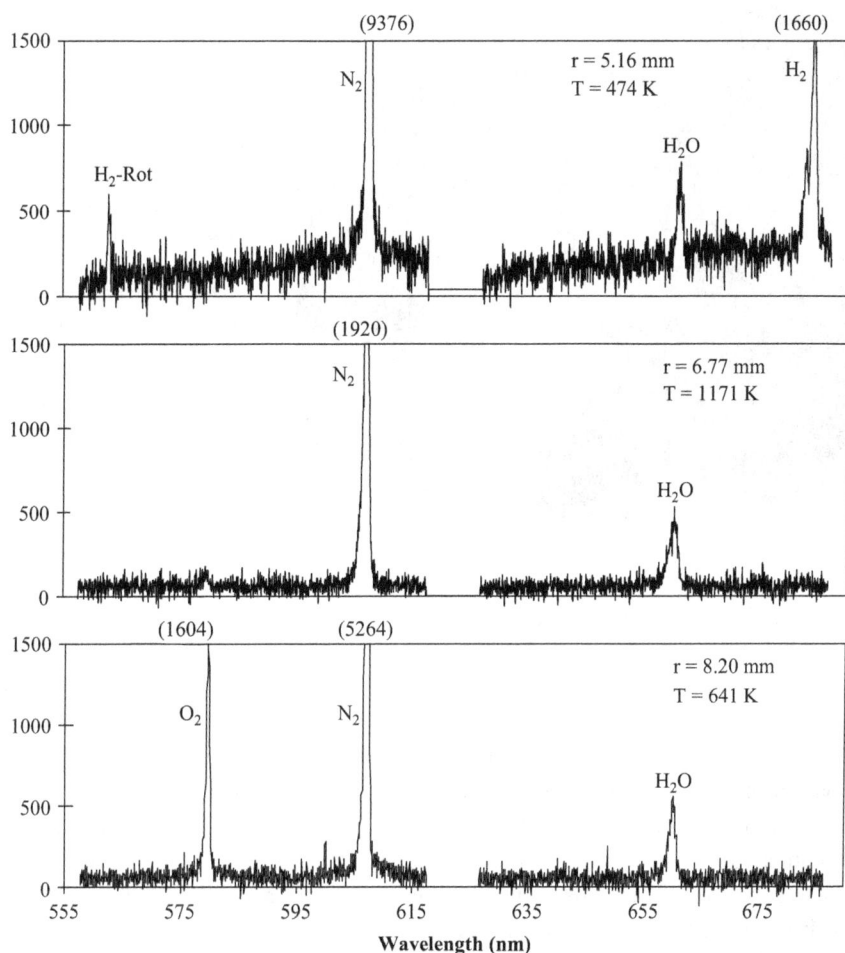

Figure 5.6. Raman spectra in a 15% H_2/85% N_2 vs air non-premixed tubular at three radial locations for $k = 91$ s^{-1}. Some peaks are truncated for clarity and their peak values are given in parenthesis. (*Source*: Hu, S. T. et al. 2007. Experimental and numerical investigation of non-premixed tubular flames. *Proceedings of the Combustion Institute* 31:1093–9. Reprinted with permission from Elsevier.)

measurement and 900 laser shots are integrated on the back-illuminated, liquid-nitrogen cooled CCD camera for each laser position. The addition of the central fuel nozzle made the Raman scattering measurements more difficult because of laser scattering from the fuel nozzle. Raman measurements were generally made in the shadow region behind the nozzle where there was much less laser scattering. Sample Raman scattering spectra from the 15% H_2/85% N_2 versus air tubular flame are shown in Fig. 5.6 for three radial locations. In general, the background noise in the Raman scattering here is much higher than in the premixed H_2–air tubular flames (see Fig. 4.2). The fuel nozzle surface is located at r = 3.2 mm. The noise background in the Raman spectra was highest closest to the fuel nozzle (r = 5.16 mm) and diminished at locations farther from the fuel nozzle (r = 8.2 mm).

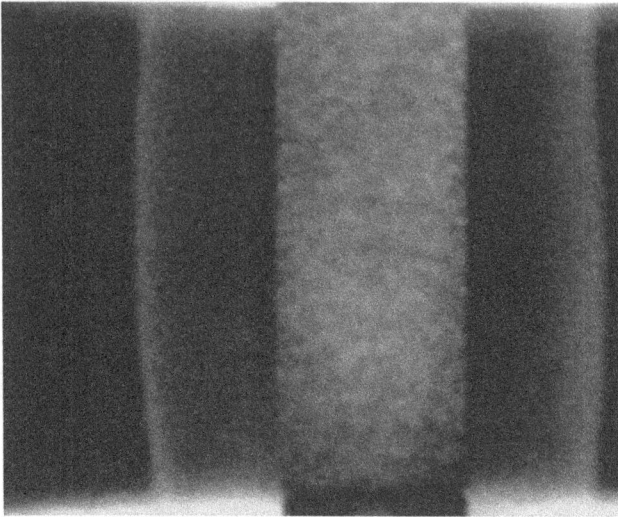

Figure 5.7. Infrared sensitive camera image of a non-premixed tubular flame showing the central porous fuel nozzle and the flame surface (15% H_2/85% N_2 vs. air; $k = 105$ s^{-1}). (*Source*: Hu, S. T. et al. 2007. Experimental and numerical investigation of non-premixed tubular flames. *Proceedings of the Combustion Institute* 31:1093–9. Reprinted with permission from Elsevier.)

The velocities of the inner and outer nozzles were matched to give the radius of the stagnation surface as $R_s = 6.5$ mm from Eq. 5.1. An image of the non-premixed 15% H_2/85% N_2 versus air tubular flame at $k = 105$ s^{-1} is shown in Fig. 5.7. The image is taken with an infrared-sensitive camera that shows the flame and the inner burner nozzle. The flame surface is symmetric with a slight amount of curvature in the vertical direction. Measurements are made in the central part of the tubular flame where the flame surface is vertical.

Raman scattering measurements of temperature are shown in Fig. 5.8 and compared with detailed numerical simulations using the modified Oppdif code.[14] The simulations are shown using chemical kinetic mechanisms by Mueller,[21] Peters[23] and GRI-Mech3.0.[24] Molecular transport is calculated using either a mixture-averaged or a multicomponent formula. Thermal diffusion of light species is included in all simulations. All of the simulations generally agree with the temperature measurements. The GRI-Mech mechanism with mixture-averaged properties predicts the narrowest reaction width and the Peters mechanism with mixture-averaged properties predicts the highest temperature. The subsequent simulations of the 15% H_2/85% N_2 versus air flames used the Mueller chemical kinetic mechanism and multicomponent molecular transport formula with thermal diffusion of light species.

The measured major species mole fractions are compared with the detailed simulations in Fig. 5.9. The measurements generally agree with the simulations. The peak temperature and H_2O mole fraction are located at r = 6.8 mm which is close to the calculated stagnation surface radius ($R_s = 6.5$ mm). At small radii, strong laser scattering from the fuel nozzle (see Fig. 5.6) leads to more uncertainty in the Raman temperature and species composition measurements.

The peak measured flame temperatures are compared with predictions in tubular flames and planar opposed jet flames in Fig. 5.10. The adiabatic equilibrium flame temperature at stoichiometric ($T_{ad} = 1180$ K) is shown by the horizontal dotted line. Preferential diffusion in the $Le_f < 1$ fuel mixture (15% H_2 diluted in N_2, $Le_f = 0.4$, $Le_o = 1$) causes the flame temperatures of the opposed-jet and tubular flames to be above the adiabatic equilibrium flame temperature. The tubular flame is negatively curved (concave toward the fuel) which weakens the flame.[15] Thus the measured and predicted temperatures of the tubular flame are lower than

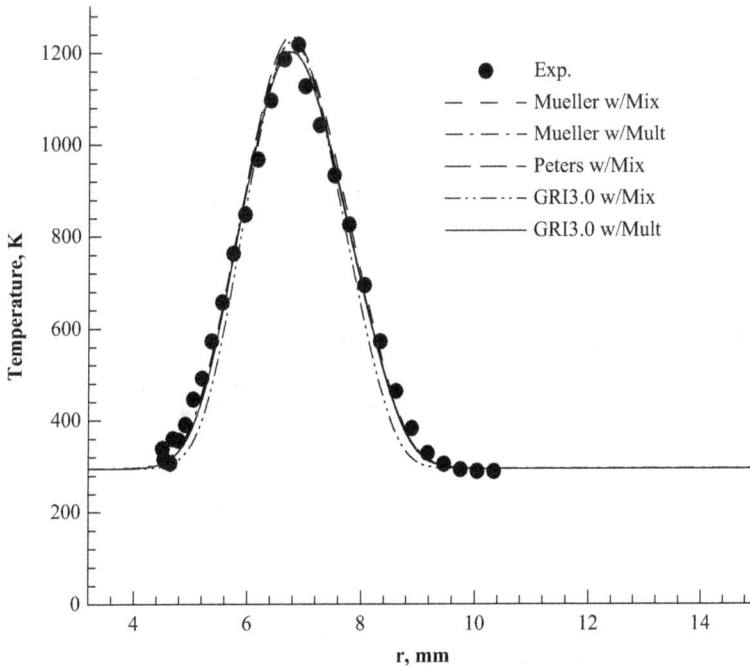

Figure 5.8. Comparison of Raman temperature measurements to detailed numerical simulations using different chemical kinetic mechanisms and either mixture-averaged or multicomponent molecular transport formula for a 15% H_2/85% N_2 vs air non-premixed tubular flame ($k = 75$ s^{-1}, $R_s = 6.5$ mm, $R_1 = 3.2$ mm. $R_2 = 15$ mm). (*Source*: Hu, S. T. et al. 2007. Experimental and numerical investigation of non-premixed tubular flames. *Proceedings of the Combustion Institute* 31:1093–9. Reprinted with permission from Elsevier.)

the planar opposed-jet flame. Since the negative curvature weakens the flame, the tubular flame extinguishes at a lower stretch rate ($k = 200$ s^{-1}) than the opposed-jet flame ($k = 420$ s^{-1}). The predicted flame thicknesses for the tubular and opposed jet flames are also shown in Fig. 5.10. Since curvature effects on tubular flames are related to the ratio of flame thickness to flame radius (δ_f/R), the decrease in the flame thickness of the tubular flame from $k = 40$ s^{-1} to 100 s^{-1} leads to a diminished effect of curvature at a high stretch rate. Above $k = 60$ s^{-1}, both the planar and tubular flame temperatures decrease with stretch rate at a similar rate due to incompleteness of combustion; this is similar to the predictions shown in Fig. 5.2.

5.4.2 HYDROCARBON–AIR NON-PREMIXED TUBULAR FLAMES

The Raman system was applied to make temperature and species concentration measurements in nitrogen-diluted hydrocarbon fuel versus air opposed tubular flames. The curvature of the tubular flames was constant at $R_s = 6.5$ mm. Since the Lewis number of air is near unity ($Le_o = 1$), the preferential diffusion effects will depend on the Lewis number of the fuel mixture. Hydrocarbon fuel mixtures were chosen with fuel (30% CH_4/70% N_2 fuel, $Le_f = 1.00$)

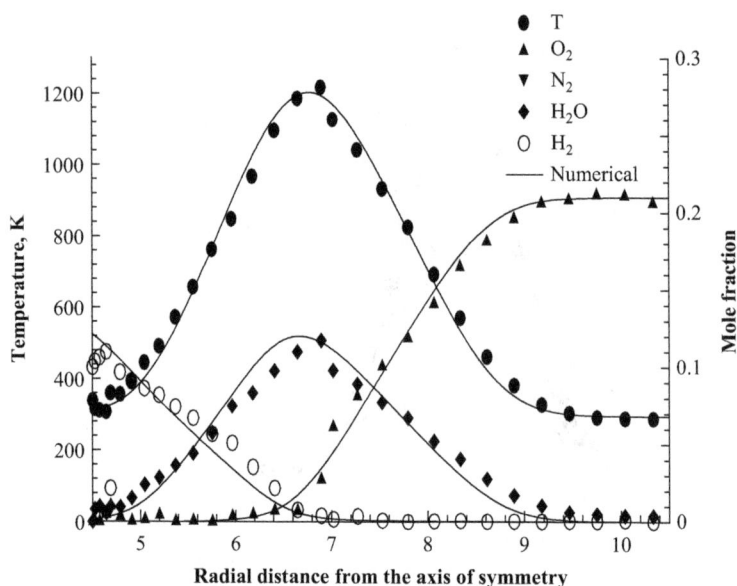

Figure 5.9. Measured and calculated temperature and major species mole fraction profiles for a 15% H_2/85% N_2 vs air nonpremixed tubular flame ($k = 75$ s^{-1}, $R_s = 6.5$ mm, $R_I = 3.2$ mm. $R_2 = 15$ mm). The radial distance is in mm. (*Source*: Hu, S. T. 2007. *Measurements and modeling of non-premixed tubular flames: Structure, extinction and instability*. Nashville, Tennessee, USA: Ph.D. Thesis, Mechanical Engineering Department, Vanderbilt University.)

that should have minimal preferential diffusion effects and with fuel (15% C_3H_8/85% N_2, Le$_f$ = 1.51) that should have significant preferential diffusion effects. All of the flames appeared blue with no visible yellow color that indicates the presence of soot.

The Raman spectra in the nitrogen-diluted methane and propane versus air flames are shown in Fig. 5.11. When compared with the hydrogen flames, additional Raman lines appear from the carbon-bearing species: CO, CO_2, CH_4, and C_3H_8. The addition of all the species increases the cross talk between Raman lines that must be taken into account (Chap. 4, Section 4.2).

A larger problem for Raman measurement is the presence of laser-induced interference in the rich flame zones. Laser-induced fluorescence (LIF) from C_2 is apparent in the 30% CH_4/70% N_2 versus air flame (Fig. 5.11b) and is very strong in the 15% C_3H_8/85% N_2 versus air flame (Fig. 5.11(f)–(g)). As will be seen later, laser-induced interferences result in an absence of interpretable temperature and species concentration data in the rich flame zones, particularly for the propane flame. The high broadband background shown in Fig. 5.11(b) was due to laser light scattering from the burner exit window that occurred at some locations.

Raman measurements of temperature and species mole fractions for the 30% CH_4/70% N_2 versus air non-premixed flame are shown in Fig. 5.12 at $k = 102$ s^{-1}. The shaded region is where laser-induced fluorescence from C_2 was so strong that the Raman spectra could not be interpreted to yield quantitative scalar data. The measurements are compared with detailed simulations using four different chemical kinetic mechanisms: C1 and C2 mechanisms,[23] GRI-Mech 3.0,[24] and Kee mechanism.[26] The simulations used multicomponent transport equations

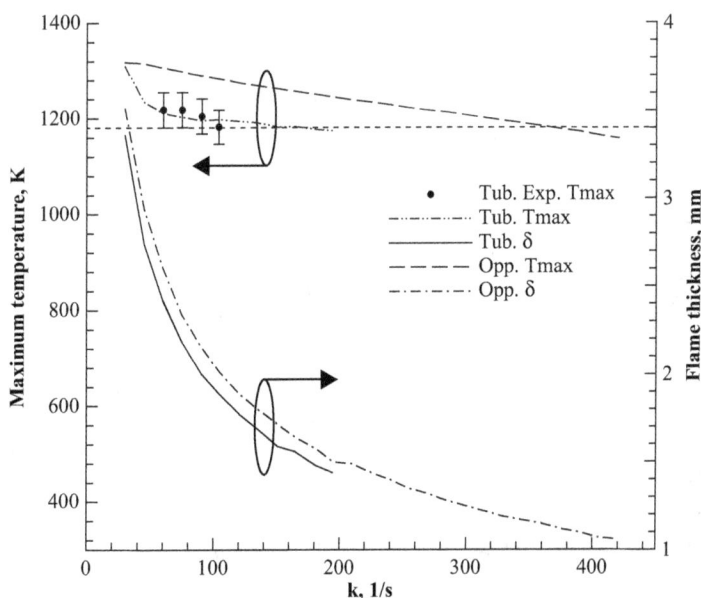

Figure 5.10. Comparison of measured peak temperatures in a 15% H_2/85% N_2 vs air non-premixed tubular flame (R_s = 6.5 mm, R_1 = 3.2 mm, R_2 = 15 mm) to predictions of non-premixed tubular and planar opposed-jet flames. The adiabatic equilibrium flame temperature (T_{ad} = 1180 K) is shown by the horizontal dotted line. The extinction occurs at the highest stretch rate shown. (*Source*: Hu, S. T. et al. 2007. Experimental and numerical investigation of non-premixed tubular flames. *Proceedings of the Combustion Institute* 31:1093–9. Reprinted with permission from Elsevier.)

and included radiation loss from CO, CO_2, and H_2O in the optically thin limit. All the simulations give nearly the same results indicating that the structure of the non-premixed tubular flame is insensitive to the particular chemical kinetic mechanism used in the simulation. This is expected because in non-premixed flames, the burning rate is controlled primarily by the rate of diffusion of the fuel and air into the flame zone. For non-premixed methane and propane flames, the experimental data are shifted 1.6 mm in the radial direction toward smaller values to better compare with the simulations.[16] This shift was thought to be caused by slight differences in the boundary conditions of the experiment and simulations.

As seen in Fig. 5.12, the temperature, N_2 mole fraction and H_2O mole fraction experimental data compare well with the simulations for the 30% CH_4/70% N_2 versus air non-premixed flame. Model–data discrepancies between the CO_2 and O_2 near the flame zone are due to the difficulty in interpreting the overlap of the CO_2 and O_2 Raman lines (e.g., Fig. 5.11(c)). Non-premixed 30% CH_4/70% N_2 versus air flame profiles were measured at stretch rates of k = 41, 61, 81, 102, and 122 s^{-1}. For $k \leq 81$ s^{-1}, the interferences were so strong that only data on the lean side of the flame could be plotted; this data is shown elsewhere.[16]

Raman measurements of temperature and species mole fractions for the 15% C_3H_8/85% N_2 versus air non-premixed flame were made for a wide range of stretch rates from k = 33 to

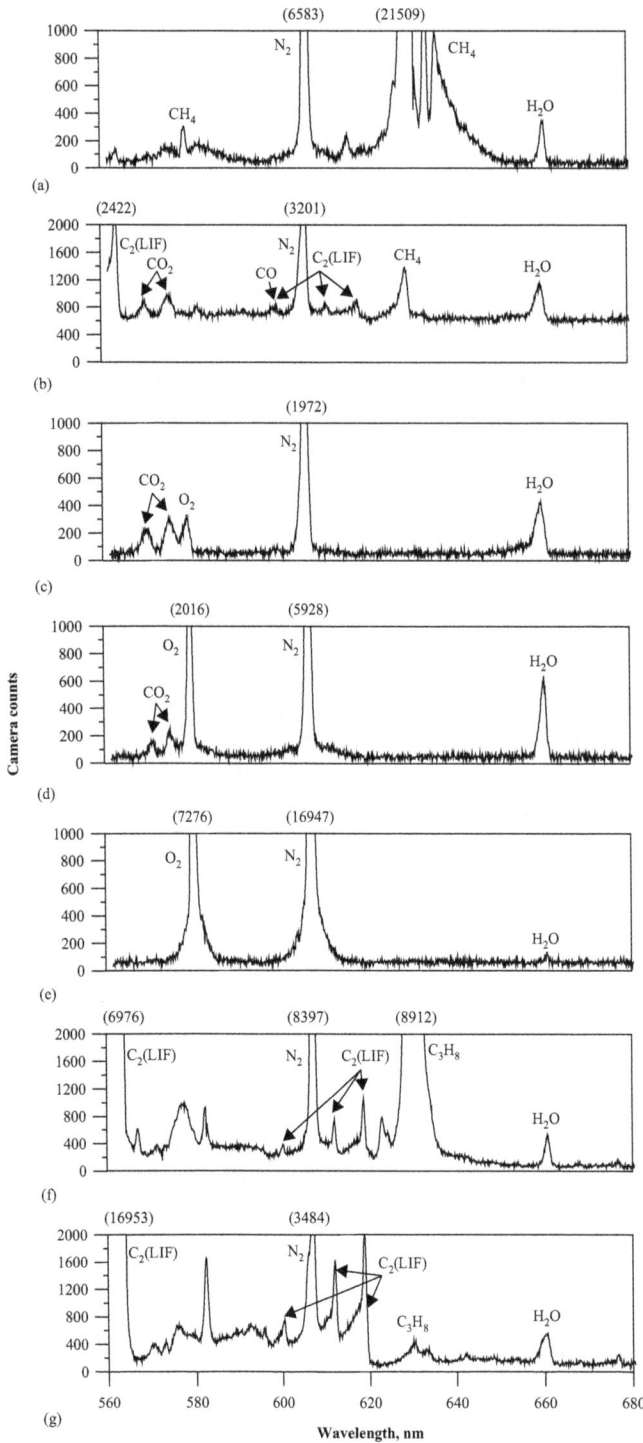

Figure 5.11. Raman spectra of the opposed tubular flame at different radial locations. (a)–(e) $k = 122$ s^{-1}, $R_s = 6.5$ mm, 30% CH_4/70% N_2 vs air flame; (f) and (g) $k = 100$ s^{-1}, $R_s = 6.5$ mm, 15% C_3H_8/85% N_2 vs air flame. (a) r = 5.2 mm, T = 563 K; (b) r = 7.0 mm, T = 1717 K; (c) r = 7.6 mm, T = 1731 K; (d) r = 8.6 mm, T = 844 K; (e) r = 10.5 mm, T = 300 K; (f) r = 5.2 mm; (g) r = 6.7 mm. (Note the different scale on y-axis in (b), (f), and (g)). (*Source*: Hu, S. T. and R. W. Pitz. 2009. Structural study of non-premixed tubular hydrocarbon flames. *Combustion and Flame* 156 (1):51–61. Reprinted with permission from Elsevier.)

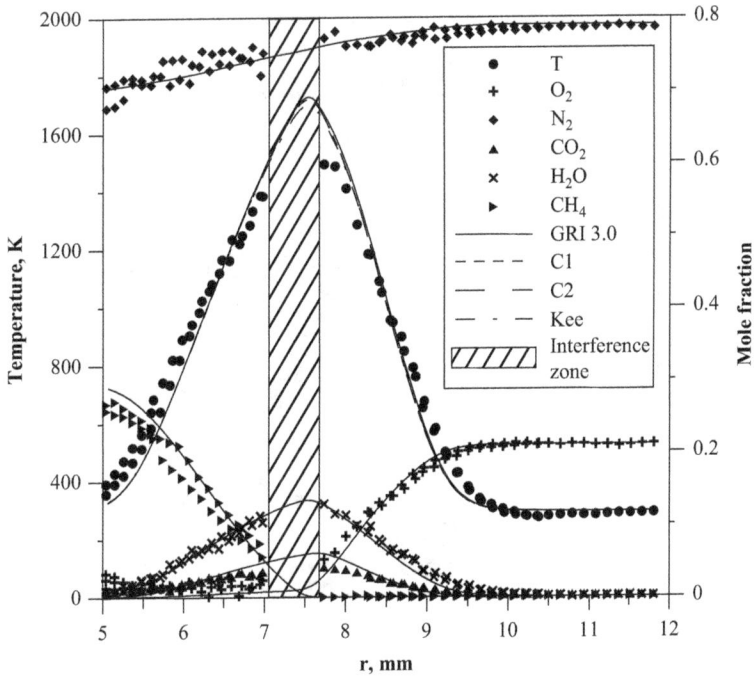

Figure 5.12. Measured and calculated temperature and major species mole fraction profiles as functions of radius for a 30% CH_4/70% N_2 vs air non-premixed tubular flame at $k = 102$ s^{-1}, $R_s = 6.5$ mm. The shaded area has strong laser-induced interference and the Raman spectra cannot be reduced to give reliable temperature and mole fraction values. (*Source*: Hu, S. T. and R. W. Pitz. 2009. Structural study of non-premixed tubular hydrocarbon flames. *Combustion and Flame* 156 (1):51–61. Reprinted with permission from Elsevier.)

100 s^{-1}. Example results for $k = 33$ s^{-1} are shown in Fig. 5.13. In all of the propane flames, the C_2 laser-induced fluorescence (LIF) and other broadband interferences were so strong in the rich region that the Raman spectra could only be interpreted to give temperature and mole fraction on the lean side of the flame. Examples of Raman spectra on the rich side of the 15% C_3H_8/85% N_2 versus air flame are shown in Fig. 5.11(f) and 5.11(g) where strong C_2 LIF interference and broadband background are evident. The non-premixed propane–air flames were simulated with the modified Oppdif code using the San Diego chemical kinetic mechanism.[27] Mixture-averaged molecular transport properties as well as adiabatic conditions (i.e., no radiation loss) were used to achieve convergence with reduced computational cost. The Raman data are compared to the simulation in Fig. 5.13 with good results, particularly for temperature, N_2 mole fraction and H_2O mole fraction. The CO_2 and O_2 mole fraction profiles departed from the simulations near the flame zone due to the difficulty in interpreting the overlap of the CO_2 and O_2 Raman lines.

To illustrate the effect of curvature on flame temperature and extinction stretch rate, simulations of convex and concave curved non-premixed tubular flames are compared with planar opposed-jet flames versus stretch rate in Fig. 5.14. All the peak flame temperatures decrease with stretch rate due to incompleteness of reaction. Since air is the oxidizer in all cases, only

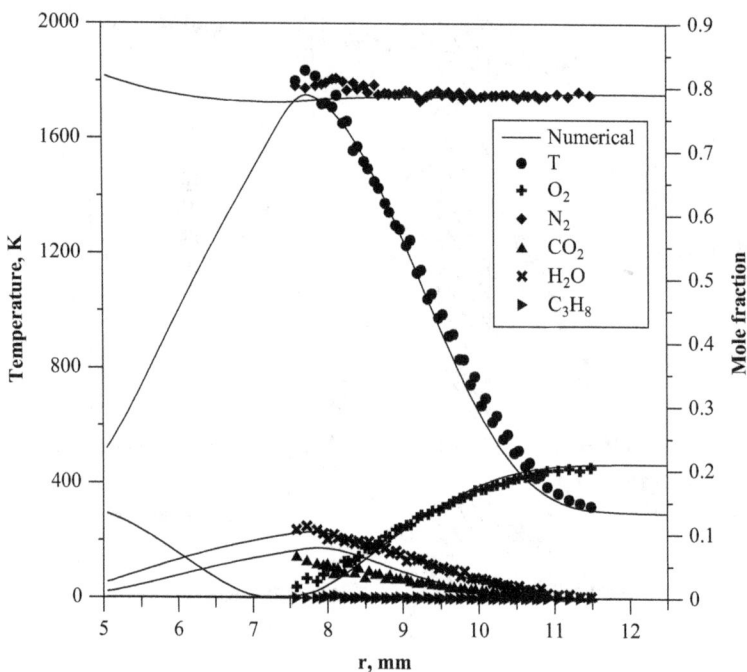

Figure 5.13. Measured and calculated temperature and major species mole fraction profiles as functions of radius for a 15% C_3H_8/85% N_2 vs air non-premixed tubular flame at $k = 33$ s^{-1}, $R_s = 6.5$ mm. The numerical simulation uses the San Diego mechanism and mixture-averaged transport properties. (*Source*: Hu, S. T. and R. W. Pitz. 2009. Structural study of non-premixed tubular hydrocarbon flames. *Combustion and Flame* 156 (1):51–61. Reprinted with permission from Elsevier.)

the fuel Lewis number affects the preferential diffusion and the flame curvature is defined with respect to the fuel side. For the 30% CH_4/70% N_2 versus air non-premixed tubular flame, the fuel Lewis number is near unity (Le$_f$ = 1) and preferential diffusion effects are small for both fuel and air. Comparing the peak temperature of the planar flame with that of the curved tubular flame, the temperature of the convex curved tubular flame is nearly the same and that of the concave curved tubular flame is slightly lower.

For the 15% C_3H_8/85% N_2 versus air tubular flame, the fuel Lewis number is greater than 1 (Le$_f$ = 1.51). In this case, convex curvature of the flame zone reduces the peak temperature and extinction stretch rate below the planar opposed-jet flame values. Conversely, concave curvature of the propane-diluted flame slightly increases the peak temperature and the extinction stretch rate above the planar opposed-jet values.

The strongest effect of curvature is seen in the 15% H_2/85% N_2 versus air tubular flames where the fuel Lewis number is below unity (Le$_f$ = 0.41). With convex curvature toward the fuel, the flame is intensified with increased flame temperature and extinction stretch rate. The opposite is true with concave curvature toward the fuel. Curvature affects the peak temperature by as much as ± 80 K. The simulations are consistent with those shown in Fig. 5.2 for more highly curved 20% H_2/80% N_2 versus air tubular flames ($R_s = 5$ mm).

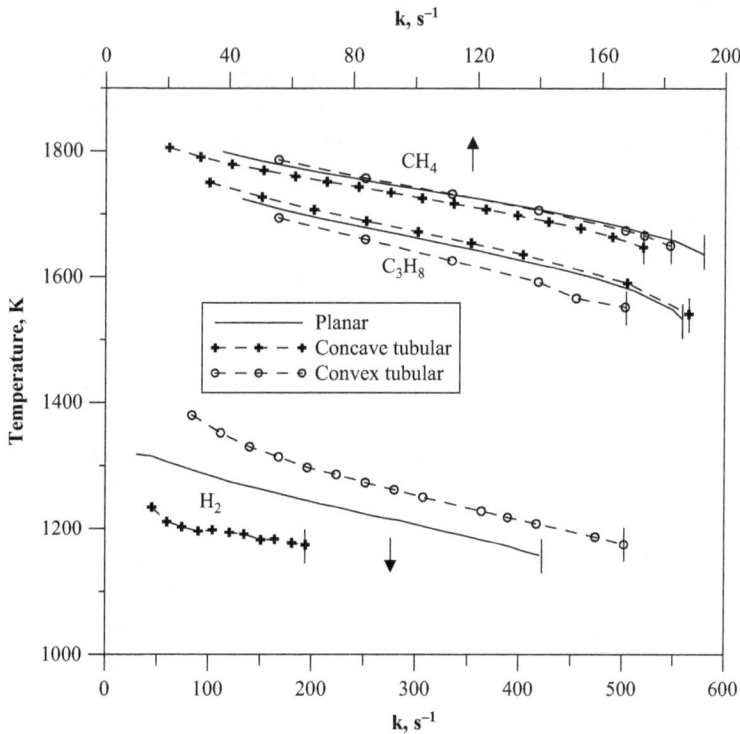

Figure 5.14. Calculated maximum flame temperature as functions of stretch rate of non-premixed opposed-jet planar and tubular flames using air versus 15% H_2/85% N_2 ($Le_f = 0.41$), 30% CH_4/70% N_2 ($Le_f = 1.00$) and 15% C_3H_8/85% N_2 ($Le_f = 1.51$) showing the different effects of curvature (Note: H_2 uses the lower x-axis and CH_4/C_3H_8 uses the upper x-axis). The curvature is fixed for the tubular flames at $R_s = 6.5$ mm. (*Source*: Hu, S. T. and R. W. Pitz. 2009. Structural study of non-premixed tubular hydrocarbon flames. *Combustion and Flame* 156 (1):51–61. Reprinted with permission from Elsevier.)

5.5 CELLULAR INSTABILITIES IN NON-PREMIXED TUBULAR FLAMES

Cellular instabilities in non-premixed tubular flames occur near extinction at low Damköhler number[a] whereas cellular instabilities in premixed tubular flames occur far from extinction (see Chap. 4, Section 5). Stability analysis of non-premixed flames finds that stable flame cells are formed at low Damköhler number and low Lewis number ($Le_f < 1$ and $Le_o \leq 1$); in addition, the region of cellularity is expanded for lean values of mixture strength ($\phi < 1$).[b, 28–30] Hu et al. first observed cellular instabilities in non-premixed tubular flames (15% H_2/85% N_2 vs air flame, $Le_f = 0.4$, $Le_o = 1$).[14]

---

[a] The Damköhler number is the ratio of the flow time to the chemical time.

[b] The initial mixture strength, ϕ, is defined as the ratio of fuel mole fraction in the fuel stream to oxidizer mole fraction in the oxidizer stream normalized by the stoichiometric fuel-to-oxidizer molar ratio.

5.5.1 CELLULAR INSTABILITIES IN DIFFUSION FLAMES

The first observation of flame cells in diffusion flames was reported by Garside and Jackson in the lifted region of a laminar jet flame of 19% H_2/81% CO_2 burning in air.[31] Dongworth and Melvin observed cells in the lifted region of a two-dimensional Wolfhard–Parker burner diffusion flame of 13% H_2/87% N_2 burning in air and attributed the cell formation to partial premixing at the flame base.[32] Ishizuka and Tsuji observed striped cells near extinction in a two-dimensional counterflow diffusion flame formed in the forward stagnation region of a porous fuel cylinder mounted in an air stream.[3] Cells appeared as stripes for H_2 fuel diluted in N_2 or Ar but not in He; they attributed the cell formation to preferential diffusion of H_2. Based on a study of two-dimensional non-premixed flames in a Wolfhard–Parker burner and analysis of previous experiments, Chen et al. concluded that cells form in non-premixed flames near extinction when the Lewis number of the "more completely consumed reactant" (either in the fuel or oxidizer stream) is less than ~0.8.[33]

In the first theoretical study of cell formation in diffusion flames, Kim et al. analyzed a one-dimensional convective diffusion flame with one-step chemistry and found striped cell formation near extinction for Lewis number less than unity; they derived an expression for the cell length that was proportional to the thermal diffusivity and showed good comparison to previously measured cell lengths.[34] Analyzing a one-dimensional diffusion flame with one-step chemistry, Cheatham and Matalon theoretically predicted cell formation for low Lewis numbers (Le_f<1 and Le_o <1) and found that the region of instability expanded for lean values of the mixture strength (ϕ<1).[28] Lo Jacono et al. experimentally confirmed the importance of initial mixture strength ϕ in their study of cell formation in an axisymmetric jet of H_2 diluted in CO_2 surrounded by a co-flow jet of O_2.[35] Cells formed at the jet flame base near extinction when Le_f<1 and $Le_o \cong 1$ and the region of cell formation was more expansive for lean values of the mixture strength (ϕ<1).

5.5.2 CELLULAR FORMATION AND EXTINCTION IN NON-PREMIXED TUBULAR FLAMES

The first observations of cellular states in non-premixed opposed tubular flames were made by Hu and are shown in Fig. 5.15.[14] For a 15% H_2/85% N_2 versus air flame (Le_f= 0.4, Le_o = 1.0) that is near extinction, the flame cells appear as stripes when viewed from the side (Fig. 5.15(a)) and short segments arranged in a circle when viewed from the top or bottom of the burner (Fig. 5.15(b)). These are flat ribbons and do not conform to the tubular surface that is concave toward the fuel. As seen in Fig. 5.14 for a 15% H_2/85% N_2 versus air flame, a stretched flame surface that is either flat or convex to the fuel has a higher temperature than a concave flame. Thus to survive near extinction, the cells have a flat flame surface. Also, adjacent to the extinction zones, the edges of the cells have a flame surface that is convex toward the fuel.

There are actually four cells in Fig. 5.15(b) but one is blocked by the inner fuel nozzle. Parts of Fig. 5.15(d) and (f) are blocked by the fuel nozzle as well. The cells are imaged from the bottom of the burner using a mirror with a slot cutout for the porous fuel nozzle as sketched in Fig. 5.5. The mirror slot and porous central fuel cylinder obscure part of the bottom view as shown in Fig. 5.16 for a non-cellular tubular flame.[18]

For a 26% CH_4/74% N_2 versus air flame (Le_f= Le_o = 1), the flame surface is a cylindrical shape and no cells are formed for a flame near extinction (Fig. 5.15(c) and (d)). Flame curvature

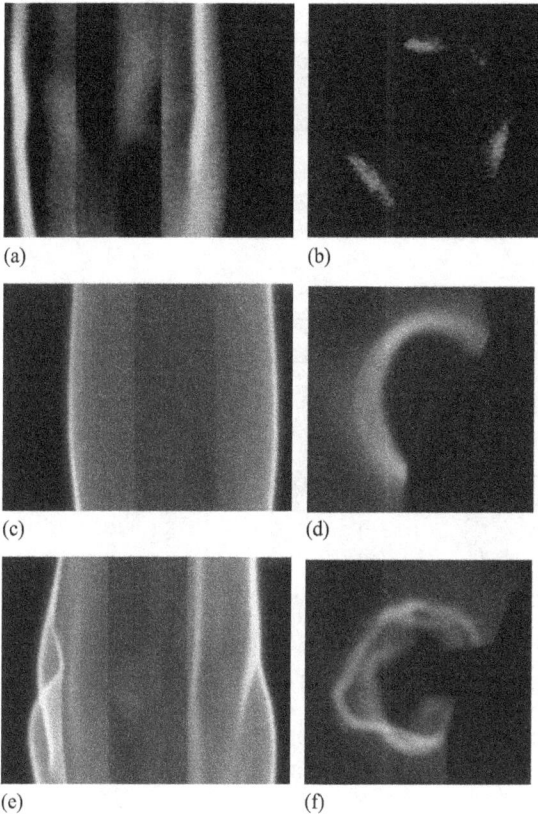

Figure 5.15. Images of non-premixed tubular flames (a) side view (b) top view of 15% H_2/85% N_2 vs air flame at $k = 210$ s^{-1}; (c) side view (d) top view of 26% CH_4/74% N_2 vs air flame at $k = 166$ s^{-1}; (e) side view and (f) top view of 15% C_3H_8/85% N_2 vs air flame at $k = 161$ s^{-1}. Top view images are partially blocked by the inner nozzle. Four cells are observed in (b) but one cell is blocked by the nozzle. (*Source*: Hu, S. T. et al. 2007. Experimental and numerical investigation of non-premixed tubular flames. *Proceedings of the Combustion Institute* 31:1093–9. Reprinted with permission from Elsevier.)

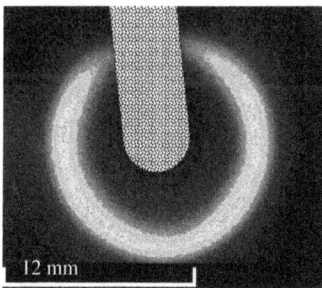

Figure 5.16. Non-cellular tubular flame image obscured by the inner fuel nozzle for 22% H_2 diluted in CO_2 vs air opposed tubular flame. (*Source*: Shopoff, S. W., P. Wang, and R. W. Pitz. 2011. The effect of stretch on cellular formation in non-premixed opposed-flow tubular flames. *Combustion and Flame* 158 (5):876–84. Reprinted with permission from Elsevier.)

has little effect on the flame temperature (Fig. 5.14) and no cellular behavior is predicted for stability theory when both Lewis numbers are near unity.[30] For the 15% C_3H_8/85% N_2 versus air flame, the fuel Lewis is greater than one ($Le_f = 1.51$) and the flame is enhanced by concave curvature toward the fuel (Fig. 5.14). Regions of concave curvature have increased luminosity in Fig. 5.15(e) and (f). However, no distinct regions of extinction and isolated cells are found.

Cellular regions and extinction were systematically studied in non-premixed tubular flames burning H_2, CH_4, and C_3H_8 with both concave and convex curvatures.[17] Images of flame cell formation in the opposed tubular flame are shown in Fig. 5.17 for an outer nozzle of air and a central fuel nozzle of H_2 diluted in N_2 ($Le_f \cong 0.4$) or CO_2 ($Le_f \cong 0.3$) The single-cell state shown in Fig. 5.17(a) was found to rotate. Other cellular states were observed up to four cells and these

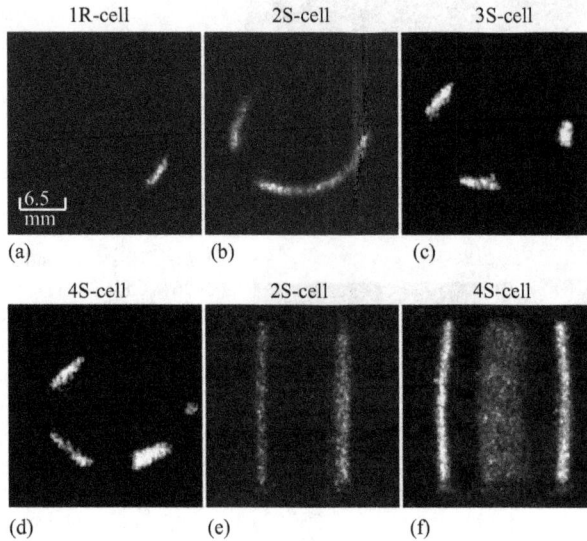

Figure 5.17. Images of the cellular structure of the diluted-H_2/ air opposed tubular concave flame (a)– (d): H_2-N_2 viewed in the axial direction; (e) and (f): H_2-CO_2 viewed in the radial direction. (a) $\phi = 0.271$, $k \approx 73$ s^{-1}, $R_s \approx 5.0$ mm; (b) $\phi = 0.281$, $k \approx 30$ s^{-1}, $R_s \approx 8.0$ mm; (c) $\phi = 0.217$, $k \approx 45$ s^{-1}, $R_s \approx 8.0$ mm; (d) $\phi = 0.236$, $k \approx 45$ s^{-1}, $R_s \approx 8.0$ mm; (e) $\phi = 0.410$, $k \approx 64$ s^{-1}, $R_s \approx 5.0$ mm; (f) $\phi = 0.398$, $k \approx 42$ s^{-1}, $R_s \approx 6.5$ mm. In the figure labels, "R" represents rotating cells and "S" stationary ones. ϕ is the initial mixture strength. (*Source*: Hu, S. T., R. W. Pitz, and Y. Wang. 2009. Extinction and near-extinction instability of non-premixed tubular flames. *Combustion and Flame* 156 (1):90–8. Reprinted with permission from Elsevier.)

were all stable. The flames with one, three, and four cells have flat surfaces but the flame with two cells (Fig. 5.17(b)) has a concave surface.

The flame extinction limits and flame cellularity regions are mapped in Fig. 5.18 as a function of initial mixture strength and stretch rate for concave and convex non-premixed tubular flames of air versus hydrogen fuel diluted in nitrogen. Concave flames are formed with an inner nozzle of diluted fuel and an outer nozzle of air (vice versa for convex flames). The extinction limits (shown by solid lines) occur at lean values of initial mixture strength. The convex curved flames are the most difficult to extinguish and exist at the leanest values of initial mixture strength, ($\phi = 0.15$ for $k = 40$ s^{-1} and $R_s = 5$ mm). A concave curved flame at the same conditions ($k = 40$ s^{-1} and $R_s = 5$ mm) will extinguish at a much higher value of initial mixture strength, $\phi = 0.25$.

The onset of cellularity is shown in Fig. 5.18 by dashed lines and the cellularity persists until extinction. Consistent with previous work, cells are found near extinction with lean values of initial mixture strength and sub-unity fuel Lewis number (Fig. 5.18). In these non-premixed tubular flames, cells occur in air versus H_2 diluted in CO_2, Ar, or N_2 for low values of fuel Lewis numbers, $Le_f \leq 0.42$, but not for H_2 diluted in He flames ($Le_f = 1.1$).[17] In Fig. 5.18, highly curved convex flames had cellularity at the leanest values of initial mixture strength ($\phi = 0.15$ for $k = 40$ s^1)

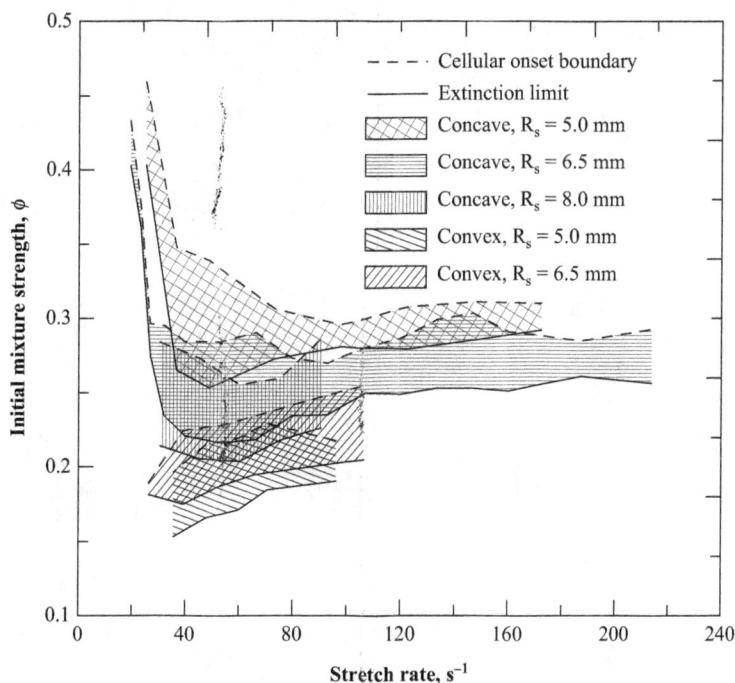

Figure 5.18. The initial mixture strength ϕ vs stretch rate indicating the regions of cellularity of opposed tubular flames with both concave and convex curvatures (fuel: H_2; diluents: N_2; oxidizer: air). The initial mixture strength ϕ is the ratio of fuel mole fraction in the fuel stream to oxidizer mole fraction in the oxidizer stream normalized by the stoichiometric fuel-to-oxidizer molar ratio. (*Source*: Hu, S. T., R. W. Pitz, and Y. Wang. 2009. Extinction and near-extinction instability of non-premixed tubular flames. *Combustion and Flame* 156 (1):90–8. Reprinted with permission from Elsevier.)

while concave flames showed cellularity at much higher values of initial mixture strength ($\phi = 0.45$ for $k = 25$ s^{-1}). Increased stretch rate reduces the extent of cellularity for concave flames.

The cellularity regions have also been mapped for tubular flames of air versus H_2 diluted in CO_2 and Ar. When compared with H_2 diluted in N_2 flames, concave tubular flames with H_2/CO_2 have an increased region of cellularity and those with H_2/Ar flames have a reduced range of cellularity with respect to initial mixture strength.[17]

A new opposed tubular burner was developed for examining tubular flames and cellular formation at higher stretch rates.[18] The burner has an outer air nozzle radius of $R_2 = 12$ mm with a height of 8 mm. The burner design has been described in detail in Chap. 4 (Section 4.5.2) and is shown in Fig. 4.25. For non-premixed flame studies, an inner fuel nozzle is added with a radius of $R_1 = 3.2$ mm and a height of 8 mm. The inner fuel nozzle has co-flows of nitrogen that match the nitrogen co-flows in the outer air nozzle.

The progress of cell formation can be seen in Fig. 5.19 as the stretch rate is increased for a 22%H_2/78% CO_2 versus air tubular flame at $\phi = 0.518$.[19] The stretch rate is increased by increasing the outer air nozzle flowrate while keeping the inner fuel nozzle flowrate constant. The non-cellular flame is shown in Fig. 5.19(a) where a part of the flame surface is masked by

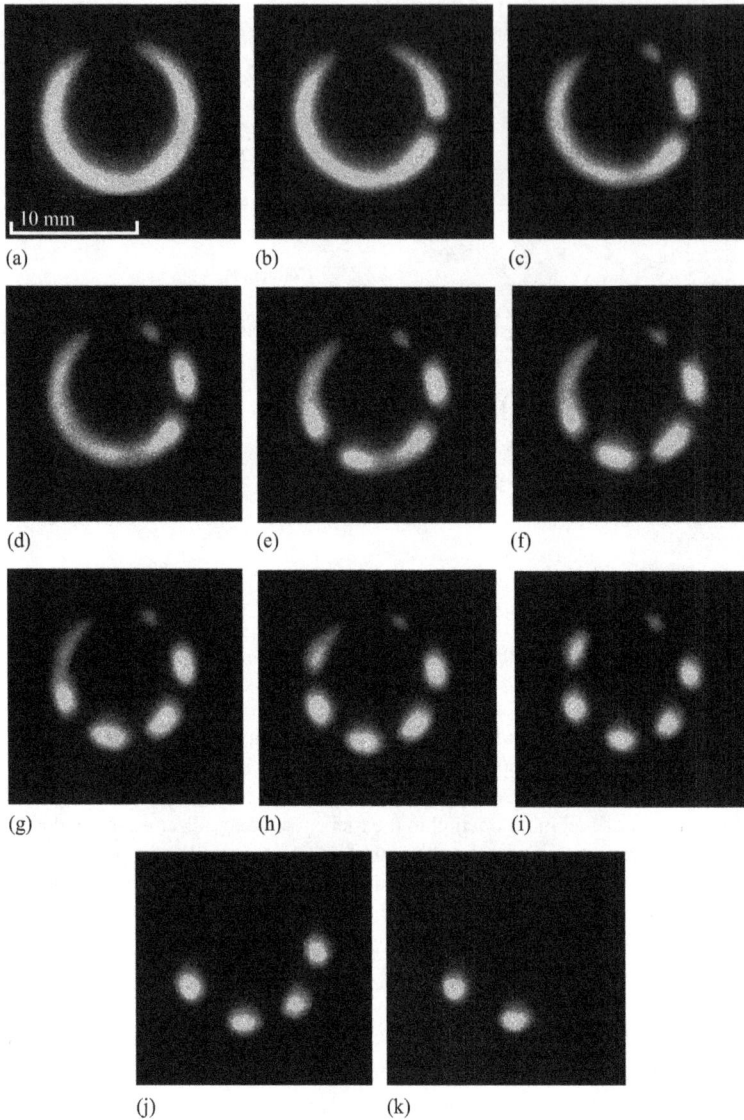

Figure 5.19. Progression of cell formation with increasing stretch rate for 22%H_2/78% CO_2 vs air opposed tubular flames (initial mixture strength $\phi =$ 0.518 and $Le_f = 0.35$). The stretch rate is changed by adjusting the air flowrate while keeping the diluted fuel flowrate constant. The backside of the flame is obscured by the inner fuel nozzle. (a) $k = 207$ 1/s, (b) $k = 238$ 1/s, (c) $k = 261$ 1/s, (d) $k = 277$ 1/s, (e) $k = 285$ 1/s, (f) $k = 292$ 1/s, (g) $k = 324$ 1/s, (h) $k = 331$ 1/s, (i) $k = 385$ 1/s, (j) $k = 402$ 1/s, and (k) $k = 417$ 1/s. (*Source*: Shopoff, S. W., P. Wang, and R. W. Pitz. 2011. Experimental study of cellular instability and extinction of non-premixed opposed-flow tubular flames. *Combustion and Flame* 158 (11):2165–77. Reprinted with permission from Elsevier.)

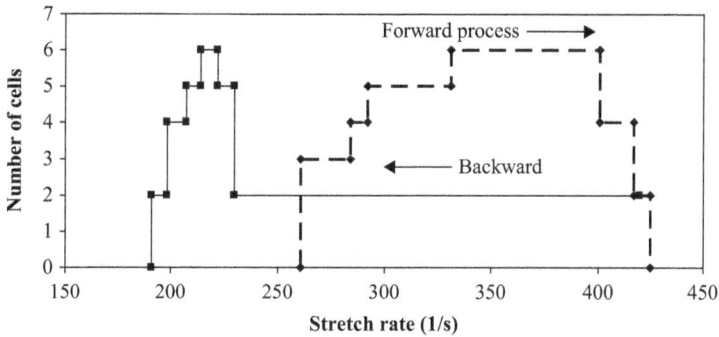

Figure 5.20. Hysteresis comparison of the number of cells vs stretch rate for the forward process (increasing stretch rate) and the backward process (decreasing stretch rate) for 22%H_2/78% CO_2 vs air opposed tubular flames ($\phi = 0.518$ and $Le_{fuel} = 0.35$). The stretch rate is changed by adjusting the air flowrate while keeping the diluted fuel flowrate constant. The point of lowest stretch rate represents the transition from non-cellular to cellular for the forward process (or cellular to non-cellular for the backward process) and the highest stretch rate represents the extinction condition. The backward process began at a condition very near the extinction condition ($k_{ext} = 425$ s^{-1}). In this test, the flame initially broke at three locations (in the forward process) but the last cellular state in the backward process was a 2-cell state. Zero cells indicate either extinction or a non-cellular flame. (*Source*: Shopoff, S. W., P. Wang, and R. W. Pitz. 2011. Experimental study of cellular instability and extinction of non-premixed opposed-flow tubular flames. *Combustion and Flame* 158 (11):2165–77. Reprinted with permission from Elsevier.)

the inner fuel nozzle. With increased stretch rate, the flame breaks in Fig. 5.19(b). The strongest chemiluminescence comes from the flame edges where the flame surface is convex to the fuel. Convex curvature increases the flame temperature for subunity fuel Lewis numbers (Fig. 5.14). Further increase in the stretch rate causes an isolated cell to form in Fig. 5.19(c). The isolated cell shows stronger reaction (increased chemiluminescence) than the rest of the flame surface due to the convex curvature of its flame edges to the diluted H_2 fuel. As the stretch rate is increased, more cells are formed. Figure 5.19(i) shows five cells with a sixth cell hidden behind the fuel nozzle; the cells are rods of reaction arranged symmetrically around the fuel nozzle. These small rods have increased temperature due to the convex curvature of their flame surface toward the fuel. As the stretch rate is increased further to $k = 417$ s^{-1}, the incompleteness of reaction is increased and most of the cells are extinguished as seen in Fig. 5.19(k). All the cells are extinguished at $k = 425$ s^{-1} (see Fig. 5.20).

Cell formation is not limited to lean mixtures of low initial mixture strength. Up to seven cells can be formed on the fuel nozzle for a rich initial mixture strength of $\phi = 1.587$ for 77%H_2/23% CO_2 versus air tubular flames.[19] Stable cells under rich conditions are less probable than stable cells under lean conditions.[29]

The hysteresis effect of increasing and decreasing the stretch rate on the cell number is shown in Fig. 5.20. The cell formation is plotted for the forward process (increasing stretch rate) and backward process (decreasing stretch rate). The cell formation for the forward process corresponds to the images shown in Fig. 5.19. Up to six cells are formed in both the forward and backward processes. The stretch rate for the maximum number of cells in the forward process is much closer to extinction (k_{ext} = 425 s^{-1}) than for the backward process.

ACKNOWLEDGMENTS

The financial support of the National Science Foundation under Grants CBET-1134268 & CTS-0314704) and the NASA Microgravity Program under Grant NNC04AA14A is gratefully is acknowledged by the author.

REFERENCES

1. Sung, C. J., J. B. Liu, and C. K. Law. 1995. Structural response of counterflow diffusion flames to strain - rate variations. *Combustion and Flame* 102 (4):481–92. DOI: 10.1016/0010-2180(95)00041-4.
2. Brown, T. M., M. A. Tanoff, R. J. Osborne, R. W. Pitz, and M. D. Smooke. 1997. Experimental and numerical investigation of laminar hydrogen-air counterflow diffusion flames. *Combustion Science and Technology* 129 (1–6):71–88.
3. Ishizuka, S. and H. Tsuji. 1981. An experimental study of effect of inert gases on extinction of laminar diffusion flames. *Proceedings of the Combustion Institute* 18:695–703.
4. Chung, S. H. and C. K. Law. 1983. Structure and extinction of convective diffusion flames with general Lewis numbers. *Combustion and Flame* 52 (1):59–79. DOI: 10.1016/0010-2180(83)90121-9.
5. Cuenot, B. and T. Poinsot. 1996. Asymptotic and numerical study of diffusion flames with variable Lewis number and finite rate chemistry. *Combustion and Flame* 104 (1–2):111–37. DOI: 10.1016/0010-2180(95)00111-5.
6. Takagi, T., Y. Yoshikawa, K. Yoshida, M. Komiyama, and S. Kinoshita. 1996. Studies on strained non-premixed flames affected by flame curvature and preferential diffusion. *Proceedings of the Combustion Institute* 26:1103–10.
7. Yoshida, K. and T. Takagi. 1998. Transient local extinction and reignition behavior of diffusion flames affected by flame curvature and preferential diffusion. *Proceedings of the Combustion Institute* 27:685–92.
8. Lee, J. C., C. E. Frouzakis, and K. Boulouchos. 2000. Numerical study of opposed-jet H$_2$/air diffusion flame–vortex interactions. *Combustion Science and Technology* 158:365–88. DOI: 10.1080/00102200008947341.
9. Ishizuka, S. 1982. An experimental study on the opening of laminar diffusion flame tips. *Proceedings of the Combustion Institute* 19:319–26.
10. Ishizuka, S. and Y. Sakai. 1986. Structure and tip-opening of laminar diffusion flames. *Proceedings of the Combustion Institute* 21:1821–8.
11. Takagi, T. and Z. Xu. 1994. Numerical - analysis of laminar diffusion flames–effects of preferential diffusion of heat and species. *Combustion and Flame* 96 (1–2):50–9. DOI: 10.1016/0010-2180(94)90157-0.
12. Takagi, T., Z. Xu, and M. Komiyama. 1996. Preferential diffusion effects on the temperature in usual and inverse diffusion flames. *Combustion and Flame* 106 (3):252–60. DOI: 10.1016/0010-2180(95)00255-3.
13. Im, H. G., C. K. Law, and R. L. Axelbaum. 1990. Opening of the Burke-Schumann flame tip and the effects of curvature on diffusion flame extinction. *Proceedings of the Combustion Institute* 23:551–8.
14. Hu, S. T., P. Y. Wang, R. W. Pitz, and M. D. Smooke. 2007. Experimental and numerical investigation of non-premixed tubular flames. *Proceedings of the Combustion Institute* 31:1093–9. DOI: 10.1016/j.proci.2006.08.058.

15. Wang, P. Y., S. T. Hu, and R. W. Pitz. 2007. Numerical investigation of the curvature effects on diffusion flames. *Proceedings of the Combustion Institute* 31:989–96. DOI: 10.1016/j.proci.2006.07.223.

16. Hu, S. T. and R. W. Pitz. 2009. Structural study of non-premixed tubular hydrocarbon flames. *Combustion and Flame* 156 (1):51–61. DOI: 10.1016/j.combustflame.2008.07.017.

17. Hu, S. T., R. W. Pitz, and Y. Wang. 2009. Extinction and near-extinction instability of non-premixed tubular flames. *Combustion and Flame* 156 (1):90–8. DOI: 10.1016/j.combustflame.2008.09.004.

18. Shopoff, S. W., P. Wang, and R. W. Pitz. 2011. The effect of stretch on cellular formation in non-premixed opposed-flow tubular flames. *Combustion and Flame* 158 (5):876–84. DOI: 10.1016/j.combustflame.2011.01.016.

19. Shopoff, S. W., P. Wang, and R. W. Pitz. 2011. Experimental study of cellular instability and extinction of non-premixed opposed-flow tubular flames. *Combustion and Flame* 158 (11):2165–77. DOI: 10.1016/j.combustflame.2011.04.007.

20. Wang, P., J. A. Wehrmeyer, and R. W. Pitz. 2006. Stretch rate of tubular premixed flames. *Combustion and Flame* 145 (1–2):401–14. DOI: 10.1016/j.combustflame.2005.09.015.

21. Mueller, M. A., T. J. Kim, R. A. Yetter, and F. L. Dryer. 1999. Flow reactor studies and kinetic modeling of the H_2/O_2 reaction. *International Journal of Chemical Kinetics* 31 (2):113–25.

22. Seshadri, K. and F. A. Williams. 1978. Laminar-flow between parallel plates with injection of a reactant at high Reynolds-number. *International Journal of Heat and Mass Transfer* 21 (2):251–3. DOI: 10.1016/0017-9310(78)90230-2.

23. Peters, N. and B. Rogg. 1993. *Reduced kinetic mechanisms for applications in combustion systems.* Chapters 1 and 5 vols, *Lecture notes in physics.* Berlin: Springer-Verlag.

24. Smith, G. P., D. M. Golden, M. Frenklach, N. W. Moriarty, B. Eiteneer, M. Goldenberg, C. T. Bowman, R. K. Hanson, S. Song, W. C. Gardiner, Jr., V. V. Lissianski, and Z. Qin. 1999. http://www.me.berkeley.edu/gri_mech/. Chicago, Illinois: Gas Research Institute.

25. Hu, S. T. 2007. *Measurements and modeling of non-premixed tubular flames: structure, extinction and instability.* Nashville, Tennessee, USA: Ph.D. Thesis, Mechanical Engineering Department, Vanderbilt University.

26. Kee, R. J., J. F. Grcar, M. D. Smooke, and J. A. Miller. 1985. *Premix: a Fortran program for modeling steady laminar one-dimensional premixed flames.* Sandia National Laboratories, Livermore, California.

27. Williams, F. A. 2005. *San Diego Mechanism,* http://maeweb.ucsd.edu/~combustion/cermech/ (15 June 2005).

28. Cheatham, S. and M. Matalon. 2000. A general asymptotic theory of diffusion flames with application to cellular instability. *Journal of Fluid Mechanics* 414:105–44. DOI: 10.1017/s0022112000008752.

29. Metzener, P. and M. Matalon. 2006. Diffusive-thermal instabilities of diffusion flames: onset of cells and oscillations. *Combustion Theory and Modelling* 10 (4):701–25. DOI: 10.1080/13647830600719894.

30. Matalon, M. 2007. Intrinsic flame instabilities in premixed and nonpremixed combustion. *Annual Review of Fluid Mechanics* 39:163–91. DOI: 10.1146/annurev.fluid.38.050304.092153.

31. Garside, J. E. and B. Jackson. 1951. Polyhedral diffusion flames. *Nature* 168:1085.

32. Dongworth, M. R. and A. Melvin. 1976. The transition to instability in a steady hydrogen-oxygen diffusion flame. *Combustion Science and Technology* 14 (4–6):177–82. DOI: 10.1080/00102207608547527.

33. Chen, R.-H., C. B. Mitchell, and R. D. Ronney. 1992. Diffusive-thermal instability and flame extinction in nonpremixed combustion. *Proceedings of the Combustion Institute* 24:213–21.

34. Kim, J. S., F. A. Williams, and P. D. Ronney. 1996. Diffusional-thermal instability of diffusion flames. *Journal of Fluid Mechanics* 327:273–301. DOI: 10.1017/s0022112096008543.

35. Lo Jacono, D., P. Papas, and P. A. Monkewitz. 2003. Cell formation in non-premixed, axisymmetric jet flames near extinction. *Combustion Theory and Modelling* 7 (4):635–44. DOI: Pii s1364-7830(03)63649-1.

CHAPTER 6

Tubular Flame Characteristics of Miniature Liquid Film Combustors

Derek Dunn-Rankin

6.1 INTRODUCTION

This chapter is inspired by the opportunity for using tubular flames in miniature liquid-fueled power systems. Recognizing that a high-performance portable power system should provide up to 1000 W for tens of hours, it is straightforward to conclude that a device based on the combustion of liquid fuel is needed.[1] The superiority of liquid fuel combustion for high-power density portable power derives from the inherent chemical energy storage density of combustible liquids. There are challenges, however, to efficiently evaporating and burning liquid fuels in small volumes. All miniature combustors suffer from an increasing surface-to-volume ratio with decreasing dimensions, and this decrease creates significant thermal management issues involving huge heat losses from the reaction zone to the burner wall.[2,3] As can be seen in the other chapters of this volume, the tubular flame often involves close wall proximity, suggesting that there may be a beneficial association between a tubular flame structure and burning liquid fuels in small volumes. One design for such an association is the miniature liquid fuel film combustor.[4] The liquid fuel film concept takes advantage of the fact that for small-scale systems a fuel film provides ample surface area for evaporation and it keeps the chamber wall cool while offering protection from heat losses and quenching. The liquid fuel film combustor relies on the realization that, at least in the sub-centimeter range which is under discussion here, liquid films can offer as high a liquid surface area for vaporization as can a vaporizing spray, particularly when considering the realistic size of droplets achievable by a microspray device. In addition, it is difficult to atomize liquid fuel without imparting high momentum to the liquid (as in typical simplex or air-blast approaches), and this momentum causes wall wetting (i.e., a wall film) in the confined spaces of miniature combustors anyway. Furthermore, the liquid film can offer protection from heat losses and/or quenching, whereas a vaporizing spray does not. With the liquid film on the solid surface, the wall temperature will not exceed the boiling point of the liquid.

This chapter explores, therefore, the relationships between tubular flames and liquid fuel film flames in order to identify potential performance enhancements that can benefit miniature power systems.

6.2 BRIEF REVIEW OF SOME KEY FEATURES OF A TUBULAR FLAME

It is prudent to first review briefly the key features of tubular flames in order to put them into context with the salient features of the liquid fuel film flame. As described throughout this volume, a tubular flame has three important practical characteristics: (1) it is nearly adiabatic, (2) it is aerodynamically stable, and (3) it is cylindrical with a long aspect ratio. The two key forms of tubular flames are *premixed* versions, often with the fuel and oxidizer injected tangentially through a slot, and *rapid-mixing* versions where the fuel and air are injected separately but in such a way as to form a mixed flame sheet. In both of these cases, the exhaust gas is ejected from the combustor along the center core region. Purely non-premixed tubular flames are not considered, in part because in the non-premixed case, there will be combustion products on both sides of the flame (i.e., between the flame and the cylinder wall and in the center of the tube) so the one-dimensional nature of the system is compromised. In practice, however, a partially premixed or partially non-premixed condition can arise and still deliver the three main features of tubular flames identified above. In fact, the 2-MW liquid kerosene-fueled combustor, described in Chap. 8, is one example of a tubular flame that is principally not premixed. In that system, evaporating liquid fuel from the wall reacts with tangentially injected air, forming a tubular flame sheet. Some of the injected kerosene evaporates, mixes, and burns in the rapid-mixing format of the tubular flame, while excess fuel spreads along the surface to evaporate downstream where it mixes and extends the tubular reaction zone. Hence, although the size scales are dramatically different, the 2-MW kerosene tubular combustor is quite similar to the miniature liquid film combustor, which is the focus of this chapter.

The qualitative similarities between miniature liquid fuel film combustors and other liquid fueled tubular combustors include: (1) swirling air flow created by tangential injection or swirl vanes, (2) tangentially injected fuel that spreads along the wall for rapid mixing with the swirling air, (3) hot exhaust gas ejected from the center of the chamber, (4) essentially adiabatic operation with heat loss to the wall avoided by the insulating layer of injected fuel and air, (5) fairly flat temperature profile across the combustor, and (6) aerodynamically stable flames with moderately wide operating limits. These similarities suggest that considering the liquid fuel film combustor as a miniature tubular flame combustor may provide useful insights into this device.

6.3 REVIEW OF THE KEY FEATURES OF A FUEL FILM COMBUSTOR FLAME

When burning liquid fuels in miniature combustors, it is logical to inject the majority of the fuel directly as a film on the solid surfaces where high heat transfer from the combustion products occurs. This film evaporation produces a fuel-rich layer in the gas near the wall. Although any small combustor burning liquid fuel will have an evaporating fuel film, cylindrical liquid fuel film combustors provide a convenient geometry for both theory and practice. The main idea (which has been described over a 10-year period in more detail in references[5-10]) is to deliver the liquid fuel into the combustion chamber in a way that allows the fuel to spread over the inside face of the chamber's wall. The film will become the fuel source of the combustion as the evaporated liquid mixes and burns with the air inside the chamber. The flame should then be located between the fuel film and the exhaust gas region in the center of the combustion chamber, as shown pictorially in Fig. 6.1.

Having a flame inside a small cylinder (which is an example of high surface-to-volume ratio combustors) means that the wall temperature can be very high since heat from the flame is transferred to the confining chamber. As mentioned earlier, however, the fuel film provides thermal protection for the wall since the heat is absorbed by the phase change of the fuel. This absorbed evaporative enthalpy is then returned to the combustion process since the gaseous fuel reaction provides the heat release. Hence, the overall heat loss from the combustor walls and their temperature can be minimal. This built-in cooling effect is one advantage of the fuel film device over droplet spray systems. Despite this potential advantage, however, it is important to confirm that there is sufficient surface area in the wall film to provide the fuel evaporation rate needed to sustain a chemical reaction at a level that is possible with a spray.

As demonstrated in several papers [e.g., Ref. 9-12], and as described somewhat more extensively in a recent review,[4] a miniature film combustor with a combustion chamber consisting of a steel tube, burning an air–heptane mixture introduced by tangential inlets, is able to give a stable flame. Figure 6.2 shows a typical fuel film combustor, indicating tangential inlets for fuel and air. The bottom of the 1 cm diameter chamber is windowed to allow visual access. Figure 6.3 shows the film combustor in operation using heptane as the fuel and the overall fuel-rich operating condition extends the flame well beyond the exit of the combustion chamber as additional air reacts with the fuel-rich exhaust products. Coherent anti-Stokes Raman spectroscopy (CARS) is employed to measure the temperature without intrusion at the chamber exit.[6] CARS was necessary because any physical probe acts as a flame holder and changes the stability characteristics. The temperature profile indicates two flames, one in the core and one attached to the rim. The values of temperature are below the stoichiometric adiabatic flame

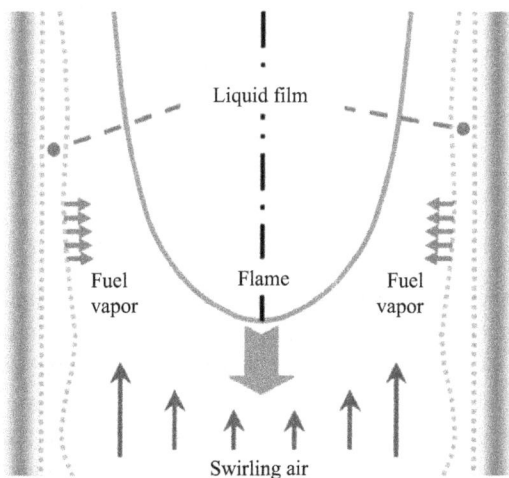

Figure 6.1. Concept drawing of the miniature liquid fuel film combustion process.[6]

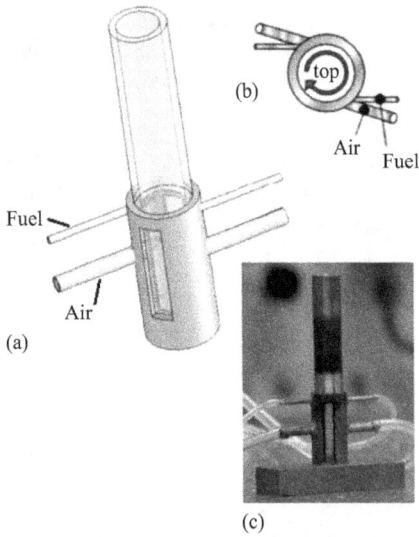

(b)

Air Fuel

Fuel

Air

(a)

(c)

Figure 6.2. Typical miniature liquid fuel film combustor; 1 cm diameter and 5–10 cm long. (a) a transparent version using a sapphire cylinder; (b) top view showing the tangential injection ports for fuel and air; (c) metal version realized in hardware.

Adiabatic flame temperatures:
$\phi = 1$, $T = 2276K$; $\phi = 1.69$, $T = 1849K$

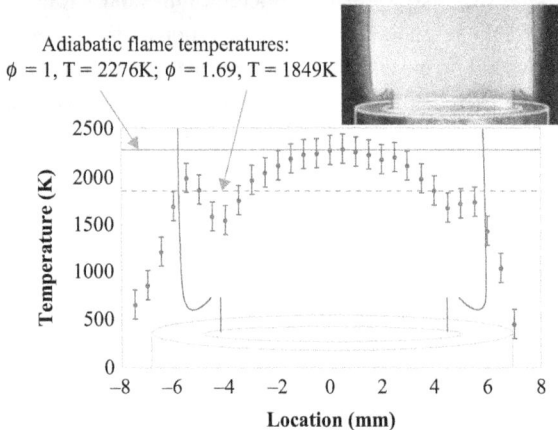

Figure 6.3. Coherent anti-Stokes Raman spectroscopy (CARS) measurements of gas temperature right at the exit of the film combustor operating on heptane and air. The upper photograph shows the experimental setup, with the CARS laser beams crossing 2 mm above the lip of the combustor. The data indicate two flames, with the local maximum near the tube wall coming from a rim-stabilized flame, and the core region exhibiting a fairly uniform high temperature.[6]

temperature, but generally above the fully premixed adiabatic flame temperature for the actual overall rich mixture. This suggests that the flame is partially premixed, with near stoichiometric conditions along the central axis and very rich regions near the walls. The core temperature has a fairly flat profile as is typical of tubular flames.

To better visualize the flame structure, Fig. 6.4 summarizes a chemiluminscent study of the film combustion process.[6,11] In this case, the chamber is sapphire to provide complete optical

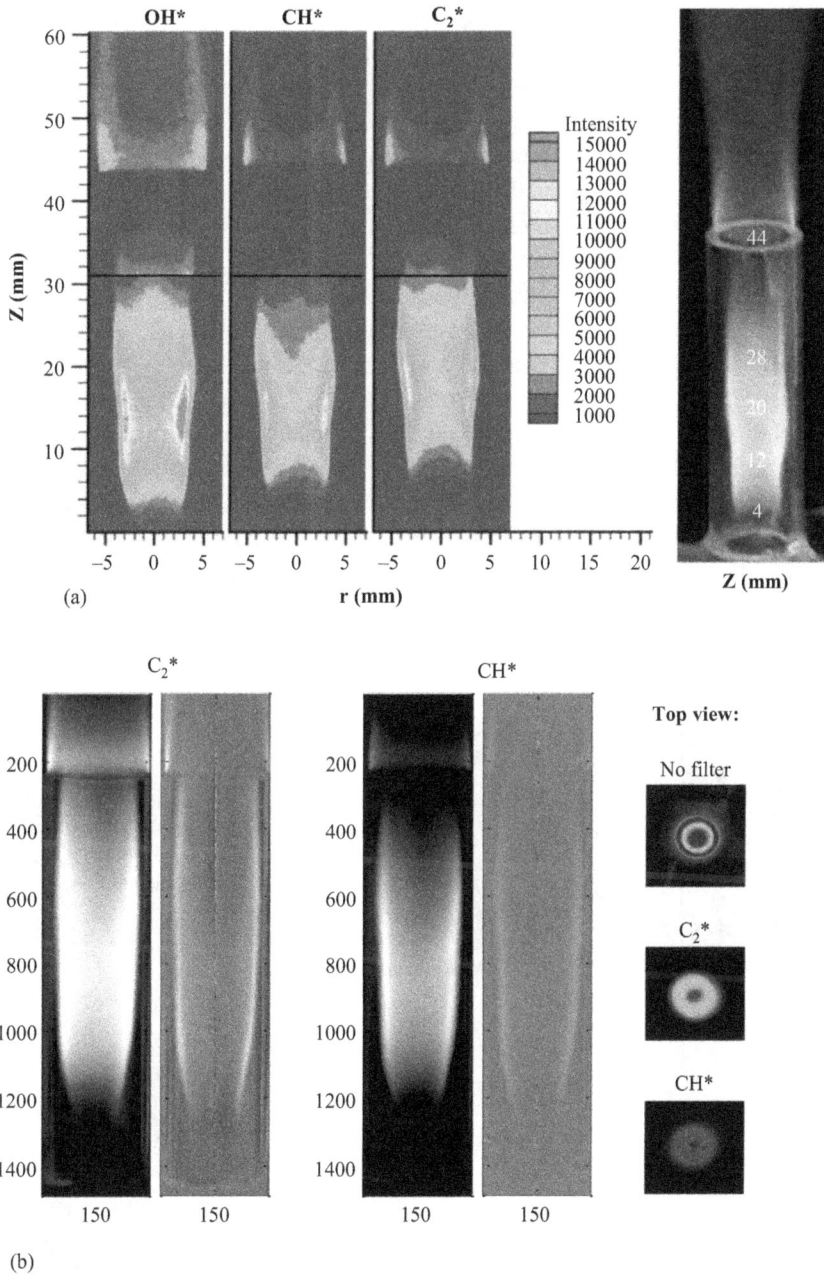

Figure 6.4. Chemiluminescence from a liquid heptane-fueled film combustor. (a) grayscale images showing the intensity of light emitted by OH*, CH*, and C_2* as recorded through appropriate notch filters in front of an intensified video camera[6]; (b) a quantitative analysis of chemiluminescence by Abel inversion of line-of-sight integrated results showing a hollow core flame shape. The top views give luminosity images as seen when observing the flame through spectral notch filters.[11]

access. The upper grayscale images are qualitative because they are line-of-sight integrated but they show clearly a separation between the internal flame and the rim flame. The lower images include Abel-inverted results to demonstrate that the flame is a cylindrical sheet near the wall.[6,11] The top view images confirm a hollow core flame structure. Figure 6.5 shows what happens when a nitrogen shroud prevents additional air from reaching the fuel-rich exhaust. From left to right the images show that as the nitrogen flow in the shroud increases, the rim flame weakens, lifts, and extinguishes. The core flame, however, is unaffected, demonstrating that its stability is independent of the rim flame.

A phenomenological explanation for the stable operation of the miniature liquid fuel film combustor that is consistent with the observed behaviors is shown in Fig. 6.6.[6] Since swirling airflows generate gradients in fuel concentration in the radial direction, fuel-rich zones near the chamber wall and fuel-lean zones near the axis of symmetry encourage the propagation of triple flames. As shown in the left-hand schematic, the central triple flame presents three branches: the fuel-rich limit, the flame limit, and the fuel-lean limit. These limits define four different regions, with (1) a fuel-rich mixture near the wall, then (2) a region with fuel excess and combustion products; on the other side of the flame (3) an air excess and combustion products zone, and (4) a last fuel-lean mixture toward the axis of symmetry. Flame stability is then established when thermal expansion behind the flame front creates flow divergence, causing the upstream flow velocity to decrease. In this manner, the flow speed matches the laminar flame speed at

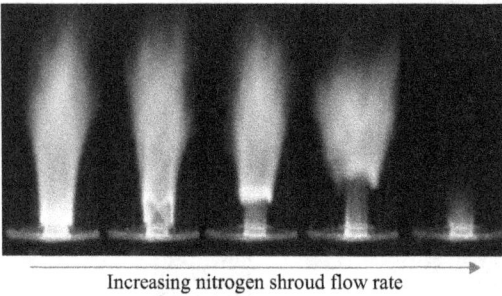

Figure 6.5. External flame plume as nitrogen shroud gas flow rate increases. The nitrogen quenches the rim flame without perturbing the core flame.

Increasing nitrogen shroud flow rate

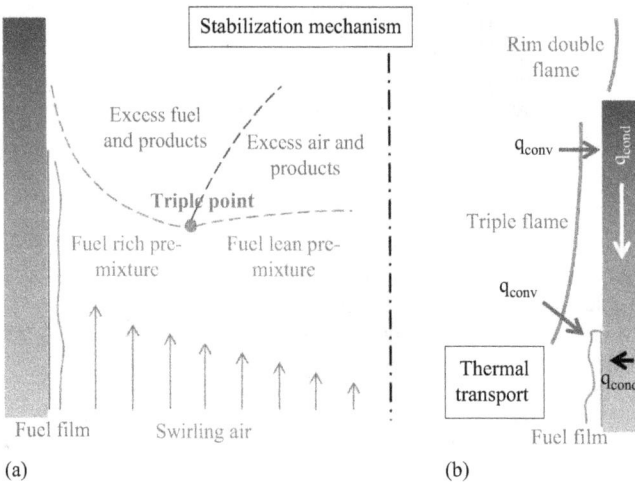

(a)

(b)

Figure 6.6. Key features of a sustained liquid fuel film flame. (a) triple flame structure that stabilizes the liquid film combustor[6]; (b) role of heat transfer in the combustion process showing the need for heat conduction to assist in fuel evaporation.

the triple-point location, providing an ignition point that is effectively immune to the impact of flows that are faster than the flame speed elsewhere in the flow field. This concept allows flame stability without recirculation or a physical flame holder. The right-hand figure indicates how heat from the flame is conducted through the chamber walls to aid in evaporating the liquid fuel.

The liquid fuel film combustor has three clear similarities to the classical tubular flame configuration: (1) fuel is injected at the wall or evaporated near the wall in the case of liquid fuel; (2) swirling air creates a cylindrical flow field with an interface between fuel and oxidizer; and (3) the exhaust gases of the combustion process exit from the center of the combustor. Interestingly, however, fuel film combustor operation has been shown to demand overall fuel-rich conditions with the flame anchored by a triple-flame structure near the liquid film dry-out region. It has also been shown that steady liquid fuel film combustion requires heat from the flame to be conducted through the chamber walls to the liquid film. This requirement means that although the system is axisymmetric, the liquid fuel film combustor cannot be one-dimensional since there is flame evolution in the axial direction. Nevertheless, the liquid fuel film flame carries many of the trademark features of the tubular flame, and recognizing the relationship between the two, can help us develop an understanding of how to improve the function of the liquid film combustor.

6.4 EXAMPLES OF TUBULAR FLAME BEHAVIORS IN A FUEL FILM COMBUSTOR

6.4.1 ORIGINAL DESIGN

As shown in Section 6.2, the original film combustor design involved a simple tangential fuel and air injection near the base of the chamber with the expectation that a fuel film would develop and evaporate into the flame zone. The experiments have shown, however, that the film does not fully cover the walls so that flame heat can conduct back upstream to the liquid through the chamber walls. This behavior requires combustor materials with reasonably high thermal conductivity (mild steel is the least conductive material that still produces a reliable film combustor). In addition, the system operates only under fuel-rich conditions to sustain the triple-flame ignition point. Despite these limitations, Fig. 6.4 shows a distinctly tubular flame structure, with the flame extending along the wall of the chamber and with a hollow central core. Unfortunately, the combustion efficiency of this tubular flame is very low because the fuel-rich operation means that it exhausts fuel-rich products along the centerline. The next section describes a new film combustor design that includes secondary air injection above the core flame to fully consume the fuel and improve the combustion efficiency.

6.4.2 SECONDARY AIR INJECTION

The secondary air injection design was described in some detail by Mattioli et al.[12] This section highlights some of the key findings of that work and relates them to a tubular flame appearance.

Figure 6.7 provides a schematic of the secondary air injection design along with typical images under operation on heptane. The images show that the combustor is stabilized by the

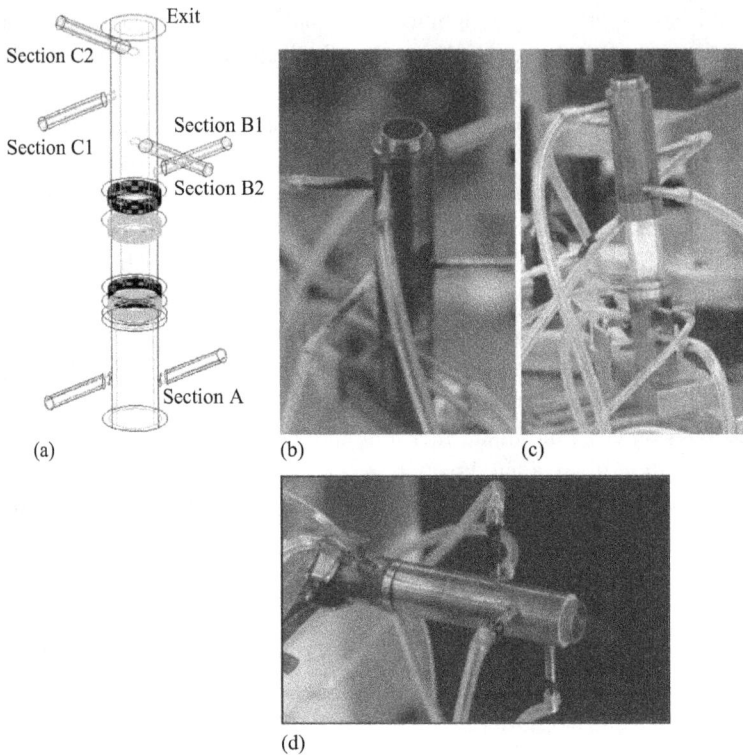

Figure 6.7. A modular film combustor design with a secondary air injection section. (a) drawing of the combustor showing the base section, the midsection, and the air injection section; (b) photograph of the combustor with a 1 cm diameter chamber; (c) combustor in operation with a windowed base section showing the triple flame anchoring mechanism; (d) film combustor operating in a horizontal direction to show that gravitational effects are minor.

same triple-flame structure as was seen without air injection and that the device can operate in any orientation with respect to gravity, so it does not rely on buoyancy for any core behaviors. Figure 6.8 shows the combustor operation with a gradually increasing secondary air injection so that the overall equivalence ratio varies from the no air-rich condition to a very lean 0.32 value. As the flow rate increases, the flame confinement likewise increases until the flame is entirely retained within the chamber. The schlieren images in the figure show how the air injection produces more rapidly mixed exit gas streams. Figure 6.9 shows that secondary air injection has an additional important role in keeping the downstream components of the combustor cool. The lower zone of the chamber is evaporatively cooled by the fuel, but the secondary air zone is dry and therefore, conduction cooling is insufficient. The figure shows that as the secondary air increases, the chamber likewise moderates its temperature.

Figure 6.10 shows the in-cylinder view of the secondary air injection flames. While not quite as distinct as the core film flame, there remains a very pronounced tubular flame appearance. It seems, therefore, that the strong swirl of the secondary air injection acts on the fuel-rich products from the triple-flame zone to create a downstream tubular flame. This concept is shown in Fig. 6.11. The reaction zone can be considered to constitute two tubular flame levels. The first is a triple flame stabilized and burns the evaporating liquid fuel in an overall rich

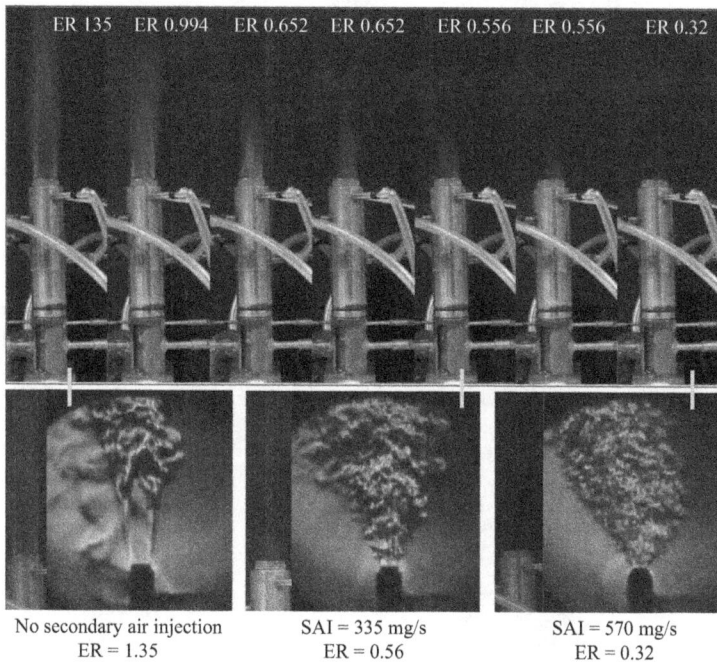

No secondary air injection
ER = 1.35

SAI = 335 mg/s
ER = 0.56

SAI = 570 mg/s
ER = 0.32

Figure 6.8. Effect of secondary air injection on tubular flame behavior and on the overall equivalence ratio (ER). Secondary air injection permits containment of the flame within the combustor. Also shown are schlieren images associated with the secondary air injection; the lines link the photograph and the appropriate schlieren image.

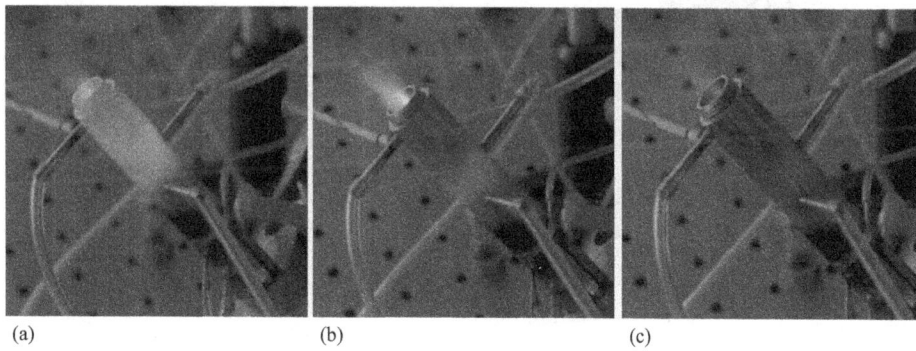

(a)   (b)   (c)

Figure 6.9. Secondary air injection cools the downstream cylinder to prevent excessive heating by the flame once the liquid film is no longer present to protect the surface by evaporative cooling. (a) Has no secondary air injection; (b) and (c) show the effects of increasing secondary air injection.

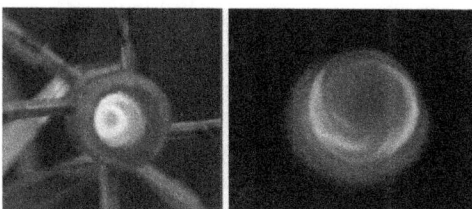

Figure 6.10. Top views of the internal film combustion flame with secondary air injection show the tubular flame nature of this process.

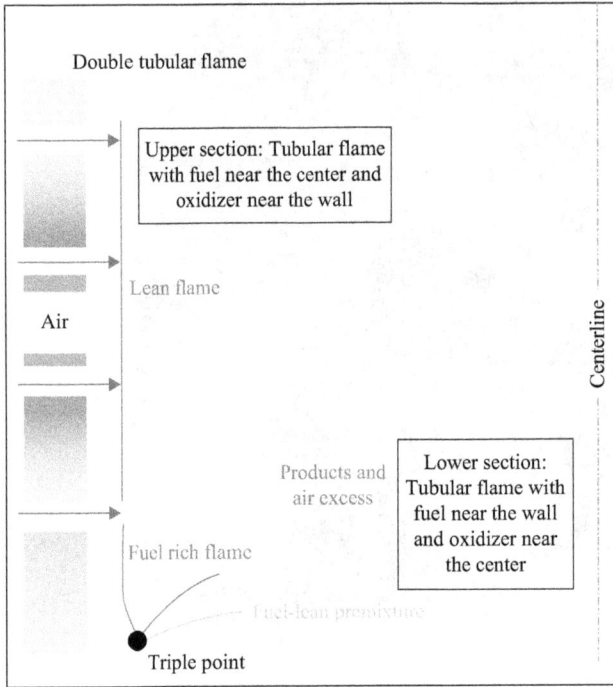

Figure 6.11. A schematic diagram showing the dual-tubular flame of a film combustor that utilizes secondary air injection. One tubular flame is anchored by a triple-flame point near the base and a second tubular flame is created by the later injected secondary air.

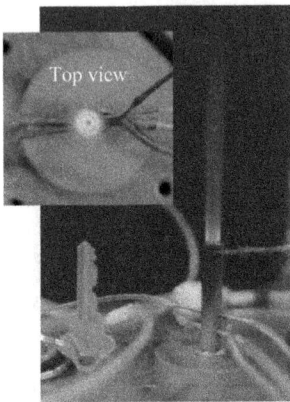

Figure 6.12. Prototype of a 0.5-cm diameter liquid fuel film combustor. This device has poorer performance characteristics than the slightly larger 1-cm diameter system.

regime. The second is a flame sheet supported by rich combustion products from the first flame and additional air injected tangentially in the downstream region.

Attempts to reduce the combustion chamber diameter below the nominal one centimeter size were only moderately successful, and in particular the injection of secondary air was not effective for smaller-diameter combustors. Figure 6.12, for example, is a 5-mm diameter film combustor. The flame still has the characteristic tubular appearance but the secondary air cannot induce the flame to remain confined within the chamber, and the operating regime is very small. As a comparison, Fig. 6.13 shows the limited stable operation zones for the original

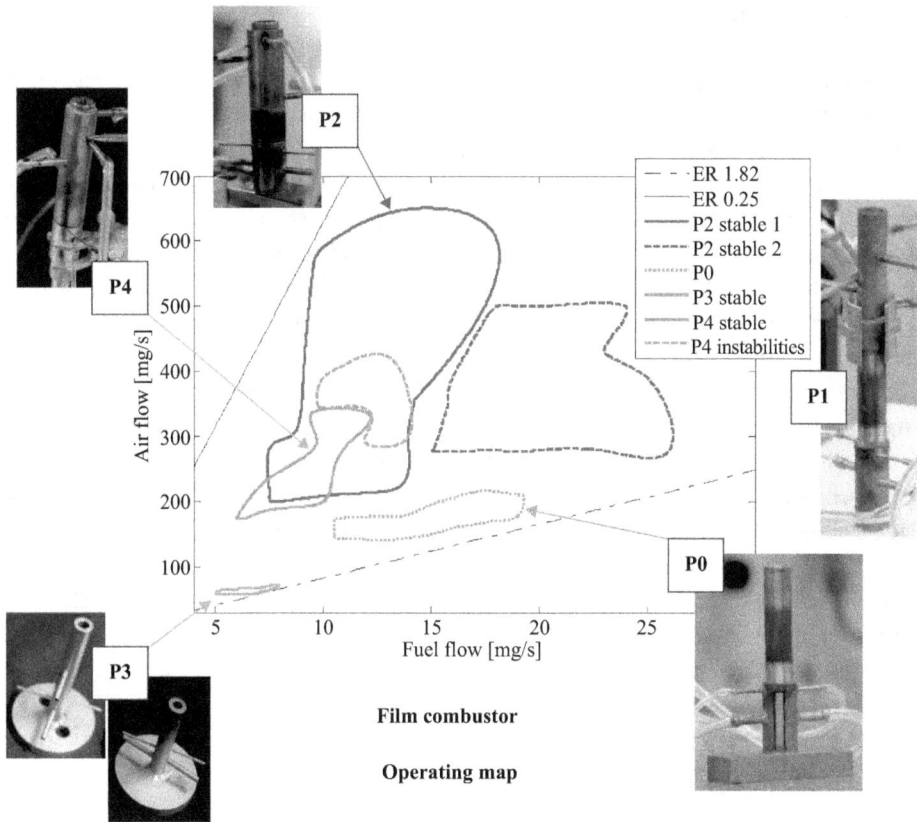

Figure 6.13. Stable operating domains of the various film combustors studied. ER, equivalence ratio.

film combustor (P0) as compared with the 5-mm design (P3), and then as compared with the improved secondary air injection systems (P1, P2, P4). The difference between these latter cases is the overall chamber length and air injection zones. Clearly, the P2 design is the best of these with the broadest operating limits. There are not yet any obvious design rules that show why this geometry performs best, but there is a trade-off between performance and chamber length. Understanding the combustion as a tubular flame can likely help with future designs.

The general conclusions of the secondary air injection design studies were that with proper design, these 1-cm diameter fuel film combustors are capable of providing about 600 W of thermal power for less than 300 g of weight, with a combustion conversion efficiency estimated at more than 90% and a total thermal efficiency between 0.5 and 0.8. The introduction of secondary air injection allows a more complete combustion and an augmentation of the stability. At the proper lean operating condition, essentially continuous burning times are achieved without flame blowout and with total flame confinement inside the chamber. However, despite the success of secondary air injection and the double tubular flame character, an ideal condition would be a strongly swirled flow to produce the rapid mixing needed for lean operation with the main airflow alone. To accomplish this, the next design concept involved a dedicated swirler at the base of the combustor.

6.4.3 SWIRLER DESIGN AND TUBULAR FLAME

The results obtained using separate injection of fuel and air through tubes oriented tangentially to the inner surface of the miniature film combustor show that this technique produces good swirl generation and good film surface spreading. On the other hand, the external tubes increase the combustor size, reducing the advantage of a small diameter. The tube requires a minimum length to introduce a uniform flow for the purpose of generating an acceptable swirling motion. Moreover, with this method, the only way to change the swirl intensity is through flow regulation, which also affects the working conditions. Another possibility to obtain reproducible swirled flow is to introduce a swirl vane. The addition of a swirl vane at the bottom of the burner has the purpose of tangentially deflecting the air flow that enters the chamber axially without any peripheral inlet tube. The swirl vane concept typically is associated only with the air injection because it is the main flow and because the fuel flow rate is small enough that its inlet tube can be reduced as desired. In this way it is possible to reduce the overall size of the burner and, above all, to allow improved control of the swirl features by somewhat decoupling the swirl strength from flow rate through swirler design. That is, with tangential air injection, the swirl level is directly linked to the axial flow velocity by mass conservation. With adjustments to the swirl vane it is possible to increase the swirl without changing axial velocities and swirl vanes can be easily interchangeable in the burner. Furthermore, a swirl vane often creates a flow path and pressure distribution that can help the flame anchor. Under some circumstances, for example, the swirler might act as a bluff body creating a recirculation zone just downstream that helps with this process. There is a trade-off, of course, between pressure drop and the creation of recirculation through a swirler, but that provides an additional optimization parameter for these small burners. Previous configurations of fuel-film combustors have considered a single swirl vane design.[10] Beginning from these past preliminary results, the challenge has been the exploration of swirl vane design.

There are not currently any clear design rules for swirl vanes at the miniature scale. The sub-centimeter dimensions and their complex features require concepts using new materials and flexible fabrication techniques. The construction challenge of building small devices with complex geometry was met by taking advantage of rapid prototyping. The approach was to take advantage of modern rapid prototyping construction methods to build swirl vanes directly into the small diameter chamber. The design was drawn using SolidWorks, a 3D CAD compatible with rapid prototyping software. The swirler is at the bottom, just below the two tangential fuel inlets. With the objective of better understanding the combustion behavior, the intention was to create a prototype in transparent material so as to allow a direct observation of the combustion process and other phenomena inside the chamber. The prototype was built by the stereolithography (SLA), a rapid prototyping technique, but when completed, we observed that the spatial resolution limits of rapid prototyping made it impossible to create a full working prototype at this combustor size using this method. In fact, the 3D stereolithography laser has 1 mm tolerance while the swirler can have features substantially smaller than this value. The laser tolerance is not sufficient to realize these small features and the resolution limits of rapid prototyping do not allow some designs at this scale. The rapid prototyping resolution issue encouraged a modular combustor design, allowing the construction of an easily replaced swirl vane.

Figure 6.14 shows the swirler-based film combustor design and a schematic of the experimental apparatus, with fuel supplied by a pair of syringe pumps. In this case, the air enters through the swirl vanes so only fuel injection ports are needed. The photograph shows that this method for obtaining air swirl creates a typical film combustion flame but it does not appear

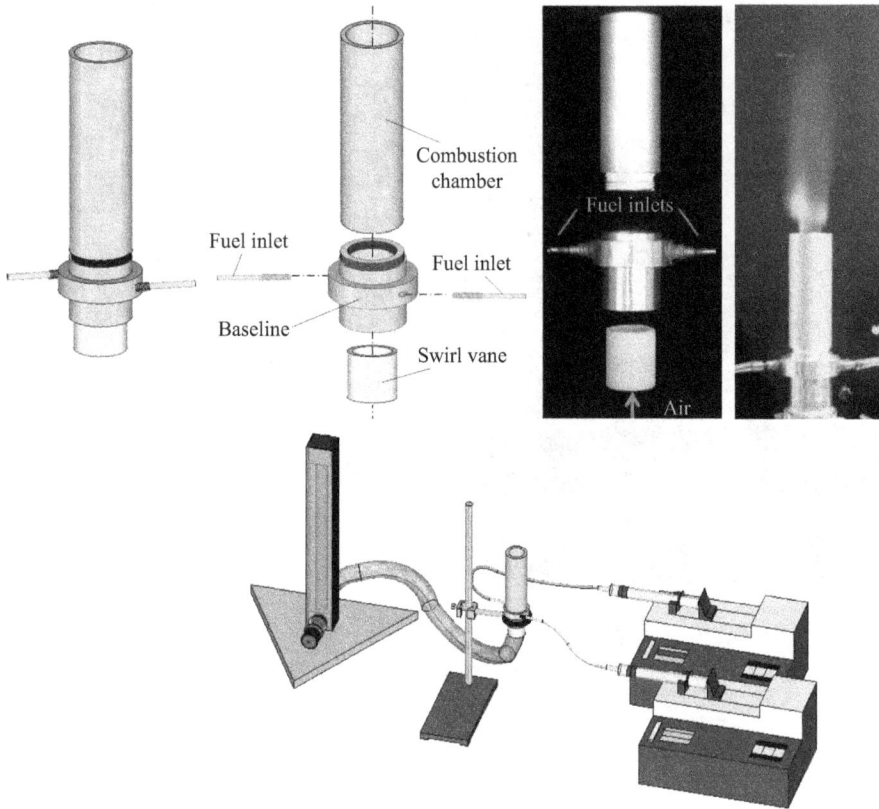

Figure 6.14. New film combustor design incorporating a swirl vane element at the base rather than tangential air injection. The rest of the design is similar to the prior ones.

superior to the flame produced by the tangential air injection approach. Unfortunately, since optimized design rules for swirlers and swirl vanes at the miniature scale are not available in the way that they are for larger scale systems, a variety of design concepts were considered. More details are described in[13,14] but the first four promising designs are shown in Fig. 6.15. All of the swirl vanes were fabricated using rapid-prototyping techniques so a wide variety of shapes and configurations were achievable. As mentioned earlier, however, the precision of the additive manufacturing methods was not always to the desired level so that experimental testing of flow restriction and swirl achieved was necessary to confirm the swirler performance.

During the characterization of swirl vanes, the stainless steel chamber became incandescent and therefore, unstable phenomena appeared after few minutes of combustion, causing quenching. In particular, the combustion chamber showed a red rim, accompanied by oscillations of the flame up and down in the chamber just prior to quenching. Hence, as with other designs, the wall material is important for stable operation and this effect was even more critical with the swirler design. To avoid this issue, an aluminum chamber tube was constructed. As shown in Fig. 6.16, while the stainless steel overheats at the exit because of poor heat transfer, the aluminum does not. The aluminum design facilitated a stable operation.

Figure 6.17 summarizes the combustion behavior of the swirl-based film combustor. The burning tests considered the four different swirl vanes and three different chamber lengths (25–75 mm). This provided several combinations to help identify the features leading to the best

Figure 6.15. Swirl vanes created via rapid prototype technology using a 3D laser stereolithographic method. Shown is a comparison between the ideal drawn design and the fabricated part to demonstrate the effect of precision limitations of the rapid prototyping approach.

Swirl I   Swirl II   Swirl III   Swirl IV

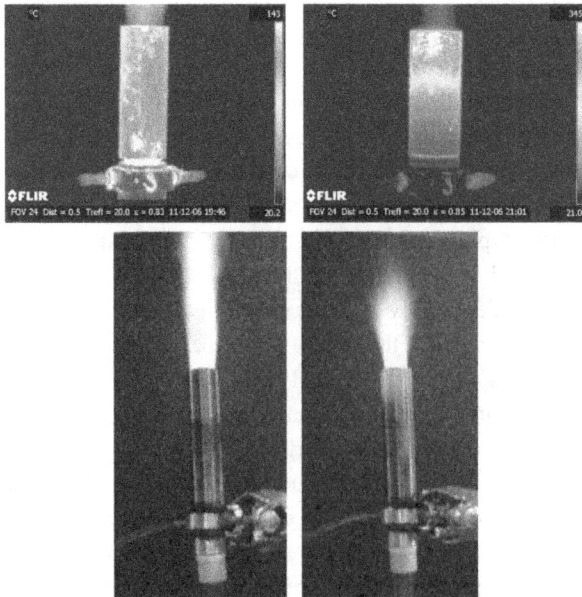

Figure 6.16. Thermal image and photograph of film combustion using an aluminum chamber (left) and a stainless steel chamber (right). The images show clearly that the low thermal conductivity of the stainless steel overheats the exit. The aluminum chamber shows fairly uniform heating throughout.

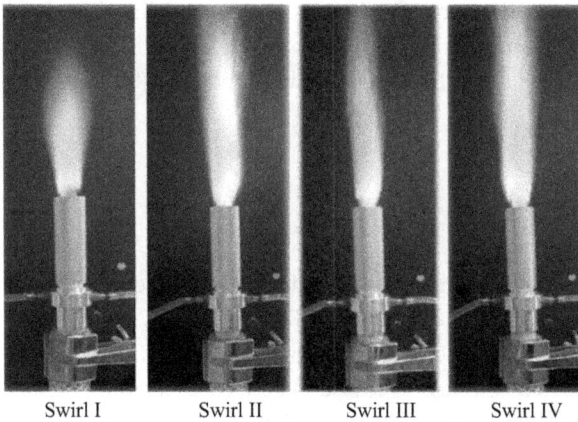

Figure 6.17. Combustion in the aluminum chamber using the four swirl vane designs.

Swirl I   Swirl II   Swirl III   Swirl IV

combustion stability and flame properties. The figure only shows the best performing 55 mm long aluminum chamber. The tests showed the superior behavior for the aluminum combustion chamber with a length of 55 mm. In this configuration there was a compromise between ignition ease and the confined flame: the former was easier in a short chamber, the latter had advantages in long cylinders. Furthermore, it is important to note that the residence time has to be sufficient for good mixing. A decrease in the chamber length would cause a decrease in the proper mixing as well as a rise in temperature at the rim. The aluminum, with its higher thermal conductivity, resulted in the best flame stability, allowing essentially unlimited run durations. This stability was independent of the swirl vane used. All of the variations had chambers with a constant inside diameter of 9.7 mm. Although the swirl vanes were made of plastic and the chamber of aluminum, neither sustained any thermal damage. The heat protection provided by the air flow and fuel evaporation, combined with thermal spreading of the load through the aluminum, maintained stable operation and moderate temperatures throughout the process despite the high combustion temperatures.

The thermal control aspect of the miniature film combustor design can be understood qualitatively by evaluating the order of magnitude values for the system when the thermal conduction down the chamber is greater than or equal to the heat transferred from the surface to the incoming air and fuel. Assuming an average wall temperature of around 500 K, and using convection coefficient $h \sim O(100)$ W/m^{2}K, chamber length $L \sim O(0.01)$ m, and wall thickness $t \sim O(0.001)$ m, a transfer balance is achieved when thermal conductivity of the chamber material $k > O(10)$ W/mK.[13,14] Looking at the thermal conductivity coefficient of different metal materials at standard conditions, aluminum, copper, chromoly, and stainless steel all have thermal conductivities above the required limit value of 10 W/mK, but aluminum and copper coefficients are an order of magnitude higher than that of stainless steel. It is not surprising, therefore, that aluminum and copper provided the best performance in past studies[9] and that aluminum provided improved performance with the new swirl vane design. This finding confirms the important relationships between chamber thermal conductivity and stability.

The results from the swirl vane design were somewhat disappointing in that they did not provide any significant improvement over the tangential air injection design. This result suggests that the triple-flame ignition is a robust feature of these systems.

Figure 6.18 shows a tubular flame from the top and slightly oblique views. As in all the previous experiments, there is a flame near the wall with a cylindrical shape. In the slightly oblique view, however, it is clear that the flame is really more of a spiral flame than a cylindrical sheet. In effect, there is a tubular structure but it its uniformity is only at the average level. Once again, there are many features similar between the tubular flame and liquid film combustor flame, but the correspondence is not exact.

6.5 CONCLUDING REMARKS

This chapter has shown that there are several important similarities between the true tubular flame and the flames arising in miniature liquid film combustors. The similarities are shown in the top views of these flames shown in Fig. 6.19. Insofar as using these miniature film flames as a radiant heat source, the differences are not significant because in both cases there is an elongated and stable flame sheet confined in a chamber near the wall. Although the fluid mechanics and the complex ignition–stability behavior of the film combustor make the fuel-film flame difficult to predict from tubular flame theory, there are clearly substantial opportunities for future overlap between these two fields of study.

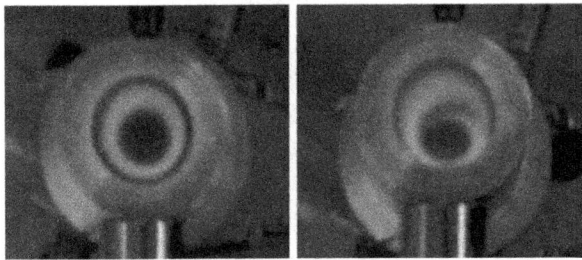

Figure 6.18. Top view of combustion using the base swirler design showing the characteristic tubular appearance. The slightly oblique view (right image) shows that the swirl remains evident in the flame structure.

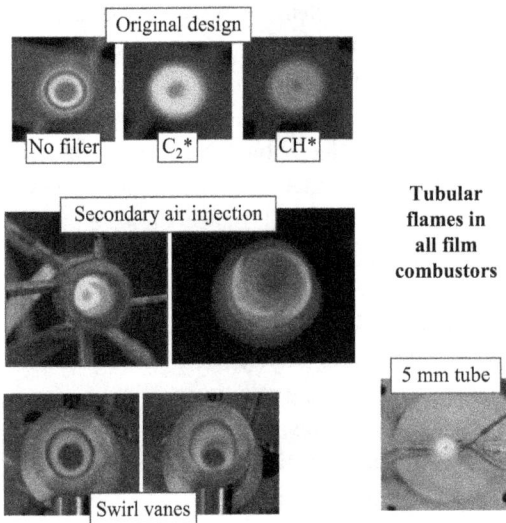

Figure 6.19. Summary images of the key tubular flame features that appear in miniature film combustors. In all cases, there is a flame with an open core that sits near the wall.

REFERENCES

1. Dunn-Rankin, D., E. Leal, and D. Walther. 2006. Personal power systems. *Progress in Energy and Combustion Science* 31:422–65. DOI: 10.1016/j.pecs.2005.04.001.

2. Fernandez-Pello, A. C. 2002. Micro-power generation using combustion: Issues and approaches. *Proceedings of the Combustion Institute* 29:883–9. DOI: 10.1016/S1540-7489(02)80113-4.

3. Maruta, K. 2011. Micro and mesoscale combustion. *Proceedings of the Combustion Institute* 33:125–50. DOI: 10.1016/j.proci.2010.09.005.

4. Dunn-Rankin, D., W. A. Sirignano, Y. H. Li, and Y. C. Chao. 2013. Miniature liquid fuel combustion. In *Micro combustion*, eds. C. Cadou, Y. Ju, and K. Maruta. Momentum Press, in production.

5. Sirignano, W. A. and D. Dunn-Rankin. 2005. Miniature liquid-fueled combustion chamber. United States Patent No. US 6,877,978 B2.

6. Pham, T. K., D. Dunn-Rankin, and W. A. Sirignano. 2006. Flame structure in small-scale liquid film combustors. *Proceedings of the Combustion Institute* 31 (2):3269–75. DOI: 10.1016/j.proci.2006.08.030.

7. Sirignano, W. A., S. Stanchi, and R. Imaoka. 2005. Linear analysis of liquid-film combustor. *Journal of Propulsion and Power* 21:1075–91. DOI: 10.2514/1.14156.

8. Li, Y.-H., Y.-C. Chao, N. Sarzi-Amade, and D. Dunn-Rankin. 2008. Miniature liquid film combustors: Double chamber and central porous inlet. *Experimental Thermal and Fluid Science* 32:1118–31. DOI: 10.1016/j.expthermflusci.2008.01.005.

9. Sirignano, W. A., T. K. Pham, and D. Dunn-Rankin. 2002. Miniature scale liquid-fuel film combustor. *Proceedings of the Combustion Institute* 29 (1):925–31.

10. Mattioli, R., T. K. Pham, and D. Dunn-Rankin. 2009. Secondary air injection in miniature liquid fuel film combustors. *Proceedings of the Combustion Institute* 32:3091–8. DOI: 10.1016/S1540-7489(02)80117-1.

11. Pham, T. K. 2006. *Flame structure and stabilization in miniature liquid film combustors*. Irvine: Ph.D. Dissertation, University of California.

12. Mattioli, R. 2009. *Liquid fuel film technology: Combustion confinement, miniaturization, and flame stabilization*. Politecnico di Milano: M. Eng. Thesis.

13. Giani, C. 2012. *Miniature fuel film combustor: Swirl vane design and combustor characterization*. Politecnico di Milano: M.Eng. Thesis.

14. Giani, C. and D. Dunn-Rankin. 2012. Miniature fuel film combustor: Swirl vane design and combustor characterization. *Combustion Science and Technology* 185:1464–1481. DOI: 10.1080/00102202.2013.804181

CHAPTER 7

Small-Scale Applications

Daisuke Shimokuri

7.1. INTRODUCTION

Developments of a small-scale combustor, the so-called "microcombustor"[1-3] have been inspired by the rapid development of the micro electromechanical system, called "MEMS." Fabrications of the MEMS were attained in the early 1990s; however, MEMS has a serious problem, namely, the lack of a suitable scale power source. Conventional chemical batteries cannot follow the downscaling of the electromechanical systems, because the specific energy (W/kg) of those batteries is quite low, and hence, the downscaling of the batteries accompanied by mass reduction leads to a shortage of the power supply.

One solution for the issue is the microcombustor, which can convert the chemical energy of hydrocarbon fuels to thermal and kinetic energies, and then, electric power. Due to the high energy density of hydrocarbon fuels (almost 100 times higher than conventional chemical batteries), the scale of the power systems is expected to be drastically reduced compared to the conventional chemical batteries, even with a low efficiency energy conversion device. For example, even if we use a micro combustor with an energy converter of 10% efficiency, the scale of the power system can be reduced one-tenth the size of conventional chemical batteries. Therefore, various scales of small combustors are being developed, for example, micro gas turbine engines, rotary internal combustion engines, Swiss roll combustors, thermoelectric systems, and so on.[1,3-6]

Although microcombustors are being developed to be suitable for MEMS, moderate-scale combustors on the orders of centimeter or millimeter (called "mesoscale combustor") have also been developed recently, because various electromechanical systems such as the cell phones, laptop computers, and humanoid robots all have the problem of power source, that is, the limitation of the running time or the shortage of the power supply. The mesoscale combustor is also expected to be a moderate-scale but portable, powerful and long time running power source (called "personal power system") for those electronic devices.[2] However, the development of the microscale and mesoscale combustors is a challenging task, because the flame shows unique behavior in narrow channels, mainly due to the thermal and chemical quenching effects by the combustor wall, which become significant in narrow channels because of its large surface-to-volume ratio, and hence, a large heat loss to heat release ratio. In most cases, the unique behavior leads to undesired, unstable combustions, and therefore, a flame stabilization technique should be adopted on the small scale combustors. However, effective flame stabilization techniques are quite limited in small combustors, for example, one way to stabilize the flame is to form a stagnant or low velocity region in the flow path, and in large-scale combustors, it can be attained with a recess wall or bluff body; however, these are ineffective in small channels due to the quenching effect by itself. Another way to stabilize the flame is to make use of the turbulent flow, in which the unburned mixture consumption rate, and hence, the heat release rate can be increased mainly due to the increase in the flame surface area, however, in micro- and mesoscale combustors, such effects cannot be expected because the turbulence is hardly induced, which can be understood by the fact that the Reynolds number of the small scale combustor is typically less than 100 because the characteristic length is less than 10^{-2} (m).

One solution for this problem is to utilize the vortex flow. Figure 7.1 shows an example of tubular flame combustion in a 2.0-mm tube. If the premixture is tangentially injected into the small-diameter tube and ignited, a flame of tubular shape can be easily established. It was found that such stable combustion can be achieved not only in the premixed case but also in the non-premixed case,[7] that is, fuel and oxidizer are separately injected into the small-diameter tube.

This chapter will describe small applications that go into establishing the tubular flame combustion. First, fundamental facts about the flame-quenching phenomena in a nonrotating flow field will be introduced, and then, the advantage of tubular flame combustion in small-diameter tubes will be described together with the recent progress of power generators with tubular flame combustion.

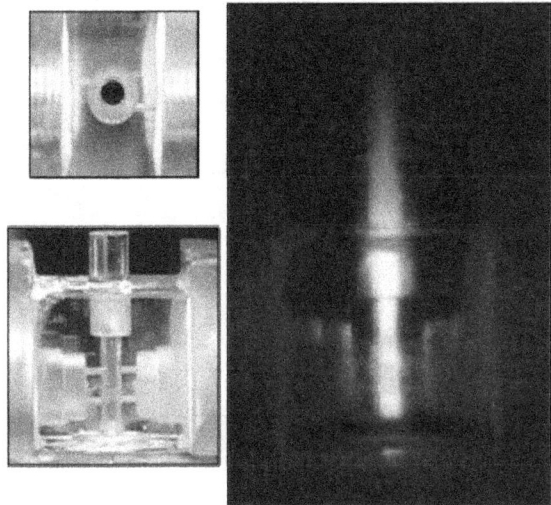

Figure 7.1. A tubular flame established in a 2 mm tube (propane with oxygen enriched oxidizer (30% O_2 + 70% N_2), Q_{oxi} = 3.0 L/min).

7.2. FLAME QUENCHING IN A NARROW CHANNEL

7.2.1 FLAME QUENCHING IN A NONROTATING FLOW FIELD

Before the initiation of the recent research on the microcombustor[1-7] and microcombustion[8-14], a scientific study on the flame-quenching phenomena in a narrow channel had been initiated by Sir Humphry Davy in 1814 from a fire-safety standpoint.[15] He found that a flame could not pass through a narrow channel and made use of this characteristic for the invention of the so-called "Davy lamp" which could prevent explosions in coal mines. Since Davy's pioneering work, extensive scientific research has been made on the flame-quenching phenomena in a narrow channel.[16-24] The critical channel size for flame propagation has been examined for various mixture systems and for various channel geometries under a variety of experimental methods. At the same time, the basic theories of flame quenching by the inert walls have also been developed.

Figure 7.2 shows an example of the behavior of a methane–air flame in a small-diameter tube ($D = 3.6$ mm). When the methane–air premixture is fed upward uniformly from the bottom of the tube and ignited, a conical shaped flame can be established when the mixture flow velocity is slightly higher than the burning velocity (i.e., rim-stabilized flame, Fig. 7.2(a)). With a decreasing flow rate, the flame propagates into the tube when the burning velocity exceeds the mixture velocity (Fig. 7.2(b)). This is a so-called "flame flashback." Figure 7.2 (c) shows the mapping of the combustion modes. The horizontal axis shows the mixture flow rate (L/min) and the mean flow velocity that can be obtained by dividing the mixture flow rate by the tube cross-sectional area. It can be seen that flame flashback occurs when the mixture velocity is decreased to less than the maximum burning velocity of the methane–air mixture (about 38 cm/s). With a further decrease in the flow rate, the flashback region is widened because the burning velocity exceeds the mixture velocity even under fuel-rich or -lean conditions.

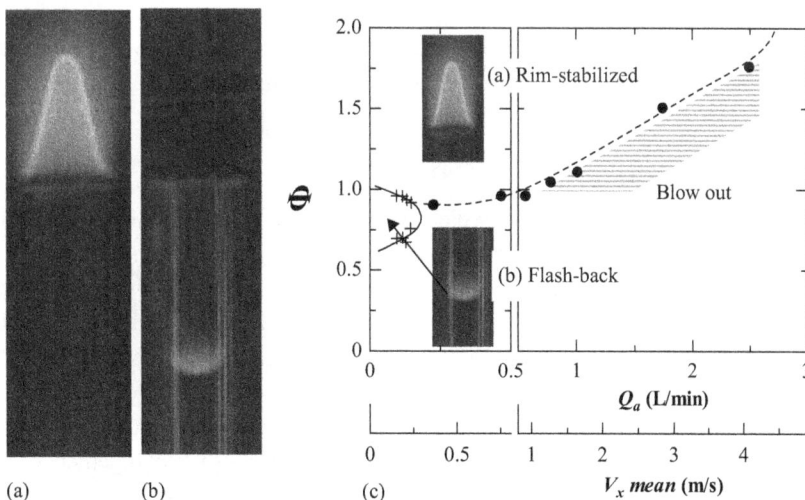

Figure 7.2. Various combustion modes in a small diameter tube (methane-air mixture, $D_g = 3.6$ mm).

It is known that if the tube diameter is decreased, the flame flashback region becomes narrow,[17] and according to Lewis and Elbe's study[17], a flame cannot enter into the tube when the tube diameter is less than 3.2 mm for a methane–air mixture even if the mixture velocity is enough lower than the burning velocity. This minimum tube diameter for the flame propagation is called the "quenching *diameter*." This critical channel size varies with its geometry. If the mixture is surrounded by two parallel plates, the critical plate separation for flame propagation is called the "quenching *distance*." A series of quenching distance values and diameters can be found in literature,[16,17] in which one can find that the quenching distance of $d_\parallel$ (2.2 mm) for a methane–air mixture is different from the quenching diameter d_0 (3.2 mm). The quenching distance is different from the quenching diameter not only for methane–air mixture but also for any other mixture. The difference between d_0 and $d_\parallel$ comes from the difference in the flame surface area that contacts the inert wall. Simon, Belles, and Spakowski[20] revealed the following relationship between the quenching distance and quenching diameter:

$$d_\parallel = A \cdot d_0 \tag{7.1}$$

where A is the geometrical factor 0.66. The value of the factor A has been obtained for various geometries of channel, such as trianglular cross section, rectangular cross section, and so on. They can be found in references.[16,21]

Based on the experimental results of the quenching distance (or diameter), numerous theoretical attempts have been made to predict the critical channel size.[16–24] Unfortunately, there was no theory that could predict the critical channel size precisely because the reaction rates or the physical properties of the unburned and/or burned gas could not be estimated precisely at that time, and the classic theories were developed based on those properties. However, the theories can help us understand the flame-quenching phenomenon in a small-diameter channel. The model proposed by Friedman[22] is the simplest and gives us fundamental insight about the flame quenching. He postulated a model in which a flame propagates through the gap, and assumed that the thermal quenching effect dominates the flame-quenching phenomena. When the gap distance attains the critical value, the heat release from the flame front should be balanced with the heat loss to the wall. As a result, he introduced the so-called quenching equation:[16,22]

$$d_\parallel = Pe \cdot \alpha / S_u \tag{7.2}$$

where Pe is a quenching Peclet number which varies with the mixture system and α is the thermal diffusivity of the gas and S_u is the burning velocity. The point of this equation is that the quenching distance is proportional to the thermal diffusivity, but inversely proportional to the burning velocity. It means that, if the thermal diffusivity of the mixture is large, the heat loss to the tube wall and hence, the quenching distance will increase. On the other hand, if S_u is large, the heat release rate will increase and the quenching distance will decrease. The experimentally obtained quenching distance follows exactly the same trends as found in the theory. In Table 7.1, the values of the quenching distance for various mixtures are shown, together with the values of α and S_u of those mixtures. For methane–air and propane–air mixtures, the quenching distance is almost the same because the values of α and S_u are almost the same. For propane–argon–oxygen mixture, the quenching diameter is smaller than that for propane–air mixture due to its high burning velocity. On the other hand, for propane–helium–oxygen mixture, the

Table 7.1. Quenching distance, burning velocity, and thermal diffusivity of several mixtures

Mixture	$d_{\parallel}$ (mm)	S_u (m/s)	α (cm²/s)
Methane–air	2.35	0.37	0.2133
Propane–air	2.07	0.42	0.1929
Propane/ (79% argon/21% oxygen)	1.04	0.94	0.1894
Propane/ (79% helium/21% oxygen)	2.53	1.35	0.9508

quenching diameter is much larger than that for the propane–air mixture although the burning velocity is high. This is because the thermal diffusivity is large.

The quenching equation also indicates that the flame quenching depends on the quenching Peclet number. An example of the quenching Peclet number dependence is the quenching distance in turbulent flow. For turbulent combustion, the heat release rate as well as the heat loss rate are greatly affected by the change in the heat and mass transport properties. Ballal and Lefebvre have found that the quenching distance can be changed if the flow is turbulent,[25] and the quenching distance in turbulent flow can be estimated by the classical quenching equation with the turbulent Peclet number correction:

$$Pe_{\text{turblent}} = Pe_{\text{laminar}} + 13 \, (Pr)^{-0.5} (u'/S_L)^{0.5} \tag{7.3}$$

where, Pe_{laminar} is the quenching Peclet number under the laminar condition, which can change with the mixture system (fuel, oxidizer, and their fraction), Pr is the Prandtl number, S_L is the burning velocity, and u' is the root mean square of the velocity fluctuation. Accordingly, the critical channel size for flame propagation can be greatly affected by the flow field, as well as the channel geometry and thermal–transport properties of the burned–unburned mixture.

7.2.2 ADVANTAGES USING SMALL-SCALE TUBULAR FLAME BURNERS

It is reasonable to think that the conventional quenching diameter, and thus, the related classical flame-quenching theories cannot be applied to the tubular flame due to its unique structure. The following are the advantages of the tubular flame combustion in a small-diameter tube.

1. Thermal advantage of the tubular flame combustion in small-scale tubes:
    Due to the axisymmetrical tubular flame shape, the temperature profile is also axisymmetrical and hence, the heat loss behind the flame is negligible. Although this fact has already been explained in previous chapters, the tubular flame has another important thermal advantage. Because the tubular flame front is covered by the unburned mixture, the tubular flame does not touch the burner wall except for the very small downstream edge part of the flame. It indicates that the heat loss from the flame front to the burner wall can be drastically reduced.

2. Aerodynamic advantage of the tubular flame combustion in small-scale tubes:

A tubular flame is aerodynamically stable according to the Rayleigh stability criterion as already explained in previous chapters. This indicates that the burned gas and unburned gas are stratified by the differences in the centrifugal forces acting upon them, and thus, a smooth tubular flame front can be fixed at the interface between the unburned and burned gas even under considerably high load conditions. The stratification effect becomes significant in small-diameter tubes because the centrifugal acceleration a is inversely proportional to the tube diameter as,

$$a = \frac{V_\theta^2}{r} \tag{7.4}$$

where V_θ is the circumferential velocity and r is the radius. Thus, the flame shape can be smoother and flame stability is increased in smaller-diameter tubes due to the strong effects of stratification.

Furthermore, the unique flow field of the tubular flame burner is by itself favorable for stable combustion in a small-diameter tube. For usual flame propagation in a nonrotating flow field, that is, in the case of the combustible mixture being axially ejected from one end of the small-diameter tube, the flame can be stabilized inside the tube only when the mixture velocity component normal to the flame front (the axial velocity for this case) is balanced with the burning velocity. Thus, the mixture velocity, and hence, the mixture flow rate should be precisely controlled to be equivalent to the burning velocity to fix the flame front position. And furthermore, because the axial flow velocity increases with decreasing tube diameters under the fixed flow rate condition, the mixture flow rate should be reduced. Therefore, the heat output of usual small-scale combustors is quite limited.

On the other hand, for tubular flame combustion, the flow field and the flame stabilization mechanisms are completely different. As the flame propagates radially outward to the unburned gas mixture, the flame is established at the position where the radial velocity V_r is balanced with the burning velocity. The profile of V_r mainly dominates the tubular flame stability.

Figure 7.3 is an example of the nonreacting (air) flow fields of a tubular flame burner obtained by the particle image velocimetry (PIV) system. The schematic and appearance of the burner used for PIV measurements are shown in Fig. 7.3(a). The burner is 30-mm in inner diameter and 150 mm in length with two tangential slits that are 2 mm wide and 70 mm long. The burner is made of quartz to be optically accessible. The swirl number of the burner is 5.7. The flow fields were obtained for the nonreacting flow, under a fixed air flow rate Q_a = 200 l/min; however, if a combustible methane–air mixture (Q_a = 200 l/min and Φ = 0.8) is ejected from the tangential slits instead of air and ignited, a tubular flame can be established in the burner as shown in the pictures on the right of Fig. 7.3(a).

Figure 7.3(b) shows the two-dimensional velocity distributions of $(r - \theta)$ plane obtained at the x = 0 mm cross section, and Fig. 7.3(c) shows the velocity distributions of $(r–x)$ plane, only for the range of x = 0–40 mm and r = 0–15 mm. As shown in Fig. 7.3(a), x and r are the axial and the radial distances from the center of the burner. In Fig. 7.3(b), it can be seen that an axisymmetrical flow field is established in the burner,

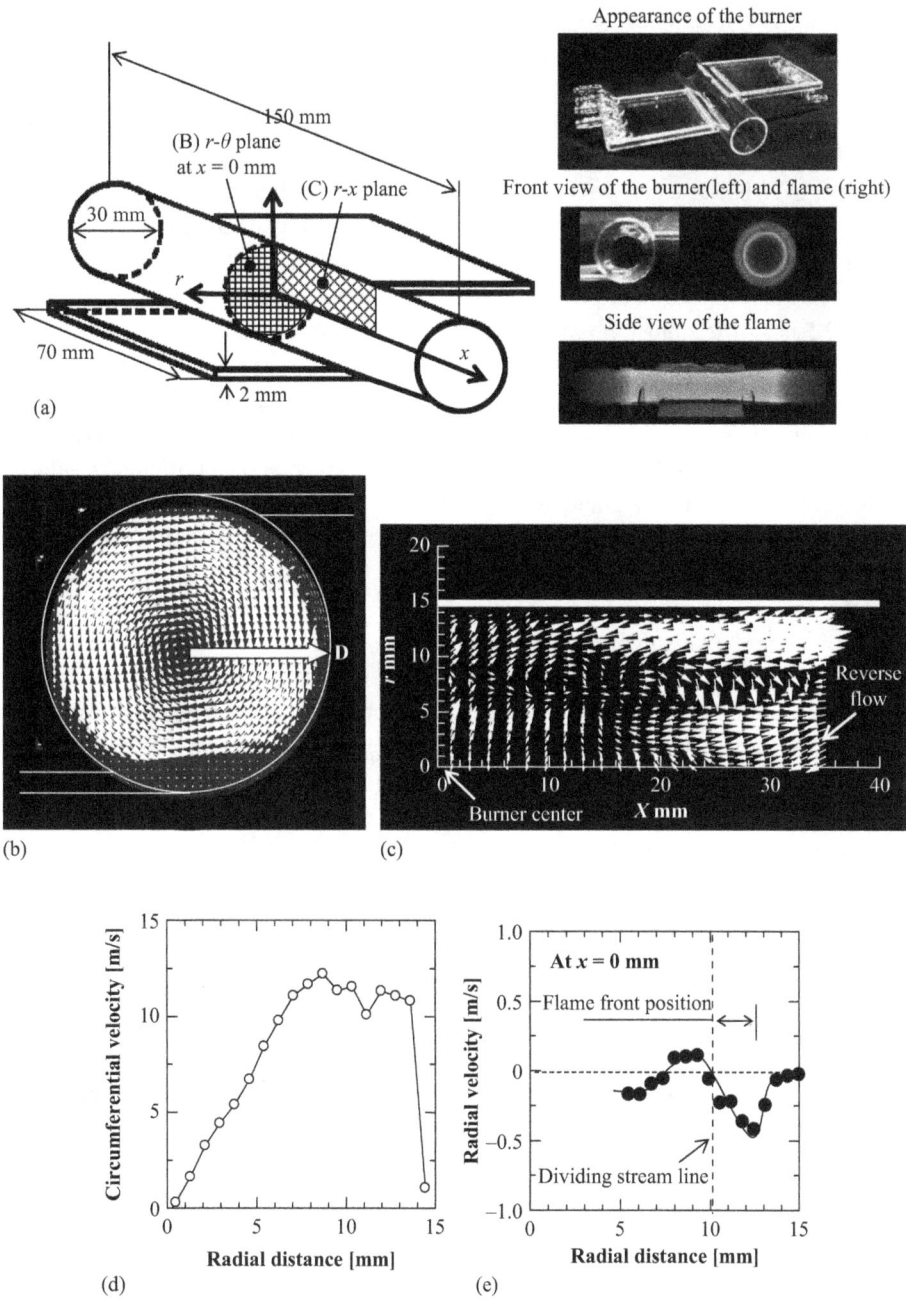

Figure 7.3. Flow field in a tubular flame burner (cold flow, $Q_a = 200$ L/min).
(a) Schematics of the burner.
(b)(c) Two dimensional vector profile obtained at r-θ and r-x plane shown in (a).
(d) Radial distribution of V_θ along line "D" in Fig. 7.3(b).
(e) Radial distribution of Vr along r axis ($x = 0$ mm) in Fig. 7.3(c).

and in Fig. 7.3(c), it can be seen that a reverse flow region is formed around the burner axis at ($r = 0$–5 mm) due to the strong swirl effect. The radial distributions of the V_θ and V_r obtained from these results are shown in Fig. 7.3(d) and (e). Note that, in Fig. 7.3(e), the positive value of V_r indicates that the gas flows radially outward (toward the burner wall) and the negative value indicates that the gas flows radially inward (toward the center axis) because the r axis is in the direction from the burner center axis to the burner wall. As the flame propagates radially outward to the unburned gas mixture, the flame can be stabilized in the region of the inward gas flow where V_r takes a negative value.

In Fig. 7.3(d), it can be seen that the maximum circumferential velocity attained is 12.5 m/s. It is almost the same value as the mean tangential ejection velocity $V_{\theta\,mean}$ 11.9 m/s, which can be obtained by dividing the air flow rate (200 L/min) by the cross-sectional area of the tangential slits (2 mm × 70 mm × 2 slits). On the other hand, in Fig. 7.3(e), it can be seen that the maximum radial velocity of the unburned mixture coming onto the flame front (the minimum value of the radial component of the gas velocity) is 0.5 m/s, which is considerably lower than V_θ by two orders of magnitude. And furthermore, as shown in Fig. 7.3(c), the axial velocity V_x is also considerably lower than V_θ around the axially center position of the burner. V_x is less than 1.0 m/s around the $x = 0$ cross section although the mean axial velocity $V_{x\,mean}$ is 2.4 m/s, which can be obtained by dividing the air flow rate (200 l/min) by the cross-sectional area of the burner exit (2 exits × $\pi/4$ × (15 mm)^2). The initial velocity components in the radial and axial directions are almost zero because the mixture is ejected only in the tangential direction, and hence, V_r and V_x are considerably lower than V_θ near the ejection slits and therefore, a tubular flame can be stabilized in a wide range of mixture flow rates and equivalence ratios even in small-diameter tubes. This is one of the unique aerodynamic characteristics of tubular flame combustion.

Additionally, in Fig. 7.3(e), the radial distributions of V_r indicate another important fact that V_r is not uniform in the radial direction, that is, V_r varies from zero to 0.5 m/s in that $r = 10$ to 12.5 mm. Because of the velocity gradient being in the radial direction, the flame front moves automatically to the point where $V_r = S_u$. Namely, a flame moves radially outward (i.e., flame diameter increases) with increases in the burning velocity, whereas a flame moves radially inward (i.e., a flame diameter decreases) with a decrease in the burning velocity. This means that a precise flow rate control is not needed to balance the V_r with S_u for tubular flame combustion.

7.2.3 TUBULAR FLAME IN A SMALL-DIAMETER TUBE

A tubular flame can be easily established in a small-diameter tube with very simple burners. An example of a small-scale burner for tubular flame combustion is shown in Fig. 7.4. The burn chamber has a 3.0-mm inner diameter. The bottom end of the burner is closed, and a quartz tube of a 3.6 mm inner diameter is installed on the top. Both the burner and combustion tube are made of quartz. The combustible mixture is tangentially injected through four injection holes which have a 0.5 mm inner diameter.

Figure 7.5 shows the appearance of methane/air flames with the burner. When the supplying mixture's flow rate is low enough $Q_{mix} = 0.17$, a flame flashback occurs just as in the

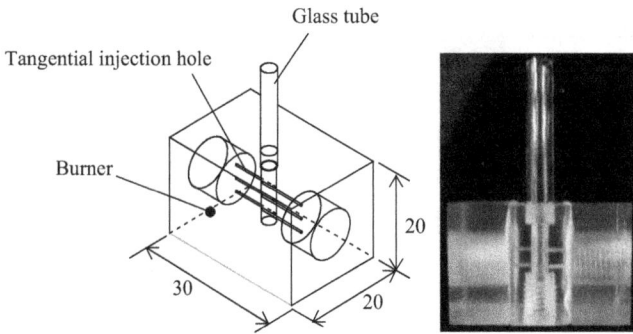

Figure 7.4. Small scale swirl combustor (tube diameter Dg is 3.6 mm, burner diameter Dc is 3.0 mm).

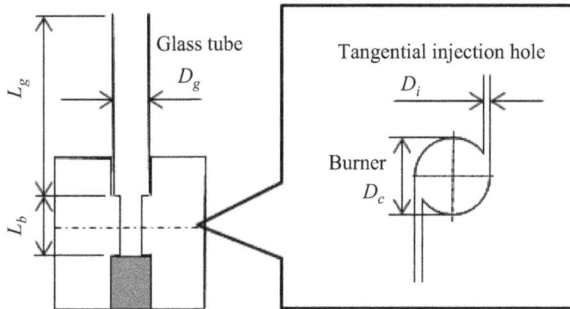

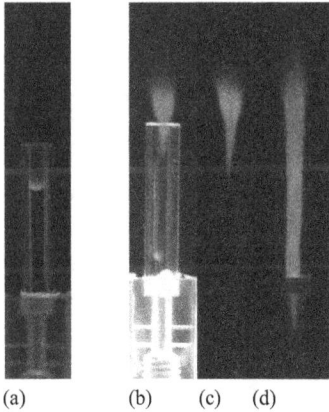

Figure 7.5. Appearance of flames in the rotating flow. (a): flashback, (b), (c): external flame, (d): establishment of the tubular flame.

nonrotating flow (see Fig. 7.5(a)) because the $V_{x \, mean}$ is enough lower (0.27 m/s) than the burning velocity although $V_{t \, mean}$ is large 3.5 m/s, where $V_{x \, mean}$ and $V_{t \, mean}$ are the mean axial and tangential velocity that can be obtained by dividing the mixture flow rate by the cross sectional area of the burner and the tangential inlets, respectively. At relatively large mixture flow rates ($Q_{mix} = 1.0$ L/min, $V_{x \, mean} = 1.6$ m/s), a flame is stabilized outside the combustion tube; however, its shape is inverted, which indicates the flame is stabilized by swirl (see Fig. 7.5(b) $\Phi = 1.0$). When the mixture flow rate is increased to 2.8 l/min ($V_{x \, mean} = 4.5$ m/s) the flame completely blows out in the nonrotating flow case, however, in the case of a rotating flow, an inverted flame is still stabilized and the flame is partially immersed into the tube for $\Phi = 0.8$ (see Fig. 7.5(c)). By varying the equivalence ratio toward the stoichiometric, the inverted flame is made to move

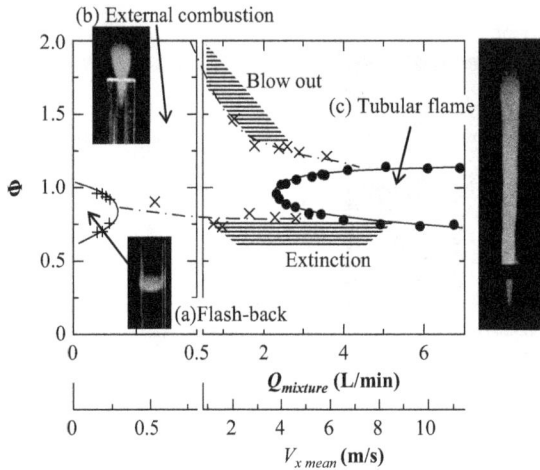

Figure 7.6. Various combustion modes in 3.6 mm tube with vortex flow (methane/air mixture, $Dg = 3.6$ mm)

further into the tube. When the flame is suddenly projected into the tube eventually to the closed end, the tubular flame can be established ($\Phi = 0.84$, Fig. 7.5(d)). The same flame propagation occurs at $\Phi = 1.05$ if the fuel flow rate is gradually decreased from the very fuel-rich condition.

Figure 7.6 shows the mapping of a stable combustion region plotted in the $Q_a - \Phi$ plane. A tubular flame can be established between two (fuel lean/rich) propagation limits. For example, under the fixed air flow rate condition of $Q_a = 2.8$ L/min (the same condition as Fig. 7.5 (c), (d)), if the fuel flow rate is increased from zero, flame propagation occurs when the equivalence ratio attains $\Phi = 0.84$. On the other hand, if the fuel flow rate is decreased from a very fuel-rich condition, the same flame propagation occurs when the equivalence ratio attains $\Phi = 1.05$. A tubular flame can be established between these two propagation limits, in the range of $\Phi = 0.84 - 1.05$.

In Fig. 7.6, it can be seen that, even under the condition of $Q_a = 7.0$ l/min, a tubular flame can still be established under a wide range of equivalence ratios $\Phi = 0.75$ to 1.2, even though $V_{x\,mean}$ is over 12 m/s, which is 30 times larger than the maximum burning velocity of a methane–air mixture of 0.38 m/s. On the other hand, if the air flow rate is decreased to less than 2.0 l/min, a tubular flame cannot be established because the tangential injection velocity decreases with decreasing the mixture flow rate, and thus, the vortex flow effect becomes weaker with a decreasing air flow rate.

As a tubular flame can be formed in such high flow-rate conditions, the heat output of the tubular flame can be much larger compared with other combustors. The heat output of combustion Q (W) can be obtained by,

$$Q = \Delta H \cdot \rho_f \cdot Q_f \tag{7.5}$$

where ΔH (J/kg) is the heat of combustion of a fuel, A_f (m^2) is the flame surface area, and Q_f (m^3/s) is the fuel flow rate consumed by the flame front. For a nonrotating flow, the heat output of the methane–air combustion in a 3.6 mm tube is calculated to be 15.2 W at most because the maximum flow rate for stable combustion in the tube is 0.24 l/min. On the other hand, for a 3.6 mm tubular flame burner, the heat output for $Q_a = 7.0$ l/min is 443.8 W, which is 30 times larger than the nonrotating flow combustors.

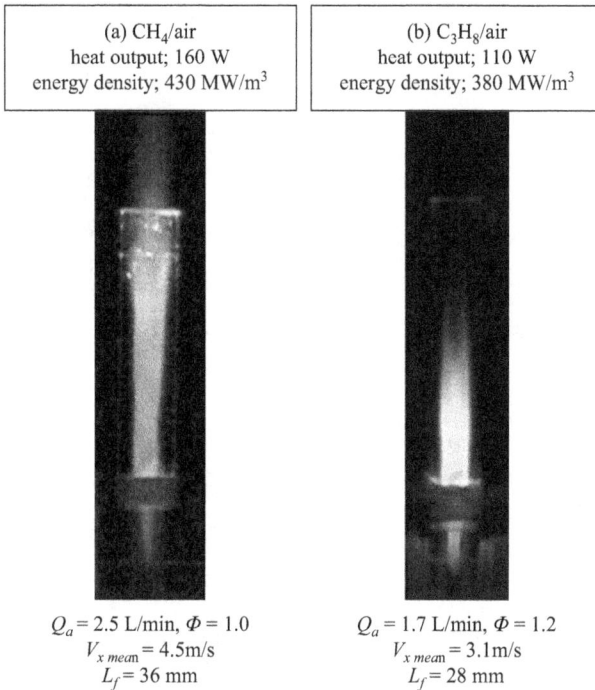

| (a) CH₄/air |
| heat output; 160 W |
| energy density; 430 MW/m³ |

(a) CH_4/air
heat output; 160 W
energy density; 430 MW/m³

(b) C_3H_8/air
heat output; 110 W
energy density; 380 MW/m³

Figure 7.7. High energy density tubular flame combustion.

Q_a = 2.5 L/min, Φ = 1.0
$V_{x\,mean}$ = 4.5m/s
L_f = 36 mm

Q_a = 1.7 L/min, Φ = 1.2
$V_{x\,mean}$ = 3.1m/s
L_f = 28 mm

Thus, energy density (W/m³) of the combustor which can be obtained by dividing the heat output of the combustor (W) by the combustor volume (m³) can also be much larger than other combustors. Figure 7.7(a) shows the appearance of a flame of methane–air mixture in a 3.6-mm tube. In this case, the tube diameter is about 3.6 mm, the flame length is 36 mm, and the methane flow rate is 0.26 l/min. Thus, the heat output is 160 W and the energy density is 430 MW/m³. For propane - air mixture, 380 MW/m³ combustion is shown in Fig.7.7(b). With oxygen-enriched combustion, the flame becomes shorter due to its high burning rate, and hence, the energy density can be increased. For example, with a mixture of propane- and oxygen-enriched oxidizer (30% O_2 + 70% N_2), a heat output 220 W and an energy density 3.5 GW/m³ can be obtained for Φ = 1.2 (the appearance is as in Fig. 7.1). Comparing the values with other combustors[26,27] (Table 7.2), a higher heat output as well as increased energy density can be obtained by the use of tubular flames.

7.2.4 EFFECTS OF TUBE SIZE ON THE TUBULAR FLAME

How do variations in tube diameter affect the tubular flame? The effects of the tube diameter on the tubular flame combustion is important from fundamental as well as practical aspects. Figure 7.8 shows variations in the stable combustion region of tubular flames with various tube diameters, 3.6, 6.0, and 10.0 mm, respectively. The stable combustion regions of methane–air tubular flame are plotted in $\Phi - V_{x\,mean}$ plane. The swirl number for each case is also shown in the figure. For large diameter tubes with D_g = 10 mm, a flame can propagate through the tube, and then, a tubular flame can be formed in a wide range of $\Phi - V_{x\,mean}$. On the other hand, for D_g = 6.0 mm, the stable combustion range is separated into two regions, a low-velocity region

Table 7.2. Comparisons of energy density and heat output of the small scale combustors

Combustor	Energy Density (GW/m³)	Heat Output (W)
Small gas turbine[26]	▼ 0.07	1000
Small catalyst combustor[27]	5.50	▼ 7
Tubular flame burner (D_g = 3.6 mm, methane–air)	0.43	160
Tubular flame burner (Fig.7.1, D_g = 2.0 mm, O_2 enriched; 30% O_2 + 70% N_2)	3.50	220

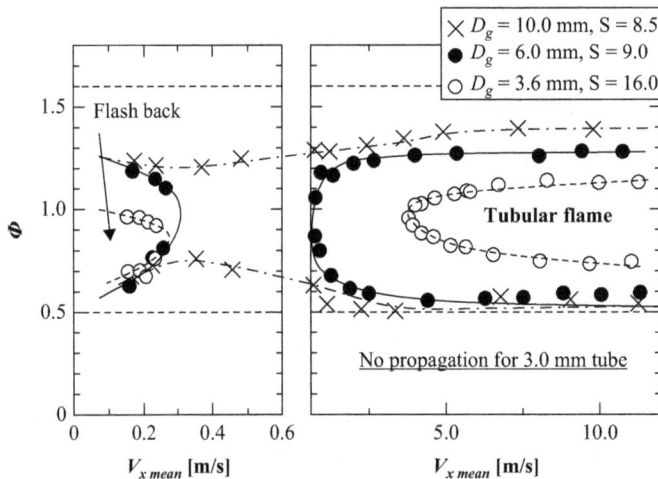

Figure 7.8. The effect of the tube diameter on the stable combustion region.

and a high-velocity region. In the low-velocity region, a thin, flat flame front propagates into the tube as shown in Fig. 7.5(a), whereas in the high-velocity region, a slender flame propagates into the tube and a tubular flame is established as shown in Fig. 7.5(d). The low-velocity region is the so-called flashback region, and the high-velocity region is the tubular flame combustion region. For D_0 = 6.0 mm, a stable tubular flame can be established for $V_{x\,mean}$ = 10.0 m/s. However, the range in Φ is 0.6 to 1.3, which is narrower than for D_0 = 10.0 mm. With a further decrease in tube diameter D_0 = 3.6 mm, the region in Φ is further decreased, and the stable tubular flame combustion range in Φ at $V_{x\,mean}$ = 10.0 m/s is Φ = 0.75 to 1.2. Stable combustion conditions were also examined with a 3.0-mm tube; however, no flames could be established. Because the swirl number is greater for smaller combustors (S = 8.5 for 10.0 mm, S = 16.0 for 3.6 mm), a decrease in the range is attributed to the tube wall effects. How does the tube wall affect the stability of the tubular flame combustion?

First, the effects of tube size can be attributed to the flow field variations. Figure 7.9 shows the comparisons of the circumferential velocity of two combustors, one small (D_g = 3.6 mm) and the other large (D_g = 10.0 mm).[13,14] Combustion tube length, injection hole diameter, and injection hole numbers of those combustors are designed to establish a similar vortex flow inside the tube. Details of the combustor configurations can be found in references.[13,14] V_θ was

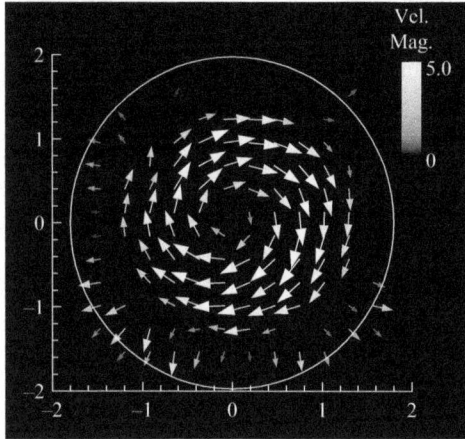

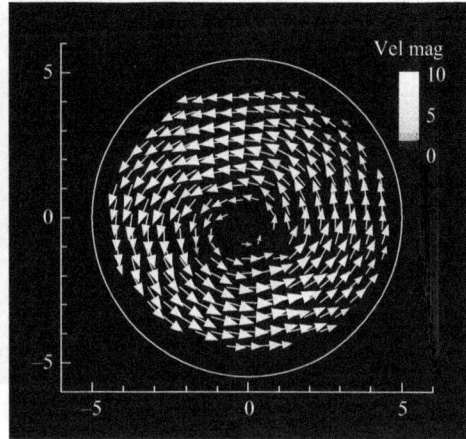

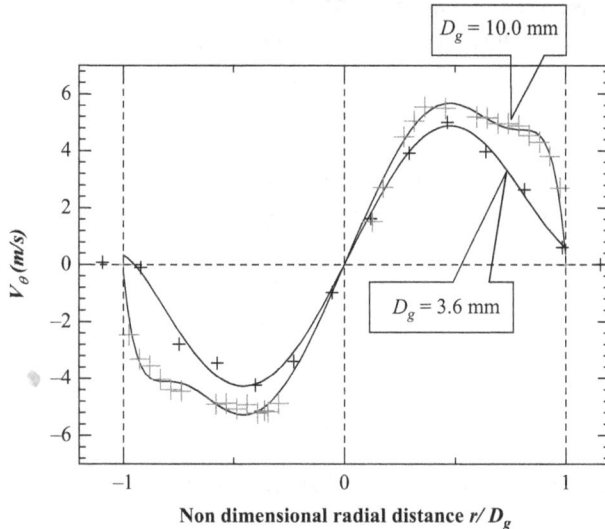

Figure 7.9. Radial distributions of circumferential velocity. Vector profile in 3.6 mm(a) and 10.0 mm(b) tube and comparisons of the circumferential velocity distributions.

obtained at the same nondimensional axial position $x/L = 0.8$, where x is the axial distance from the closed end of the combustor, and L is the tube length. In both combustors, almost the same maximum tangential velocity ($V_{\theta max}$ = 5 m/s) can be obtained at the almost same normalized radial position ($r/R = 4.0$, where R is the tube radius). This indicates that a similar vortex flow is formed in those tubes; however near the tube wall, circumferential velocity drastically decays in 3.6-mm tubes. This is due to the boundary layer effect, that is, in small-diameter tubes, the boundary layer thickness is comparable to the tube diameter, and thus, its effect becomes more evident than in large-diameter tubes. This indicates that, a strong swirl is needed in small-diameter tubes.

Second, in small-diameter tubes, even though the tubular flame front is established away from the tube wall, the thermal and chemical quenching effects by the tube wall can be more

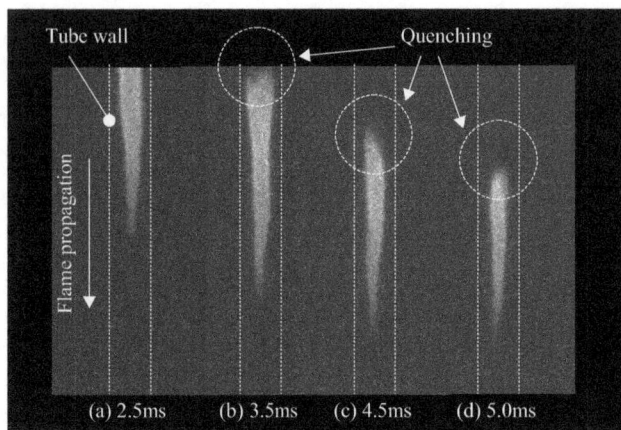

Figure 7.10. Sequence images of the initial step of flame propagation near the flame propagation limit (propane / air mixture, $Q_a = 3.0$L/min, $\Phi = 1.05$). Time after ignition is shown below each figure.

evident than in the case of large-diameter tubes. The quenching mechanism can be understood by looking at Fig. 7.10, which is a sequence of pictures obtained at the exit of a 3.0-mm tube, in which no tubular flame can be established. Immediately after ignition, a flame kernel is formed at the exit of the tube and then the flame propagates along the tube axis by the so-called vortex bursting mechanism (Fig. 7.10(a)). However, in Fig. 7.10, it can be seen that the flame quenching occurs near the tube exit after ignition (Fig. 7.10(b)–(d)), and eventually the whole flame is extinguished. This indicates that the flame propagation cannot be sustained due to the flame quenching immediately after the ignition, and thus, a stable tubular flame cannot be established. Due to this fact, although the flame stabilization mechanism is quite different from that of the nonrotating flow field, critical tube diameter in the rotating flow field exists in the 3.0–3.6 mm range for the methane–air mixture, which is almost the same as the quenching diameter of the nonrotating flow field ($d_0 = 3.2$ mm).

7.2.5 CRITICAL TUBE DIAMETER FOR A ROTATING FLOW FIELD

If one can eliminate the quenching effect immediately after ignition, a flame can be stabilized in further smaller-diameter tubes. Figure 7.11 shows a small-scale swirl burner which was designed to eliminate these effects. The tube diameter varies as D_c, D_0, and D_g from the bottom of the burner. D_c and D_0 are smaller than the conventional quenching diameter d_0 whereas D_g is larger than d_0. At first, a flame kernel is formed at the exit of the tube immediately after the ignition and then, a flame propagates through the channel of D_g. Because D_g is larger enough than the quenching diameter, the quenching effects by the tube wall can be eliminated, thus a flame tip can propagate into the smaller diameter D_0. Figure 7.12 shows the stable combustion ranges. The stability limits are plotted in the D_0 and Φ plane. Without the quenching effect in the initial step, a flame can be stabilized in a 2.4-mm channel for a methane–air mixture, and in a 2.0-mm channel for a propane–air mixture. The minimum channel diameter is about 1.0 mm smaller than the conventional quenching diameter for a methane–air mixture, and is about 1.5 mm smaller than the quenching diameter for a propane–air mixture, suggesting that the minimum tube diameter for tubular flame combustion can be much less than the conventional quenching diameters.

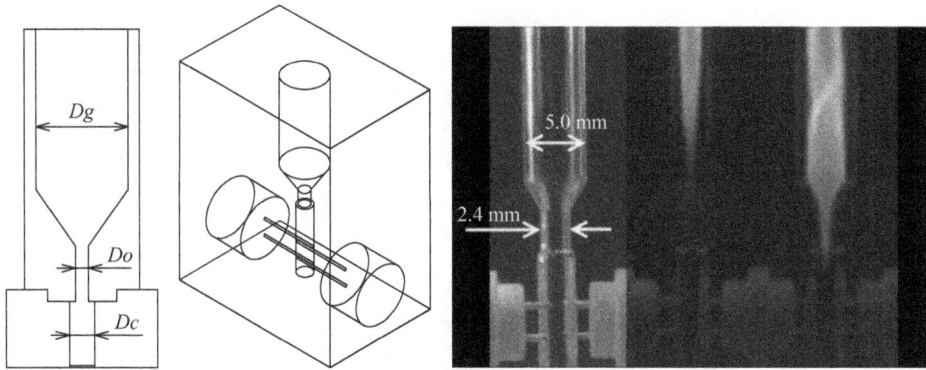

Figure 7.11. Schematics of the swirl burner to examine the quenching diameter in a rotating flow field.

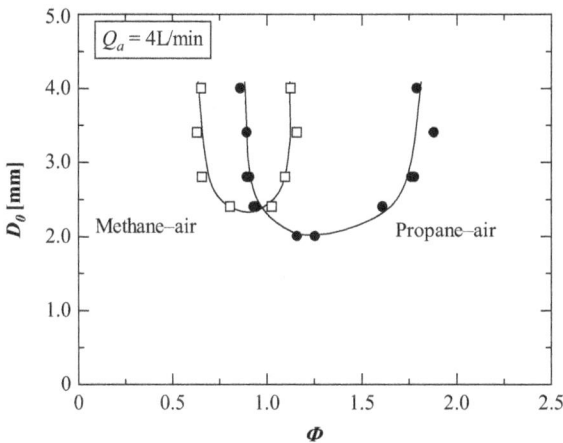

Figure 7.12. Stable combustion rage in D_0–Φ plane for methane–air and propane–air mixture.

7.3 DEVELOPMENT OF SMALL POWER SOURCES USING A TUBULAR FLAME

Very recently, these advantages of the small scale tubular flame combustion are being applied to the practical usages. In the development of micro combustor, two critical issues should be considered not only for the tubular flame but also other micro combustors, those are (1) the mobility of the system and (2) an energy conversion device (or system). On the first point, the scale of the oxidizer and fuel source should be considered, namely, even if the combustor is small enough, the system cannot be truly portable if the source of the fuel and oxidizer are too large. For the tubular flame burner, this issue has already been solved by using liquid fuels[28–30] as explained in previous chapter, or using liquefied gaseous fuels. Figure 7.13 shows an autonomous tubular flame burner with a fuel tank and a tubular flame established in the 10 mm burner.[31] The fuel tank is a conventional butane cartridge, which can be found in any DIY shop. The oxidizer is the surrounding air which is entrained into the burner by the fuel jet. The fuel tank is not so small but still portable, and with this system, 100 W heat output can be obtained over 24 hours.

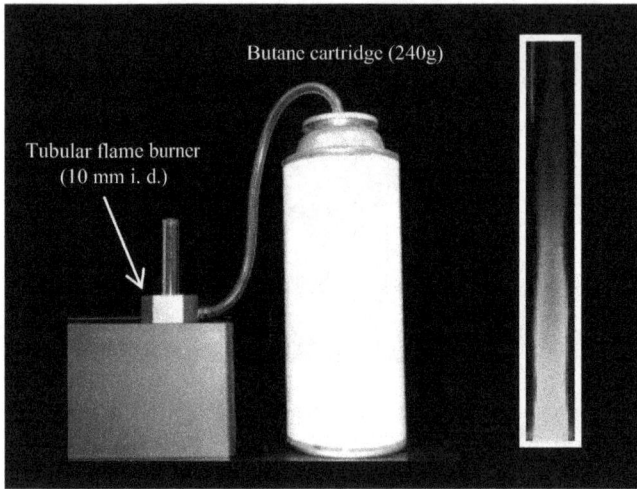

Figure 7.13. A portable tubular flame burner with butane cartridge and an appearance of the flame.

As for the second critical point, the energy conversion device, development of small-scale power sources using tubular flames is currently in progress and some challenging work has been done by Li et al.[32] They are attempting to adopt a thermo photovoltaic (TPV) power generation system, which directly converts radiation into electrical power. They have found that radiation for the TPV system can be increased by using a quartz tube with metal-oxide deposits as an emitter. They have also found that by using a "*reverse tube*," the illumination of the emitter can be made uniform; furthermore, CO and NO_x emissions can be reduced.

A promising energy conversion device is the thermoelectric device although its efficiency is *currently* quite low (less than 5%).[27] One of them, the Peltier device, is light and thin and its *output is presently* several times higher than conventional chemical batteries. However, the device cannot endure high temperatures. The maximum temperature tolerance is typically less than 300°C. Because the device cannot withstand the high temperatures of the burned gas, a heat medium should be used. At present, development of small-scale power sources with a tubular flame using the Peltier device is in progress. A prototype is shown in Fig. 7.14. The system has a heat medium into which the combustor is directly incorporated. The device is 60 mm × 60 mm × 60 mm in size and is made of 410 g of duralumin, which consists of part A, part B and two cover plates (Fig. 7.14(a)). Figure 7.14(b) shows the top side and the bottom side of the part A and B. As shown in the figure, narrow channels, 5.0 mm deep and 5.0 mm wide, are fabricated on the surface of the medium. In Fig.7.14(c), a cross sectional image of the channel is shown. A fine hole of 0.5 mm is drilled at the gas inlet position, and this part is the "embedded" tubular flame burner. The cross section of the channel is not round; however, if a combustible mixture is injected into the channel through the 0.5-mm hole and ignited, a tubular flame is established at the gas inlet position. Figure 7.14(d) shows the appearance of a flame in which the flame is visualized by covering the medium with a quartz plate. As shown in the figure, a stable tubular flame can be established at the gas inlet position, and the hot burned gas flows around the heat medium through the channels, of which details are shown in Fig.7.14(b), from gas inlet, channel 1 to 18, and to outlet.

Figure 7.15 shows the time history of the surface temperature of the medium. Because the heat output of the tubular flame is large (see Table 7.2), the surface of the medium heats up

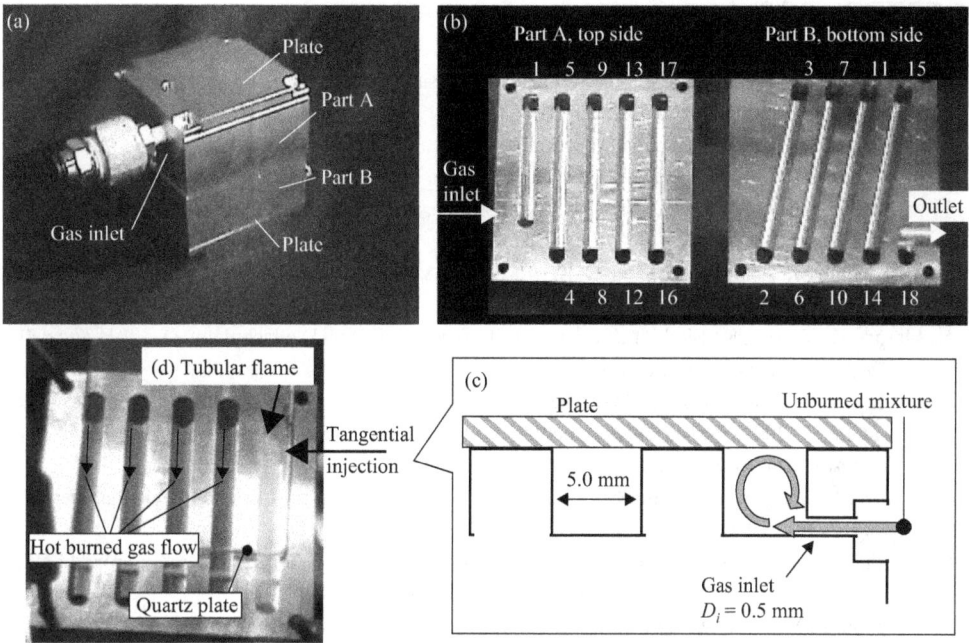

Figure 7.14. Heat media for power generation using a Peltier device with a tubular flame burner embedded inside.

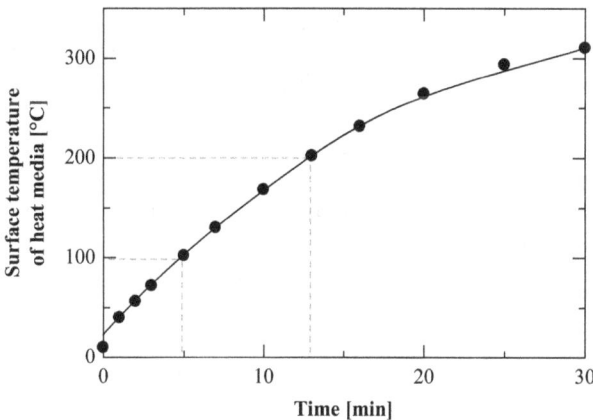

Figure 7.15. Time history of the media surface temperature.

quickly. The surface temperature reaches 100°C within 5 min and 200°C in 15 min. Ideally, electric power of 12 W (3.0A by 4.0V) can be obtained with a commercial Peltier device (60 mm × 60 mm, 200°C on the hot side); hence, about 60 W can be obtained in 15 min using five devices attached to the available sides. The quantity of fuel used in this test (propane) was 1.9 liters for a 15-min burn time, which is 3.4 g of mass. At present, the device weighs 413.4 g (410 g + 3.4 g) and takes 15 min to start up. With improvements in operation time and the system scale in terms of both volume and mass, it can become a very powerful mobile power source.

REFERENCES

1. Fernandez-Pello, C. A. 2002. Micropower generation using combustion: Issues and approaches. *Proceedings of the Combustion Institute* 29:883–99.
2. Dunn-Rankin, D., E. M. Leal, and D. C. Walther. 2005. Personal power systems. *Progress in Energy and Combustion Science* 31:422–65.
3. Maruta, K. 2011. Micro and mesoscale combustion. *Proceedings of the Combustion Institute* 33: 125–50.
4. Spadaccini, M., Zhang, X., Cadou, C. P., Miki, N., I. A. Waitz. 2002. *MEMS 2002 IEEE International Conference*, 20–25.
5. Maruta, K., Takeda, K., Sitzki, L., Ronney, P. D., Wussow, S., and Deutschmann, O. 2001. *Third Asia-Pacific Conference on Combustion*, 24–27.
6. Fu, K., A. J. Knobloch, F. C. Martinez, D. C. Walther, A. C. Fernandez-Pello, A. P. Pisano, and D. Liepmann. 2001. *Proceedings of the 2001 International Mechanical Engineering Congress and Exposition (IMECE)*, 875–880.
7. Wu, M., Y. Wang, V. Yang, and R. A. Yetter. 2007. Combustion in meso-scale vortex chambers. *Proceedings of the Combustion Institute* 31:3235–42.
8. Akram, M. and S. Kumar. 2011. Experimental studies on dynamics of methane–air premixed flame in meso-scale diverging channels. *Combustion and Flame* 158:915–24.
9. Kurdyumov, V. 2011. Lewis number effect on the propagation of premixed flames in narrow adiabatic channels: Symmetric and non-symmetric flames and their linear stability analysis. *Combustion and Flame* 158:1307–17.
10. Richecoeur, F. and C. D. Kyritsis. 2005. Experimental study of flame stabilization in low Reynolds and Dean number flows in curved mesoscale ducts. *Proceedings of the Combustion Institute* 30:2419–27.
11. Maruta, K., T. Ktaoka, N. Kim, S. Minaev, and R. Fursenko. 2005. Characteristics of combustion in a narrow channel with a temperature gradient. *Proceedings of the Combustion Institute* 30:2429–36.
12. Xu, B. and Y. Ju. 2007. Experimental study of spinning combustion in a mesoscale divergent channel. *Proceedings of the Combustion Institute* 31:3285–92.
13. Shimokuri, D., Y. Honda, and S. Ishizuka. 2011. Flame propagation in a vortex flow within small-diameter tubes. *Proceedings of the Combustion Institute* 33:3251–8.
14. Shimokuri, D., Y. Karatsu, and S. Ishizuka. 2013. Effects of inert gases on the vortex bursting in small diameter tubes. *Proceedings of the Combustion Institute* 34:3403–10
15. Davy, H. 1816. On the fire-damp of coal mines, and on methos of lightning the mines so as to prevent its explosion. *Philosophical Transactions* 105:1–22.
16. Potter, A. E. 1960. *Program Combustion Science and Technology* 1:145–81.
17. Lewis, B. and G. Elbe. 1987. *Combustion flames and explosions of gases*, 341. Orland, Florida: Academic Press, 242, 341.
18. Harris, M., J. Grumer, G. Elbe, and B. Lewis. 1949. Burning velocities, quenching, and stability data on nonturbulent flames of methane and propane with oxygen and nitrogen. *Proceedings of the Combustion Institute* 3:80–9.
19. Blanc, M. V., P. G. Guest, G. Elbe, and B. Lewis. 1949. Ignition of explosive gas mixtures by electric sparks. *Proceedings of the Combustion Institute* 3:363–7.
20. Simon, D., F. Belles, and A. Spakowskp. 1953. Investigation and interpretation of the flammability region for some lean hydrocarbon-air mixtures. *Proceedings of the Combustion Institute* 4:126–38.
21. Berlad, A. and A. E. Potter. 1955. Prediction of the quenching effect of various surface geometries. *Proceedings of the Combustion Institute* 5:728–35.
22. Friedman, R. 1949. The quenching of laminar oxyhydrogen flames by solid surfaces. *Proceedings of the Combustion Institute* 3:110–20.

23. Elbe, G. and B. Lewis. 1949. Theory of ignition, quenching and stabilization of flames of nonturbulent gas mixtures. *Proceedings of the Combustion Institute* 3:68–79.

24. Jensen, R. and A. Putnam. 1949. Application of dimensionless numbers to flash-back and other combustion phenomena. *Proceedings of the Combustion Institute* 3:89–98.

25. Ballal, D. and A. Lefebvre. 1977. Flame quenching in turbulent flowing gaseous mixtures. *Proceedings of the Combustion Institute* 16:1689–98.

26. Sakurai, T., S. Yuasa, M. Murakami, and K. Isomura. 2007. *National Symposium on Power and Energy Systems in Japane* 12:45–8.

27. Takahashi, S. and K. Wakai. 2009. *Proceedings of 45th AIAA/ASME/SAE/ASEE Joint Propulsion Conference*, 5246.

28. Sirignano, W. A., T. K. Pham, and D. Dunn-Rankin. 2002. Miniature-scale liquid-fuel-film combustor. *Proceedings of the Combustion Institute* 29:925–31.

29. Pham, T. K., D. Dunn-Rankin, and W. A. Sirignano. 2007. Flame structure in small-scale liquid film combustors. *Proceedings of the Combustion Institute* 31:3269–75.

30. Mattioli, R., T. K. Pham, and D. Dunn-Rankin. 2009. Secondary air injection in miniature liquid fuel film combustors. *Proceedings of the Combustion Institute* 32:3091–8.

31. Kumagai, K., S. Ishizuka, H. Taketomi, H. Nakajima, Y. Iino. 2007. *Proceedings of the 44th Combustion Institute*, 220–1, Japan.

32. Li, Y.-H., T.-S. Cheng, Y.-S. Lien, and Y.-C. Chao. 2011. Development of a tubular flame combustor for thermophotovoltaic power systems. *Proceedings of the Combustion Institute* 33:3439–45.

CHAPTER 8

LARGE-SCALE APPLICATIONS

Satoru Ishizuka

This chapter provides information on large-scale use and applications of tubular flame. Most of the tubular flame burners discussed in this chapter are large in diameter, from 2 to 12 inches, and large in heat output, from a few kilo Watts to 2 MW. In the first section, a variety of developed burners are classified into four groups, and general characteristics of flame diameter and length, and fundamental knowledge on rapidly mixed tubular flame combustion are given. In subsequent sections, the four classified burners are specifically introduced.

8.1 INTRODUCTION

As briefly mentioned in Chap.1, tubular flame has thermal and aerodynamic advantages in practical use and application. Close attention had been paid to tubular flame first by engineers in steel works in Japan.

The first demand was to make a large heat output, hence, a large-area tubular flame for the industrial processes. This was easily achieved by injecting premixture tangentially into a tube of large diameter. Initially, the tube used was 4 inches in diameter, which was then extended to 12 inches. As far as the author knows, the largest one used so far was 30 inches in diameter. The second demand was non-premixed combustion, since premixed combustion is dangerous due to the occurrence of flame flashback; it becomes more and more dangerous as the burner size is increased. This had been achieved by adopting rapidly mixed-type combustion, that is, by injecting fuel and air separately into a tube.

Thereafter, a research project "A Burner System of Next Generation (Development of an Advanced Tubular Burner System)" proposed by JFE steel company had been adopted by NEDO (New Energy and Industrial Technology Development Organization, Japan) as a Fundamental Technology Research Facilitation Program (Private Sector Fundamental Technology Research Support Scheme) and a variety of tubular flame burners have been developed during the four and a half years from September 2002 to March 2007. Besides, very unique tubular flame burners have been developed for their own practical purposes as well.

8.1.1 CLASSIFICATION

Up to now, a variety of tubular flame burners have been designed for use. Although most of them are still under development, they are currently classified into four groups and are summarized in Table 8.1.

1. Wide flammable range: Due to the merits of the wide flammable range, a tubular flame burner was used to burn blast furnace gas (BFG) in steel works, the combustion heat of which is just one-twentieth of that of propane. It was successfully burned with a blue tubular flame in a 12-inch diameter rapidly mixed-type burner. Also, tubular flame burners were used to determine the peak concentrations for various extinguishers as mentioned in Chap.1.
2. Fuel diversity: A tubular flame burner can burn various kinds of fuels, which include gaseous, liquid, and solid fuels as well. The gaseous fuels are many and used in several ways: [1] methane, propane, and hydrogen are used as fuel in tubular flame burners in laboratory-scale experiments; [2] city gas (13A) is used in tubular flame burners installed in water heaters, super-heated steam generators, and boilers; [3] by-products gas fuels (BFG, Linz–Donawitz converter gas (LDG), and coke oven gas (COG)) are used in tubular flame burners in steel works; [4] a biomass gas made by methane bacterium (60% CH_4 + 40% CO_2) can be used in a tubular flame burner installed in a Stirling engine. Liquid fuels can be burned by being tangentially sprayed into a tube. Kerosene, banker-A, and banker-C can be successfully burned in tubular flame burners. Although only one case, a solid fuel of biomass wooden powders can be burned in a tubular flame burner. Biomass powders are supplied with air and injected tangentially into a tube. At appropriate conditions, self-sustained combustion can be achieved without any aid of auxiliary gaseous fuels such as propane.
3. Compactness: Tubular flame burner is simple in structure. It consists of a tube and tangential slits. Thus, the burner is compact. A small burner of 30 mm in inner diameter and 40 mm in length was installed into a fuel-processing system for polymer electrolyte fuel cell (PEFC), which needed a heat source to maintain its temperature around 600°C to optimize the performance of catalysts. Very low emissions of NOx and CO were achieved due to its uniformity in temperature distribution. Also, instead of an electrical heater, a very compact, detachable tubular flame burner was developed for heating a hollow fasten bolt; the burner was used to fasten a high-pressure vessel of steam turbine with a shrinkage fitting method. In this case, homogeneity and completeness of burned gas were required to maintain the quality of the fattening bolt. Tubular flame burners were also installed into a superheated steam generator and a water heater. A portable microtubular flame burner was developed, which used a liquefied fuel of a gas lighter, while air was supplied into the burner using the ejector effect of the fuel injection.
4. Geometry: Tubular flame has a peculiar geometry; its cross-sectional shape is circular and the inside is a hot burned gas of uniform temperature and homogeneous composition. Thus, a tubular flame can be used to stabilize a flame in a high-velocity stream and also to assist the combustion of PET resin, which needs to burn swiftly not to spoil the wall of

a furnace. Tubular flame burners can be installed in industrial processes such as a heating process of a gas stream, a melting process of glass-making powders, and a chemical treatment process of a waste, exhaust gas. A fire place is designed by putting a fuel into the inner burned gas region, yielding a yellow luminous flame. Very recently, because of the advantage of its cylindrical shape, a tubular flame burner has been used to heat the head of a Stirling engine. Some of these burners will be specifically introduced in the Sections 8.2–8.5.

8.1.2 FLAME DIAMETER AND LENGTH

In industries, most types of combustion adopted are turbulent combustion whether in a premixed mode or a non-premixed mode. Laminar premixed flame is quite limited in use, be it household or small-scale manufacturing or industrial processes. This is partly because it is very hard to obtain a large laminar flame for the requirements in industries.

As well known, a laminar flame can be obtained with a Bunsen burner, which was invented by Robert Wilhelm Bunsen of Heidelberg University in 1885 for laboratory-scale experiments in chemistry. Due to its special device, a laminar flame can be stabilized quite easily on the burner. It is, however, very difficult to achieve a large flame.

Table 8.1. Classification of tubular flame burners

Advantage		Applications
Wide flammable range	Blast furnace gas, Flammability determination	
Fuel Diversity	Gases	By-product fuel gases in steel works, City gas (13A), LPG
	Liquids	Kerosene, Banker-A, Banker-C
	Solids	Biomass powder
Compactness	Uniform temperature	Fuel-processing system for PEFC
	Convenience	Detachable burner for heating a hollow fastening bolt
	Little space	Superheated steam generator
	Multistaged	Water heater
	Handy	Portable microburner
Geometry	Pilot flame	Flame stabilization in a high speed stream
	Assist	PET resin combustion
	Industrial process	Heating gas, Melting powder, Chemical treatment
	Radiation	Fire place
	Ring shape	Heating the head of a Stirling engine

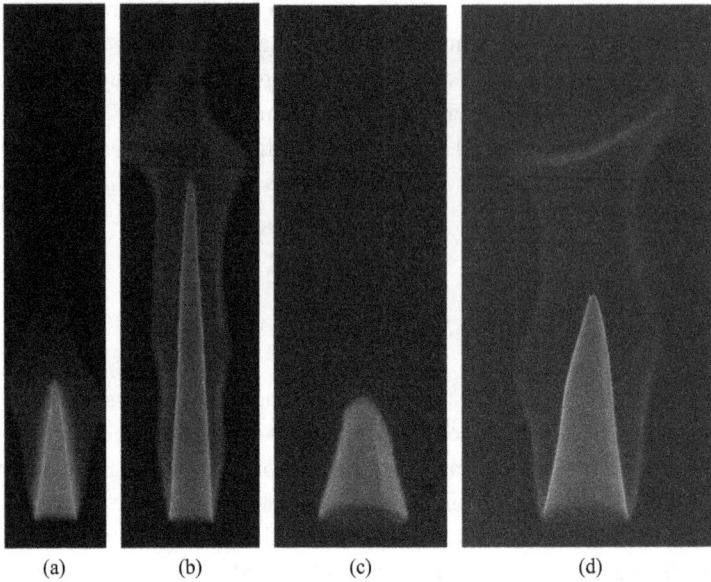

Figure 8.1. Photographs of methane–air flames (inner diameter, equivalence ratio, and mean velocity: (a) 14 mm, 1.0, 4 m/s; (b) 14 mm, 1.4, 4.2 m/s; (c) 30 mm, 0.9, 0.86 m/s; (d) 30 mm, 1.4, 0.9 m/s).

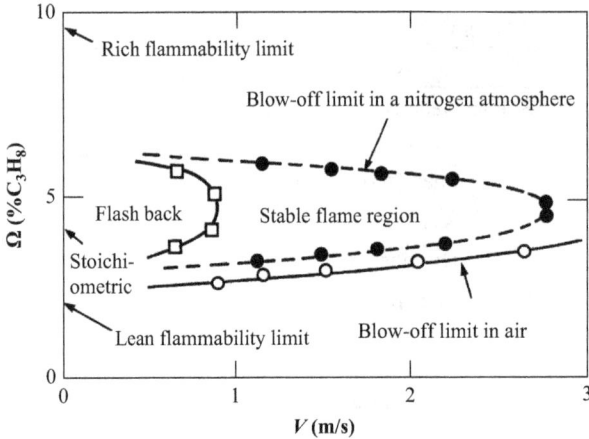

Figure 8.2. Stability diagram of a flame of propane–air mixtures anchored on a 14-mm inner diameter nozzle in air and nitrogen atmospheres.

Figure 8.1(a) shows a picture of the stoichiometric methane–air premixed flame, stabilized on a 14-mm inner diameter nozzle, which is designed to give a uniform velocity distribution at the exit. The flame is almost conical in shape. One way to increase the flame is to increase the mixture flow rate. However, as shown in Fig.8.2, the flame blows off quite easily at higher velocities with lean and stoichiometric mixtures. With rich mixtures, which are common in Bunsen burners, the flame is anchored at the rim by a diffusion flame, which is formed between the excess fuel and the ambient air. Thus, the fuel-rich flame can be stabilized for higher velocities, and as shown in Fig.8.1(b), the flame height is increased as the mixture becomes more fuel-rich, and accordingly the burning velocity decreases. However, the flame is always flickering, which could be seen from the corrugation of the outer diffusion flame although the picture

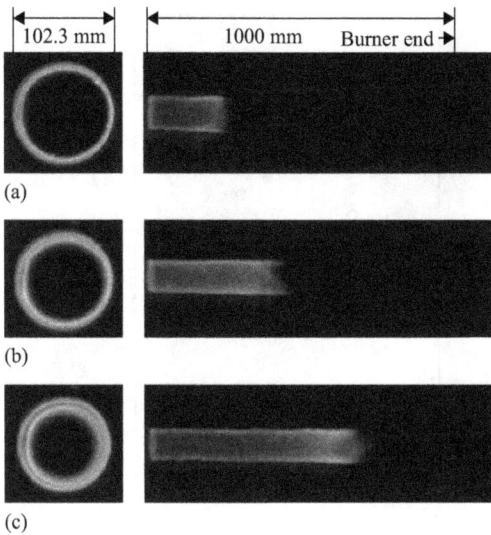

Figure 8.3. Appearance of flame (equivalence ratio $\Phi = 0.6$; (a) $Q_{air} = 60$ m³/h (stp), (b) $Q_{air} = 100$ m³/h (stp), (c) $Q_{air} = 140$ m³/h (stp), Q_{air}: Air flow rate). (*Source*: Hagiwara, R., Okamoto, M., Ishizuka, S., Kobayashi, H., Nakamura, A., Suzuki, M., Combustion Characteristics of a Tubular Flame Burner for Methane, Transaction of JSME 66(2000)3226–3232.[3] Used by permission of JSME.)

was taken at a shutter speed of $1/250$ s⁻¹. Furthermore, the flame becomes turbulent when the Reynolds number of the combustible stream exceeds around 3000 in this burner (the critical Reynolds number is 2300 for the Hagen–Poiseuille flow in a pipe[1]). Thus, an increase in the flow rate will not yield a large laminar flame.

Another way to increase the flame area is to increase the burner diameter. Figure 8.1(c) and (d) shows pictures of methane–air premixed flames obtained with a 30-mm inner diameter nozzle. When the tube diameter becomes larger, another type of instability appears at the flame front. When the mixture is fuel lean, the so-called polyhedral-type instability[2] appears at the flame front (Fig.8.1(c)). As known clearly, polyhedral flames are formed on the tube mouth for combustible mixtures, whose Lewis number is less than unity; rich heavy hydrocarbon–air mixtures and lean hydrogen–air and lean methane–air mixtures are used in this case. This instability is damped in rich methane–air mixtures, because the Lewis number of the deficient, limiting oxygen is larger than unity, as shown in Fig.8.1(d). The flame, however, is strongly disturbed by a buoyant force, which works on the hot gas column. Flame flickering occurs as well. Thus, an increase in the tube diameter will not yield a large laminar flame.

On the contrary, a large-area laminar flame can be successfully obtained with a swirl-type tubular flame burner. It should be noted that in the case of a nonrotating, counterflow-type porous cylinder burner, the range of fuel–air flow rates for uniform laminar flames is limited; first, the injected flow is quite disturbed when the velocity exceeds about 30 cm/s owing to the porous material, and second, a buoyancy force distorts the flame when the burner is horizontally mounted, which can be seen in Fig.1.8, or lifts and blows out the flame when the burner is mounted vertically.

Figure 8.3 shows the variations of the tubular flame configuration of a lean methane–air mixture with the air flow rate in the swirl-type burner introduced in the Section 1.5.1 (Fig.1.35).[3] With an increase in the flow rate, the flame length is increased while the flame diameter is slightly reduced.

Figure 8.4 shows the variations of flame diameter in the tubular flame burner of 4-inch in diameter introduced in the Section 1.5.1 at two representative conditions of the total flow rates

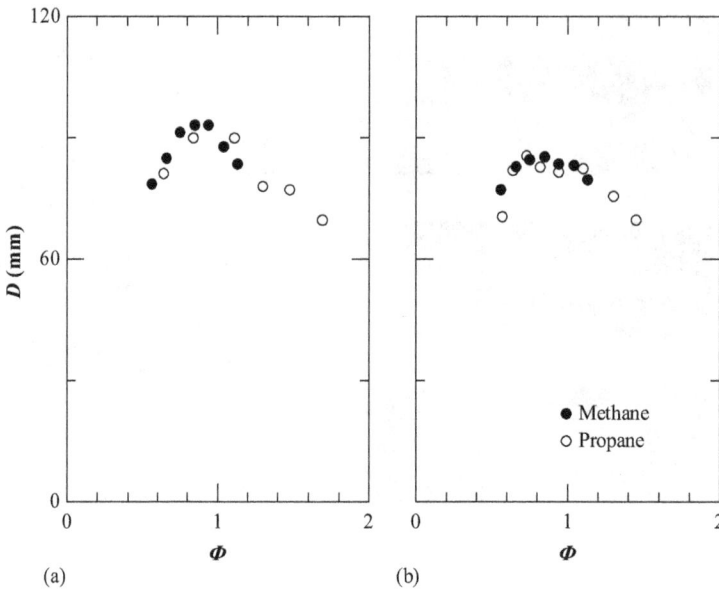

Figure 8.4. Variations of the flame diameter D with the equivalence ratio Φ ((a) Q_{air} = 80 m³/h (stp), (b) Q_{air} = 140 m³/h (stp)). (*Source*: Hagiwara, R., Okamoto, M., Ishizuka, S., Kobayashi, H., Nakamura, A., Suzuki, M., Combustion Characteristics of a Tubular Flame Burner for Methane, Transaction of JSME 66(2000)3226–3232.[3] Used by permission of JSME.)

of 80 and 140 m³/h (stp).[3] Around the stoichiometric condition, the flame diameter takes its maximum and gradually decreases as the mixture becomes leaner or richer. Accordingly, the flame increases in length as shown in Fig.8.5.[3] This is because the tubular flame is stabilized at a position where the inward radial velocity and the burning velocity of the mixture balance each other.

Under a constant condition of the mixture flow rate, the diameter becomes largest and the length smallest for the mixture whose burning velocity becomes fastest. Comparing the results of the two flow rates, flame diameter becomes smaller for a larger flow rate. This is due to a decrease in the mixture burning velocity through flame stretch, which increases as the flow rate increases.

Figure 8.6 shows the variations of flame area with the equivalence ratio.[3] The flame area takes its minimum around the stoichiometric composition, 0.06 m² at 80 m³/h (stp), which is increased almost linearly with the mixture flow rate to 0.1 m² at 140 m³/h (stp). The maximum flame area is about 0.3 m², which corresponds with the inner surface area of a 100-mm diameter and 1000-mm long tube.

To design a burner, the diameter and length of the burner should be decided. The heat outputs are about 0.15 MW (low caloric value) for the stoichiometric methane and propane–air mixtures at the air flow rate of 140 m³/h (stp). Thus, the burner diameter recommended is 4 inches for the order of 0.2 MW heat output. Since the heat output is proportional to the flame area, the burner diameter should be increased with an increase in the square root of the heat output. Thus, it becomes 12 inches for 2 MW, for example. Once the burner diameter is fixed, the flame length can be estimated. If the flame diameter, the flame length, the burning velocity, and the mixture flow rate are denoted by D_f, L_f, S_u, and $Q_{mixture}$ respectively, a following simple relation holds.

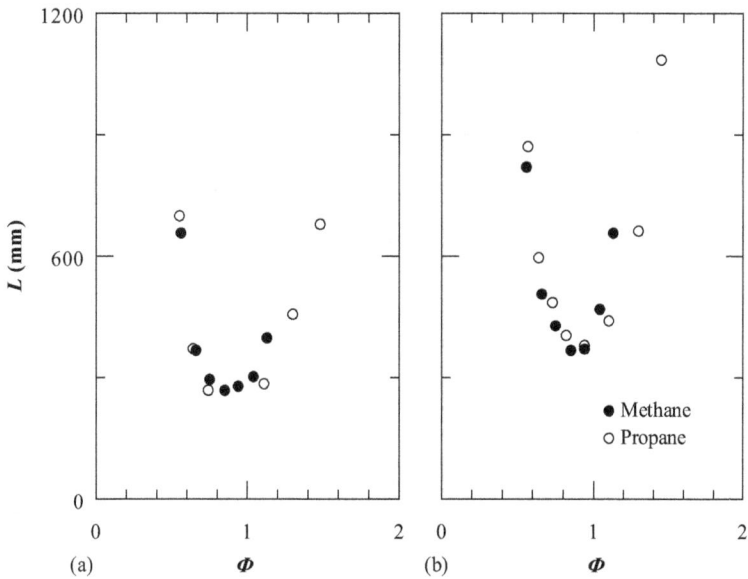

Figure 8.5. Variations of the flame length L with the equivalence ratio Φ ((a) Q_{air} = 80 m³/h (stp), (b) Q_{air} = 140 m³/h (stp)). (*Source*: Hagiwara, R., Okamoto, M., Ishizuka, S., Kobayashi, H., Nakamura, A., Suzuki, M., Combustion Characteristics of a Tubular Flame Burner for Methane, Transaction of JSME 66(2000)3226–3232.[3] Used by permission of JSME.)

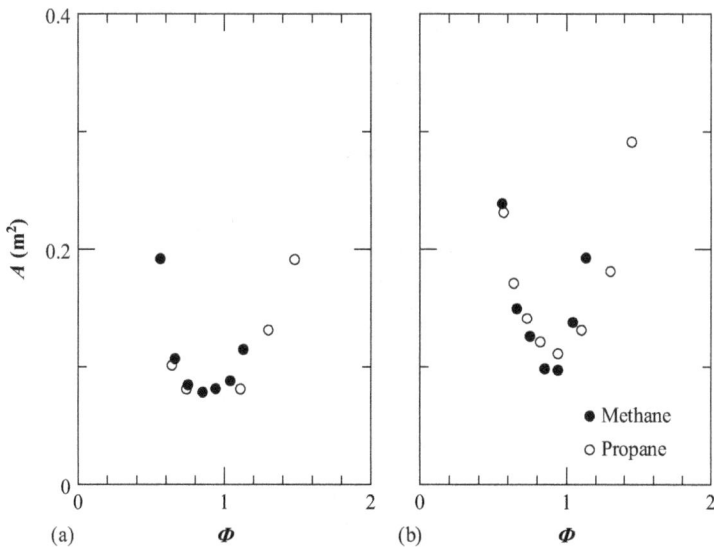

Figure 8.6. Variations of the flame area A with the equivalence ratio Φ ((a) Q_{air} = 80 m³/h (stp), (b) Q_{air} = 140 m³/h (stp)). (*Source*: Hagiwara, R., Okamoto, M., Ishizuka, S., Kobayashi, H., Nakamura, A., Suzuki, M., Combustion Characteristics of a Tubular Flame Burner for Methane, Transaction of JSME 66(2000)3226–3232.[3] Used by permission of JSME.)

$$\pi D_f L_f S_u = Q_{mixture}. \tag{8.1}$$

In a practical situation, most of the combustible mixtures are near stoichiometric and the volume flow rates are far from the blow-off limit. Thus, the flame expands to the diameter just inside the tangential slit, that is, the available diameter D_a, which is given by $D_{tube} - 2W$, where D_{tube} is the burner inner diameter and W is the slit width. The burning velocity is close to the unstretched adiabatic value S_u^0. Then, the flame length can be roughly given as:

$$L_f = \frac{Q_{mixture}}{\pi D_f S_u} \cong \frac{Q_{mixture}}{\pi D_a S_u^0}. \tag{8.2}$$

If $S_u^0 = 40$ cm/s and $D_a = 100 - 3.7 \times 2 = 92.6$ mm are substituted into Eq. 8.2, we obtain $L_f = 211$ and 369 mm for the air flow rates of 80 and 140 m^3/h (stp), respectively.[4]

At the lower flow rate ($Q_{air} = 80$ m^3/h), the flame diameter determined is close to the available diameter (see Fig.8.4) and the burning velocity seems close to the adiabatic burning velocity. Thus, the estimated flame length should be close to the flame length obtained with the burner. However, the determined length is 263 mm, about 1.25 times more than the estimated value.

At a higher flow rate ($Q_{air} = 140$ m^3/h), the flame shrinks in diameter; the maximum diameter is reduced to 85 mm in Fig. 8.4. Accordingly, the burning velocity should be reduced. These will result in a longer flame length in estimation. However, the determined flame length is 358 mm, which is in good accordance with the estimated value of 369 mm. Thus, there seem large discrepancies between the real and estimated flame lengths. Equation 8.2 just yields measure of the flame length as the first approximation.

8.1.3 RAPIDLY MIXED TUBULAR FLAME COMBUSTION

As the burner sized is increased, the occurrence of flame flashback becomes more serious. To avoid explosion hazard of premixed combustion, an inherently safe technique of rapidly mixed combustion has been proposed. To elucidate the validity of this technique, experiments have been conducted with optically accessible burners of different diameters and different slit widths,[5] which is introduced in Section 1.5.2 (Fig.1.38).

Figure 8.7 shows the appearance of flames of premixed and rapidly mixed combustion of propane fuel.[5] The burner inner diameter is 50 mm and the width and length of the four tangential slits are 3 and 100 mm, respectively. The equivalence ratios in the rapidly mixed combustion are those assuming complete mixing of the fuel and the air injected.

In the premixed combustion (Fig.8.7(a)), the flame luminosity is uniform for rich and stoichiometric mixtures, but the luminosity becomes weaker as the tangential slit is approached for lean mixtures. This is due to the Lewis number effect. Namely, the mass diffusivity of the deficient component, propane, is less than the thermal diffusivity of the mixture, and hence, the flame is weakened due to a stretch around the injection area.

In the rapidly mixed combustion, the nonuniform luminosity of the lean mixture becomes more apparent; as seen in the bottom picture of Fig.8.7(b), flame luminosity cannot be detected around the tangential slit area. For the rich mixtures also, the flame luminosity becomes nonuniform; as seen in the upper picture of Fig.8.7(b), the luminosity is intensified as the tangential slit is approached. The nonuniformity seems to be enhanced with incomplete mixing between propane and air. Incompleteness of mixing can be seen also in the case of stoichiometric mixture, because the combustion is completed inside the tube in the premixed combustion (Fig.8.7(a), middle), whereas the flame extends outside the combustion tube (Fig.8.7(b), middle).

When the burner diameter is increased, flame configurations are more complicated. Figure 8.8 shows the flame appearance in the 76-mm burner. The slit width is 3 mm and the

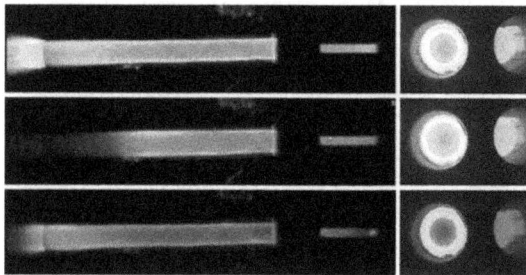

(a)

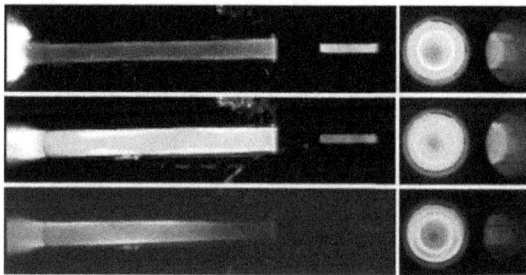

(b)

Figure 8.7. Appearance of flame (a) premixed combustion; (b) rapidly mixed combustion; burner diameter, 52 mm; upper, $\Phi = 1.2$; middle, $\Phi = 1.0$; lower, $\Phi = 0.7$; $Q_{air} = 60$ m^3/h (stp); fuel, propane, Φ equivalence ratio. (*Source*: Ishizuka, S., Motodamari, T., Shimokuri, D., Rapidly-Mixed Combustion in a Tubular Flame Burner, Proceedings of the Combustion Institute 31(2007)1085–1092.[5] Used by permission of Elsevier.)

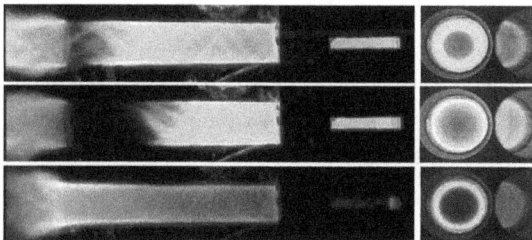

(a)

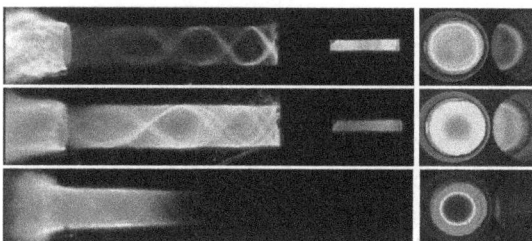

(b)

Figure 8.8. Appearance of flame (a) premixed combustion; (b) rapidly mixed combustion; burner diameter, 76 mm; upper, $\Phi = 1.5$; middle, $\Phi = 1.0$; lower, $\Phi = 0.7$; $Q_{air} = 80$ m^3/h; (stp) fuel, propane, Φ equivalence ratio. (*Source*: Ishizuka, S., Motodamari, T., Shimokuri, D., Rapidly-Mixed Combustion in a Tubular Flame Burner, Proceedings of the Combustion Institute 31(2007)1085–1092.[5] Used by permission of Elsevier.)

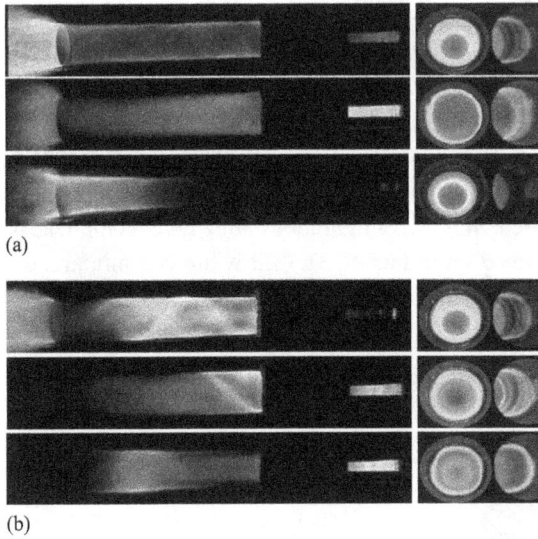

Figure 8.9. Appearance of flame (a) premixed combustion; (b) rapidly mixed combustion; burner diameter, 102 mm; upper, $\Phi = 1.5$; middle, $\Phi = 1.0$; lower, $\Phi = 0.6$; $Q_{air} = 100$ m³/h (stp); fuel, propane, Φ equivalence ratio. (*Source*: Ishizuka, S., Motodamari, T., Shimokuri, D., Rapidly-Mixed Combustion in a Tubular Flame Burner, Proceedings of the Combustion Institute 31(2007)1085–1092.[5] Used by permission of Elsevier.)

fuel is propane. In the premixed combustion, the flames are smooth in luminosity. In the rapidly mixed combustion, however, nonuniform flames with spiral-shaped stripes are formed for rich and stoichiometric mixtures. These stripes seem to be caused by poor mixing of fuel and air along the concave wall, since the centrifugal force working on different density gases may prevent the mixing. For lean mixtures, flames are relatively smooth in luminosity for both the premixed and rapidly mixed combustion. However, in the premixed combustion, a very faint luminosity can be seen around the tangential slit area, whereas flame luminosity cannot be detected around the tangential slit area in the rapidly mixed combustion.

Figure 8.9 shows the flame appearance in the 102-mm burner. The slit width is 3 mm and the fuel is propane. When the burner diameter is further increased, the nonuniformity in flame luminosity is rather suppressed. Instead, a yellow luminous zone appears at the center for both the premixed and rapidly mixed combustion types. It is interesting to note that for lean mixtures, a luminous flame is not formed around the tangential slit area in the premixed combustion, whereas in the rapidly mixed combustion, a flame with strong blue-green luminosity that is peculiar to rich propane–air flames is formed. At the center, a yellow luminous zone can be seen. Although in the 2-inch diameter tubular flame burner, a PIV measurement has shown that a hot stagnant gas column occupies the center region.[6] This kind of hot gas column may assist combustion. Hence, it is interesting to note that a small flame is separated from the main flame and it can survive near the closed end wall even when the overall equivalence ratio becomes smaller than the lean flammability limits.

Figure 8.10(a) and (b) are the mappings of stable combustion regions obtained with the 76- and 102-mm burners, respectively.[5] In the 102-mm burner (Fig. 8.10(b)), the residual flame region is located below the lean flammability limit. If this residual flame region is ignored, the lean limit occurs around the lean flammability of $\Phi = 0.5$ in both the 76- and 102-mm diameter burners. On the rich side, a luminous yellow zone appears around the center axis in the 102-mm burner, and its diameter becomes larger with an increasing equivalence ratio; the yellow flame approaches the blue-green flame zone, while the luminosity of the blue-green main flame becomes weaker. For reference, the limits at which the blue-green flame zones seem

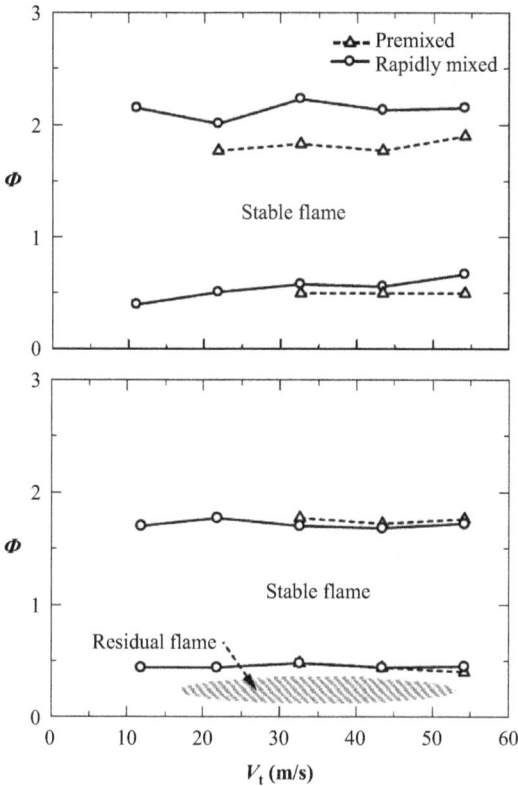

Figure 8.10. Mapping of stable combustion region in the (a) 76- and (b) 102-mm diamter burner. Fuel, propane. (*Source*: Ishizuka, S., Motodamari, T., Shimokuri, D., Rapidly-Mixed Combustion in a Tubular Flame Burner, Proceedings of the Combustion Institute 31(2007)1085–1092.[5] Used by permission of Elsevier.)

to disappear are plotted in Fig. 8.10(b). With a further increase in the equivalence ratio, the luminosity of the yellow zone becomes weaker, but recirculation of a hot burned gas, which is made at the exit of the burner in a form of diffusion burning with ambient air, becomes remarkable. However, a dark, hot zone still exists beyond the rich flammability limit. This situation is the same as the rich combustion in a swirl-type tubular flame burner of diameter 21 mm for fundamental study (see Fig. 38 of Ref.[7]); that is, the rich limit is dependent on the amount of recirculating gas, which supplies heat and oxygen inside the burner. Therefore, the rich limit has not been determined in Fig.8.10(b).

From the results with different burner diameters and slit widths, it is found that various flame configurations are formed in the rapidly mixed combustion. Because the rapidly mixed combustion is of practical importance, its success or failure is judged whether tubular-shaped flame combustion is obtained for equivalence ratios from 0.7 to 1.3 around the stoichiometry. Table 8.2 summarizes the results.[5] In this table, the characters "S" and "F" denote the success and failure of establishment of tubular flame combustion, respectively. Since the rotational strength seems to be important for its establishment, the swirl number was calculated for each burner of different diameters and slit widths. The swirl number is given approximately by the equation,

$$S_w \cong \frac{\pi D_e D_0}{4 A_t} \tag{8.3}$$

in which D_e is the exit throat diameter, approximately given by extracting the slit width from the exit diameter, D_0 is the burner diameter, and A_t is the tangential slit area.[5]

Table 8.2. Summary of the establishment of rapidly mixed tubular combustion. (*Source*: Ishizuka, S., Motodamari, T., Shimokuri, D., Rapidly-Mixed Combustion in a Tubular Flame Burner, Proceedings of the Combustion Institute 31(2007)1085–1092.[5] Used by permission of Elsevier.)

Diameter (mm)	Slit	Width	
	3 mm	6 mm	9 mm
52	S(3.45)	F(1.72)	F(1.15)
76	S(7.63)	S(3.81)	F(2.54)
102	8(13.5)	S(6.77)	S(4.52)

Abbreviations: S, success; F, failure; ( ), swirl number.

It is seen that although the flames are not exactly the same, stable tubular flame combustion could be obtained with the rapidly mixed combustion at swirl numbers greater than about 5. It is known that a recirculating flow occurs for swirl numbers larger than 0.6.[8] Thus, to achieve rapidly mixed combustion, swirl numbers of one order of magnitude larger than those for recirculating flows are needed. In addition to large swirl numbers, large injection velocities are also required to establish rapidly mixed combustion.

There are two pints, which should be noted. One is flame laminarization in the rotating flow. When the velocity of a flow increases, the flow becomes turbulent. For example, the Reynolds number of the air flow which flows in a 3-mm diameter duct at 20 m/s is about 3750, while the flame established in the burner is laminar as seen in Fig.8.7. Strong rotation makes the flame laminar and, accordingly, the burning velocity remains at the same level as the laminar burning velocity.[9,10] The second point to be noted is the swirl number. Usually, the swirl number is a measure of angular momentum relative to linear momentum and it is generally accepted that when the swirl number exceeds about 0.6, a toroidal recirculating flow is induced; hence, the flow is called a strong swirl flow.[8] In the rapidly mixed tubular flame combustion, however, the swirl number has a different physical meaning, which will be explained below.

Figure 8.11 shows a two-slit tubular flame burner, in which air is injected from the upper left slit, while methane is injected from the lower bottom slit into a Pyrex tube of 36 mm inner diameter and 160 mm length. The slit width and length are 3 and 80 mm, respectively. The flow rates of air and methane are 8.54 and 2.66 l/min, respectively. It is seen in the upper right picture that a yellow luminous flame whorl is formed in the burner. In the lower picture, taken through the Pyrex tube, there can be seen many fine horizontal lines, showing a cigarette-like flame roll-up.

Figure 8.12 is a development of a four-slit tubular flame burner, showing illustratively the geometry of mixing of two streams, which are individually injected through two slits apart from $\pi/2$ radian at the periphery. The burner diameter, the slit length, and the slit width are denoted by D, L, and w, respectively. If only air is injected at the flow rate of Q_{air}, and a cylindrical symmetry is assumed, the axial and tangential velocities, $V_{axial,air}$ and $V_{tangential,air}$ are simply given as

$$V_{axial,air} = \frac{Q_{air}}{(\pi/4)D^2}, \quad V_{tangential,air} = \frac{Q_{air}}{w \times L} \tag{8.4}$$

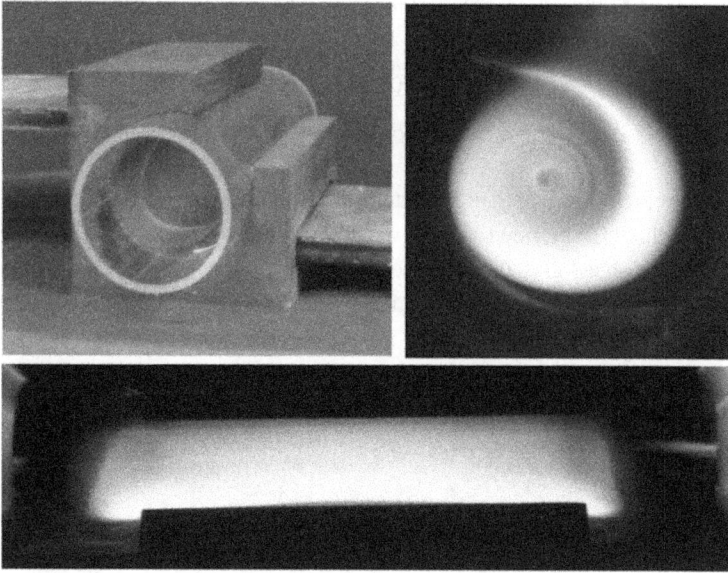

Figure 8.11. Flame whorl (Pyrex tube; 36 mm inner diameter and 160 mm long, slits; 3 mm wide and 80 mm long, upper left; air, 8.544 L/min, lower right; methane, 2.66 L/min).

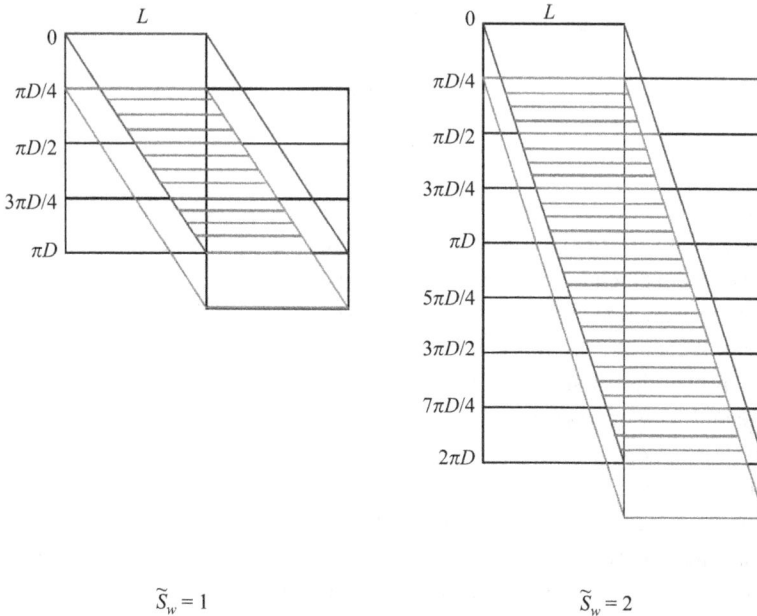

$$\widetilde{S}_w = 1 \qquad\qquad \widetilde{S}_w = 2$$

Figure 8.12. Development of a four-slit tubular flame burner, showing illustratively the geometry of mixing of two streams, injected individually through two slits, which are apart from $\pi/2$ radian at the periphery.

Then, with the use of Eq.8.3, the tangential to axial velocity ratio becomes

$$\frac{V_{\text{tangenttial, air}}}{V_{\text{axial, air}}} = \frac{Q_{air}}{w \times L} \div \frac{Q_{air}}{(\pi/4)D^2} = \frac{(\pi/4)D^2}{w \times L} = \frac{\pi D^2}{4A_t} \cong S_w \tag{8.5}$$

Namely, the velocity ratio is approximately equal to the swirl number of the burner.

The time Δt_1 which the air needs to turn one revolution is given by $\pi D/V_{\text{tangential, air}}$. During this time, the air goes ahead in the axial direction by a distance $V_{\text{axial, air}} \times \Delta t_1$. If this distance is less than the slit length L, the whole tube of diameter D is wrapped with the air stream of width L. This condition is given as

$$V_{\text{axial, air}} \times \frac{\pi D}{V_{\text{tangentail, air}}} \leq L \tag{8.6}$$

From this equation, the following relation is obtained.

$$\frac{\pi D}{L} \leq \frac{V_{\text{tangentail, air}}}{V_{\text{axial, air}}} \cong S_w \tag{8.7}$$

If the swirl number S_w is equal to $\pi D/L$, the whole tube of diameter D is wrapped with the air stream of width L. We define this wrapping state of $S_w = \pi D/L$ as the single wrapping state. Furthermore, we introduce a normalized swirl number $\tilde{S}_w$, which is defined by dividing the swirl number S_w by that of the single wrapping state $\pi D/L$. Then, a following relation holds.

$$\tilde{S}_w \equiv \frac{S_w}{\pi D/L} \cong \frac{\pi D^2/4wL}{\pi D/L} = \frac{D}{4w} \tag{8.8}$$

Here, $\tilde{S}_w$ is the normalized swirl number. For a fixed D, the wrapping state can be varied through the slit width w. If the width w is reduced to half as much as that of the single wrapping state, the tangential velocity is increased to twice while the axial velocity remains unchanged for a constant air flow rate. Thus, at $\tilde{S}_w = 2$, the whole tube of diameter D is wrapped with a double layer of the air stream of width L. If $\tilde{S}_w = 3$, the whole tube of diameter D is wrapped with a triple layer of the air stream of width L. That is, the normalized swirl number $\tilde{S}_w$ corresponds to the multicity of wrapping state such as the single, double, and triple wrapping states, and so on.

Next, for simplicity, we further assume that the fuel stream also goes axially and tangentially at the same velocities as those of air, that is, $V_{\text{axial, air}}$ and $V_{\text{tangential, air}}$, respectively. Then, the overlapped area of the two streams will be those shown in Fig.8.11. In the case of $\tilde{S}_w = 1$, the air particle initially at a position (azimuthal, axial) = (0,0) reaches at (πD, L), while the fuel particle initially at ($\pi D/4$, 0) reaches at ($5\pi D/4$, L) when the fluids proceed at a distance L. In the case of $\tilde{S}_w = 2$, the air particle at (0,0) reaches at ($2\pi D$, L), while the fuel particle at ($\pi D/4$, 0) reaches at ($9\pi D/4$, L).

From a simple consideration, the ratio of the overlapped area to the original area Ψ and the overlapped area Σ for $\tilde{S}_w$ are given as

Ratio of the overlapped area:
$$\Psi = \left(1 - \frac{1}{4\tilde{S}_w}\right)^2, \qquad\qquad (8.9)$$

Overlapped area:
$$\Sigma = \tilde{S}_w \left(1 - \frac{1}{4\tilde{S}_w}\right)^2 (\pi D L), \qquad\qquad (8.10)$$

for the present $\pi/2$ shift streams.

When the normalized swirl number is unity ($\tilde{S}_w = 1$), the overlapped area ratio is just 56.25%; however, it becomes about 90.25% for $\tilde{S}_w = 5$. The overlapped area is increased from $0.5625\pi DL$ for $\tilde{S}_w = 1$ to $4.5125\pi DL$ for $\tilde{S}_w = 5$. Then, a premixed-like tubular flame can be established for large swirl numbers such as 5, recommended by the previous work.[5]

It is important to note that the swirl number S_w can be rewritten in term of the normalized swirl number $\tilde{S}_w$ such that

$$S_w = \frac{\pi D}{L}\tilde{S}_w = \frac{\pi D \tilde{S}_w}{L}. \qquad\qquad (8.11)$$

In other words, the swirl number means how many revolutions ($\tilde{S}_w$) and, resultantly, how long peripheral distance ($\pi D \tilde{S}_w$) the flow makes during when the flow proceeds a distance of the slit width L. Thus, it should be kept in mind that the physical meaning of the swirl number in the rapidly mixed-type tubular flame burner is different from that in the conventional swirl burner; the swirl number in the rapidly mixed combustion means how extent the fuel and oxidizer streams are overlapped in geometry, whereas in the conventional swirl combustors, the swirl number means the intensity of swirling motion, and swirl numbers larger than 0.6 means the occurrence of a toroidal recirculating hot burned gas flow,[8] which stabilizes combustion, yielding high-intensity combustion.

Once the overlapping of the fuel and air streams is enough, an intermixing between the two streams becomes the next concern. Concerning this, an interesting result has been very recently obtained. Figure 8.13 is a schematic of a four-slit burner, in which methane and oxygen are injected separately from the two opposed slits and the other two opposed slits, respectively.[11] The inner diameter is 16 mm and the slit length is 8 mm. Two slit width (W), 2 and 1 mm were used, which gave the swirl numbers of 6.28 and 12.56, respectively.

Figure 8.14 shows the appearance of flames with the 2- and 1-mm slit width burners, in which Φ is the overall equivalence ratio assuming complete mixing.[11] When the slit width is large, diffusion flame are anchored at the exit of the fuel slits because the velocity at the exit of the fuel slits is much lower than that of the oxidizer slits. Once diffusion flames are anchored, mixing of fuel and oxygen is inhibited, resulting in the failure of rapidly mixed tubular flame combustion. On the other hand, when the slit width becomes smaller, tubular flame combustion becomes possible for small equivalence ratios of 0.12 and 0.15.

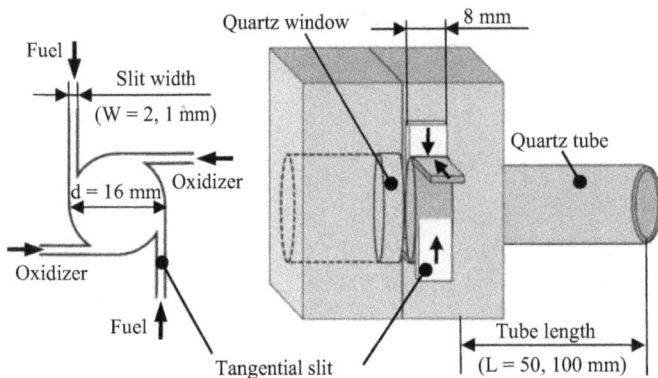

Figure 8.13. Schematic of a four-slit burner with a quartz window. (*Source*: Shi, B., Shimokuri, D. and Ishizuka, S., Methane/oxygen combustion in a rapidly mixed type tubular flame burner, Proceedings of the Combustion Institute 34(2013)3369–3377.[11] Used by permission of Elsevier.)

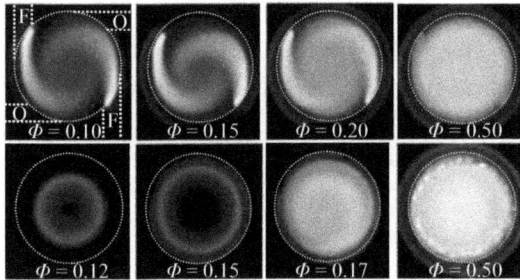

Figure 8.14. Appearances of methane/oxygen flames (oxygen flow rate: 4 m³/h, upper: W = 2 mm; lower: W = 1 mm). (*Source*: Shi, B., Shimokuri, D. and Ishizuka, S., Methane/oxygen combustion in a rapidly mixed type tubular flame burner, Proceedings of the Combustion Institute 34(2013)3369–3377.[11] Used by permission of Elsevier.)

As a criterion for the establishment of tubular flame combustion, the Damköhler number, which is the ratio of the mixing time τ_m to the chemical reaction time τ_r defined in the following was proposed.

$$D_a = \tau_m / \tau_r. \tag{8.12}$$

The reaction time was calculated by a relation

$$\tau_r = \delta_L / S_u, \tag{8.13}$$

in which δ_L is laminar flame thickness and S_u is laminar burning velocity. The value of δ_L was determined by computing the temperature profile and using the following equation

$$\delta_L = \frac{T_b - T_u}{\max\left(\left|\partial T / \partial x\right|\right)}, \tag{8.14}$$

in which T_u and T_b are unburned and burned gas temperatures, respectively, and $\partial T / \partial x$ is the temperature gradient.[12] The laminar burning velocity and temperature profile were calculated using the Chemkin Premix code.[13,14]

On the other hand, the mixing time τ_m was calculated by dividing the thickness of the fuel and oxidant streams, δ_{CH_4} and δ_{O_2} by the binary diffusion coefficient $D_{CH_4-O_2}$ in a form as

$$\tau_m = \frac{\delta_{O_2}\delta_{CH_4}}{D_{CH_4-O_2}}, \text{ or } \tau'_m = \frac{\delta_{CH_4}\delta_{CH_4}}{D_{CH_4-O_2}}, \tag{8.15}$$

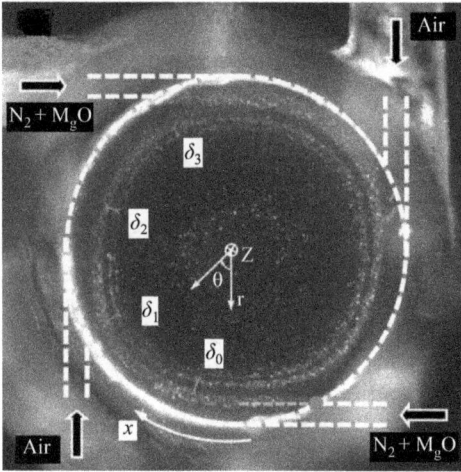

Figure 8.15. Flow visualization inside the 1-mm slit width burner (the flow rate: 0.048 m³/h). (*Source*: Shi, B., Shimokuri, D. and Ishizuka, S., Methane/oxygen combustion in a rapidly mixed type tubular flame burner, Proceedings of the Combustion Institute 34(2013)3369–3377.[11] Used by permission of Elsevier.)

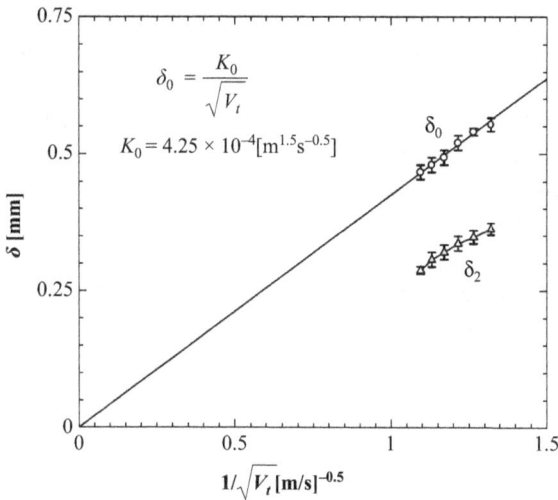

Figure 8.16. Variations of the thickness δ_0 and δ_2 with the injected tangential velocity V_t. (*Source*: Shi, B., Shimokuri, D. and Ishizuka, S., Methane/oxygen combustion in a rapidly mixed type tubular flame burner, Proceedings of the Combustion Institute 34(2013)3369–3377.[11] Used by permission of Elsevier.)

in which τ'_m is a modified mixing time that is described later.

Figure 8.15 shows the flow visualization inside the 1-mm slit burner.[11] The broken lines show the positions of the tangential slits and the broken circle shows the position of the burner wall. From the two slits of upper right and lower left, air is injected, while from the two slits of upper left and lower right, nitrogen is injected with seeded MgO particles. By denoting the width of the upper right air stream by δ_0, δ_1, δ_2, and δ_3 as shown in Fig.8.15, the variations of δ_0 and δ_2 with the tangential injection velocity are shown in Fig.8.16. It is seen that δ_0 follows the thickness of a boundary layer-type flow, that is, the thickness is inversely proportional to the square root of the approach velocity.

Based on the results, the thicknesses of the fuel and oxygen streams at high velocities are estimated, and the resulting mixing times and Damköhler numbers are listed in Table 8.3. Since the injection velocity of the fuel is much smaller than that of the oxidizer under very lean conditions, the thickness of methane seems to be larger than that of oxidizer. Diffusion through the

Table 8.3. Damkohler number

Φ	τ_r (μs)	Boundary Layer-Type Flow						Tubular Flame Establishment
		δ_m-O_2 (μm)	δ_m-CH_4 (μm)	τ_m (μs)	τ'_m (μs)	D_a	D'_a	
0.12	19,000	51	208	462	1885	0.024	0.099	Success
0.15	2070	51	186	413	1508	0.199	0.725	Success
0.17	981	51	175	388	1331	0.395	1.356	Failure
0.50	32	51	102	226	452	7.020	14.049	Failure

thicker layer seems to be a rate determining process, and hence, the modified mixing time τ'_m is also taken into consideration.

It is seen that in the cases where a tubular flame is established, that is, $\Phi = 0.12$ and 0.15, the modified Damköhler numbers based on the modified mixing times τ'_m are less than unity; the mixing time is shorter than the reaction time, and hence, mixing is completed before the onset of reaction, resulting in tubular flame combustion of a premixed type. In the cases where the tubular flame cannot be established, that is, $\Phi = 0.17$ and 0.5, the modified Damköhler numbers are larger than unity; it can be understood that the mixing is incomplete, resulting in the formation of a diffusion flame between the two streams.

Thus, the condition for the establishment of tubular flame combustion in the rapidly mixed-type tubular flame burner can be well argued based on two parameters, the swirl number and the Damköhler number. The swirl number is a measure of degree of overlapping between the fuel stream and the oxidant stream. The Damköhler number is a measure of degree of mixing between the fuel and the oxidant, which indicates whether the mixing is completed or not before the onset of reaction.

8.2 WIDE FLAMMABLE RANGE

As briefly introduced in Sections 1.2.1 and 1.2.2, the flammable range of tubular flame is very wide. Thus, in tubular flame burners, stable combustion can be obtained for a wide fuel concentration range from the lean to rich flammability limits and up to a large peak value for inert gas dilution. In addition, stable combustion can be also obtained for a wide range in combustible mixture flow rate. As an example, a burner for BFG is introduced. The burners for the determination of flammability limits or peak concentration can be found in Refs.[15–17]

8.2.1 BFG BURNERS

In steel works, by-product fuel gases are generated in various processes. Typical fuels are blast furnace gas (BFG), Linz–Donawitz converter gas (LDG), and coke oven gas (COG). Table 8.4

Table 8.4. Typical properties of by-product fuel gases in steel works. (*Source*: Ishioka, M., Okada, K., Ishizuka, S., Development of Tubular Flame Burner for By-Product Fuel Gases in Steel Works, Journal of the Combustion Society of Japan 48–145(2006)250–256.[18] Used by permission of the Combustion Society of Japan.)

By-Product Fuels		BFG	LDG	COG
Compositions	H_2	5.5	1.0	55.0
	N_2	47.0	11.5	4.8
	CO	24.0	77.5	8.0
	CO_2	23.0	10.0	3.0
	CH_4	–	–	26.0
	C_2H_4	–	–	2.0
	C_2H_6	–	–	1.0
	O_2	0.5	–	0.2
Lower heating value (MJ/m^3(stp))		4.0	8.8	19.3
Stoichiometric air–fuel ratio (m^3(stp) air/m^3(stp) fuel)		0.78	1.6	4.6
Stoichiometric burned gas–fuel ratio (m^3(stp) air/m^3(stp) fuel		1.6	2.3	5.3
Density (kg/m^3(stp))		1.36	1.31	0.49
Adiabatic flame temperature (air excess ratio: 1.0, dry air) (K)		1656	2269	2264
Adiabatic flame temperature (air excess ratio: 1.2, dry air) (K)		1573	2168	2096

Abbreviations: BFG, blast furnace gas; COG, coke oven gas; LDG, Linz–Donawitz converter gas.

summarizes their properties;[18] composition, lower heating value, stoichiometric air–fuel ratio, stoichiometric burned gas/fuel ratio, density, and adiabatic flame temperatures at the air excess ratio of 1.0 and 1.2. Since the compositions are daily varied, the average values are listed in Table 8.4.

It is seen that BFG has the lowest heating value of 4 MJ/m^3(stp) among three. The lower heating values of typical gaseous fuels of hydrogen, methane, and propane are 10.8, 35.8, and 91.2 MJ/m^3(stp) respectively, and hence, the lower heating value of BFG is very small, below one-twentieth that of propane. Usually, BFG is burned with an aid of a pilot flame, or mixed with other fuels of larger lower-heating value and burned. In a tubular flame burner, however, BFG can be burned without the aid of any auxiliary fuel.

To elucidate the wide flammable range of a tubular flame burner, a combustion test had been made with a 12-inch diameter tubular flame burner, which had four rectangular, tangential slits, two for fuel and two for air; hence rapidly mixed-type combustion had been attempted. The burner is schematically shown in Fig.8.17.[19] The slit length was 300 mm, while the slit width could be varied in a step wise, 6, 8, 16, and 24 mm by inserting plates of different thicknesses into the slit room in order to vary the swirl intensity.

The swirl number is defined as the ratio of the axial flux of angular momentum to the flux of axial momentum, which is made dimensionless by using the representative radius of

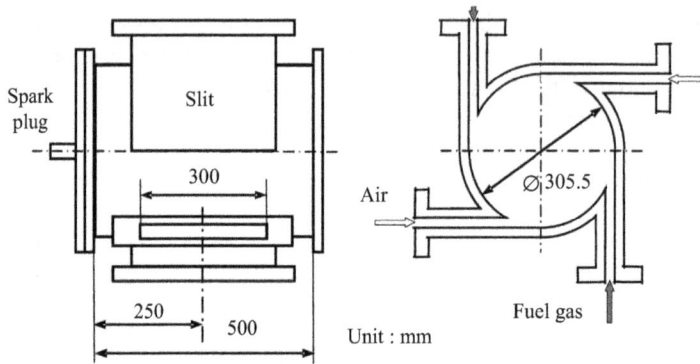

Figure 8.17. Schematic of a tubular flame burner. (*Source*: Ishioka, M., Okada, K., Ishizuka, S., Development of Tubular Flame Burner for By-Product Fuel Gases in Steel Works, Journal of the Combustion Society of Japan 48-145(2006)250–256.[19] Used by permission of the Combustion Society of Japan.)

the burner. In a tubular flame burner, the swirl number can be roughly calculated from its geometry as

$$S_w \equiv \frac{G_a}{G_x \cdot R_b} = \frac{\left(a\rho_a Q_a w_a + \rho_f Q_f w_f\right)}{\left(\rho_a Q_a + \rho_f Q_f\right) \cdot \dfrac{Q_a + Q_f}{A_b}} \cdot \frac{1}{R_b} \tag{8.16}$$

in which G_a is the angular momentum flux, G_x is the axial momentum flux, R_b is the representative radius of the burner, ρ is the density, w is the injection velocity through a tangential slit, Q is the volumetric flow rate flow, A_b is the cross-sectional area of the burner, and the subscripts, a and f, denote air and fuel, respectively. Based on this equation the swirl numbers are calculated for different slit widths under conditions of the air excess ratio of 1.2 and room temperature. BFG and its several mixtures diluted with nitrogen at different extents are listed in Table 8.5.[20] The swirl number slightly changes with the mixture composition; however, it decreases as the slit width is increased. For example, the swirl number for BFG decreases as 11, 7.5, 3.7, and 2.4 as the slit width is increased as 6, 8, 16, and 24 mm.

Ignition tests had been carried out by attaching the burner to a furnace that was 6 m long, 3.2 m wide, and 2.5 m high for 2 MW combustion. For comparison, a conventional burner was also tested with this furnace. Figure 8.18 shows schematically the setup of the burners.[20] The conventional burner was for burning BFG and LDG fuels, and the burning was always assisted with a 30-kW pilot flame of COG. LDG or BFG was mixed with the inner and outer swirling air streams and injected through six holes placed at the periphery of a center plate. The tubular flame burner was designed for 2 MW and the flame became long for large heat outputs; hence, a 500- or 1000-mm long combustion tube was inserted between the burner and the furnace. The mixture was ignited by an electric spark of 10 KVA discharge with a duration time of 10 s. Several positions were examined for the spark such as around the middle or periphery of the end plate, upstream and downstream of the tangential slit on the burner wall and so on. However,

Table 8.5. Swirl numbers for fuel gases (air–fuel ratio, 1.2; air, room temperature). (*Source*: NEDO Progress Report No. 03002451-0.[20] Used by permission of NEDO.)

Slit size, length × width (mm)		300×6	300×8	300×16	300×24
Number of slits		4	4	4	4
Fuel Gas* by volume	Blast furnace gas (BFG)	11	7.5	3.7	2.4
	BFG : N_2 = 9:1*	-	7.7	3.8	2.5
	BFG : N_2 = 8:2*	-	7.9	3.9	2.5
	BFG : N_2 = 7:3*	-	8.1	4.0	2.6
	BFG : N_2 = 6:4*	-	8.6	4.2	2.7
	Linz–Donawitz converter gas	12	9.0	4.4	2.8

Source: "NEDO Progress Report No. 03002451-0, Table 2-4, March 2004," published by the New Energy and Industrial Technology Development Organization (NEDO).

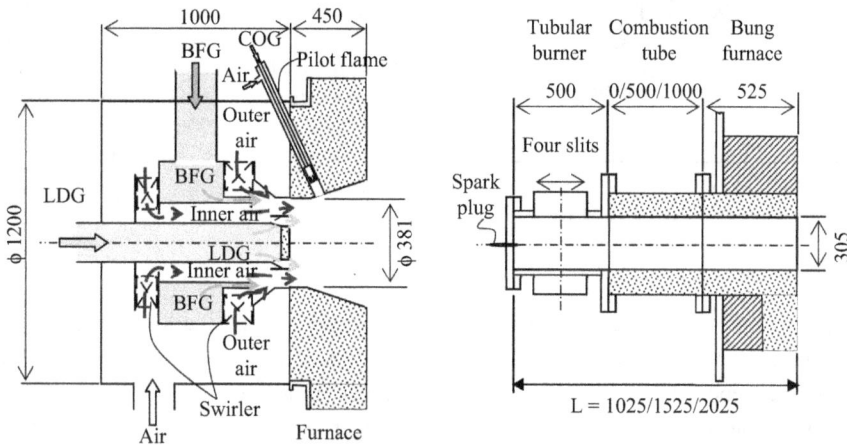

Figure 8.18. A conventional LDG burner (left) and a 12-inch tubular flame burner (right) BFG, blast furnace gas, LDG, Linz–Donawitz converter gas. (*Source*: NEDO Progress Report No. 03002451-0, March 2004.[20] Used by permission of NEDO.)

it is found that ignition at the center of the end plate ensures ignition and smooth transition to stable combustion, hence adopted. This is because rapid flame propagation along a vortex axis occurs due to the so-called vortex bursting.[21] In the conventional burner, the pilot flame of COG was always working at 30 kW.

Figure 8.19 shows the appearance of flames of BFG mixtures after ignition. In the conventional burner, six round flames were stabilized around the base plate. In the tubular flame burner, an almost constant diameter tubular flame was formed. Although the fuel and air were separately injected, the intensity of flame luminosity seemed quite uniform, suggesting less heat loss condition due to symmetry of temperature distribution.

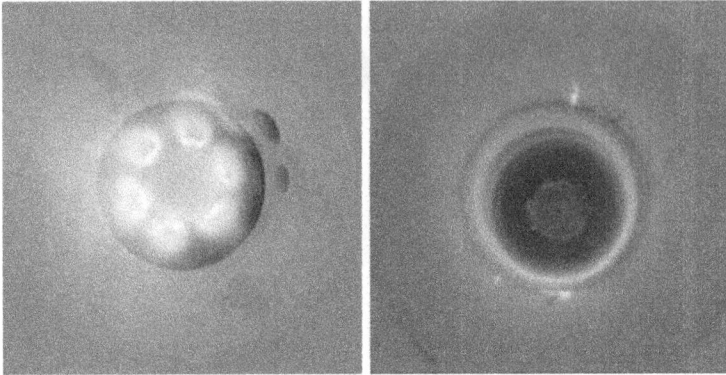

Figure 8.19. Appearance of flames in the conventional burner (a) and a tubular flame burner (b). (*Source*: Courtesy of JFE Steel Corporation.)

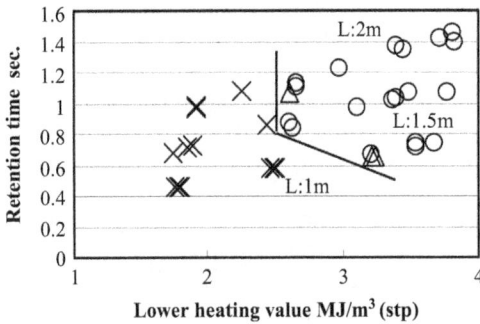

Figure 8.20. Mapping of a stable combustion region in a plain of retention time and lower heating value. (*Source*: Ishioka, M., Okada, K., Ishizuka, S., Development of Tubular Flame Burner for By-Product Fuel Gases in Steel Works, Journal of the Combustion Society of Japan 48-145(2006)250–256.[19] Used by permission of the Combustion Society of Japan.)

Figure 8.20 shows the relation between the retention time and the lower heating value of the BFG mixtures for ignition being the combustion tube length L as a parameter.[19] This test was made under conditions of 0.2 MW and the air excess ratios between 1.13 and 1.27 at room temperatures. The open circles denote success, the open triangles denote success for ignition but failure for transient to stable combustion, and the crosses denote failure for ignition. It is seen that ignition and transition to stable combustion is possible if the lower heating value exceeds 2.6 MJ/m³(stp) for 1.5 and 2.0 m combustion tubes, although larger lower-heating values are needed when the combustion length is shorter. When the lower heating value is 2.6 MJ/m³(stp), the mixture consists of 65% BFG and 35% nitrogen. Thus, it is found that stable combustion is possible in the tubular flame burner, even if BFG is diluted with nitrogen in the extent of about 54% by volume.

Figure 8.21 shows the limits for ignition and transition to stable combustion,[19] in which the swirl number is plotted against the lower heating value. Independently of the swirl number, ignition is possible when the lower heating value exceeds above 2.6 MJ/m³.

Finally, Figure 8.22 shows a comparison between the conventional BFG burner and the tubular flame burner, in which the burning load is plotted against the lower heating value.[20] In the conventional burner, which is used for combustion of BFG and LDG fuels, a pilot flame of COG is always operated at 30 kW. However, when BFG is diluted with nitrogen by 10%

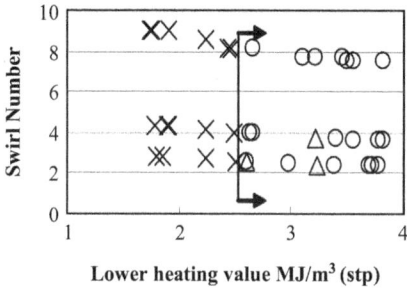

Figure 8.21. Mapping of a stable combustion region in a plain of swirl number and heating value. (*Source*: Ishioka, M., Okada, K., Ishizuka, S., Development of Tubular Flame Burner for By-Product Fuel Gases in Steel Works, Journal of the Combustion Society of Japan 48-145(2006)250–256.[19] Used by permission of the Combustion Society of Japan.)

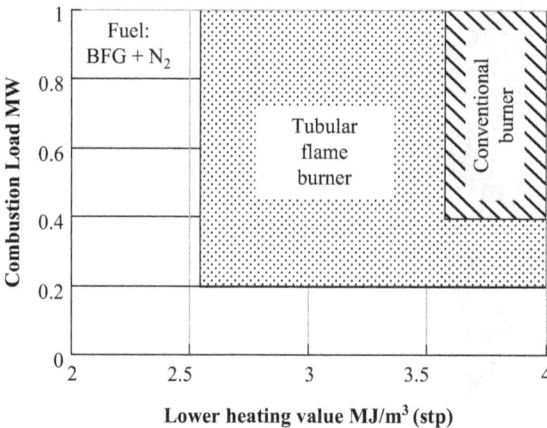

Figure 8.22. Mapping of a successfully ignitable region in a plain of combustion load and heating value. BFG, blast furnace gas. (Extracted from "Fig. 2-5 of NEDO Progress Report, "A Burner System of Next Generation", No.03002451-0, Figure 2-5, March 2004,"[20] published by the New Energy and Industrial Technology Development Organization (NEDO).)

by volume, stable combustion becomes impossible. On the contrary, stable combustion is possible in the rapidly mixed tubular flame combustion without any aid of pilot flame, if the lower heating value exceeds 2.6 MW/m^3(stp) and the combustion load is above 0.2 kW up to 2 MW, which is the upper limit of the present test facility.

8.3 FUEL DIVERSITY

A variety of fuels can be burned using tubular flame burners. In this section gaseous, liquid, and solid fueled burners are separately introduced.

8.3.1 GASEOUS FUELS

As briefly introduced in 8.2.1, there are many gaseous fuels that can be burned in the tubular flame burners. They are methane, propane, and hydrogen in laboratory-scale burners, city gas (13A), by-products gas fuels in steel works (BFG; LDG; COG), and a biomass gas produced using methane bacterium (60% CH_4 + 40% CO_2) in proto-type tubular flame burners. The largest heat output is 2.3 MW, which is for a boiler with a city gas fuel (13A). A 12-inch tubular flame burner with propane as fuel is briefly introduced in the following text.

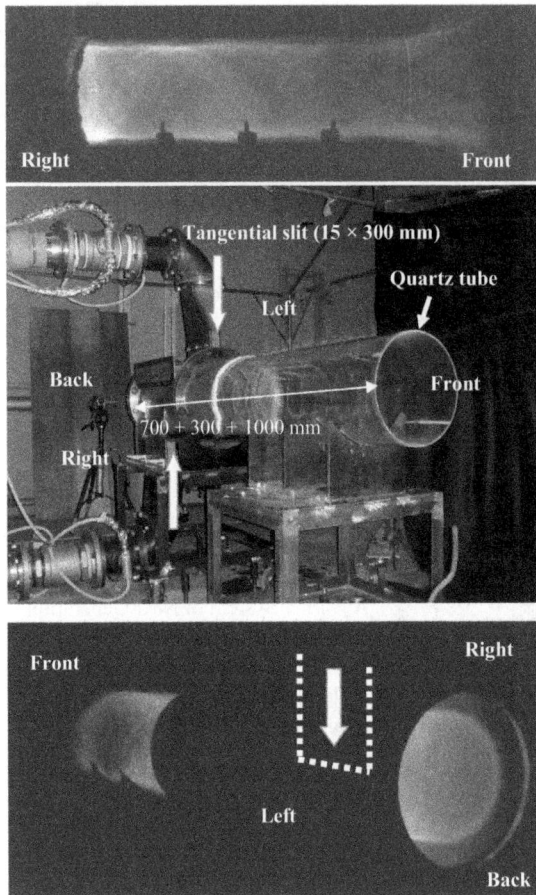

Figure 8.23. Appearance of a flame when a propane–air mixture is injected through the two slits in a 12-inch diameter tubular flame burner.

The burner is 12 inches in diameter with two rectangular slits that are 300 mm long and 15 mm wide, which gives a swirl number of 8.1. The length of the main body is 700 mm, attached with a 1500-mm long steel combustion tube, or with a 300-mm long steel tube and a 1000-mm long quartz tube for photographing.

Figure 8.23 shows the appearance of a flame when a premixed gas of propane and air is injected through the two slits. This picture was taken at a lean condition close to extinction limit; hence, the flame is long and extended outside the quartz tube. Even when the tube diameter is extended to 12 inches, a large laminar flame with uniform flame front can be obtained as in BFG fuel (see Fig.8.19).

However, at large volume flow rates of air and under near stoichiometric or slightly fuel-rich conditions, strong vibratory combustion abruptly occurs. Figure 8.24 shows pressure fluctuation spectra of the 12-inch tubular flame burner[22] obtained at the air flow rate is 1400 m^3/h (stp) and the equivalence ratios are 0.7 and 1.2.

When the equivalence ratio is 0.7, the combustion is rather stable and the pressure fluctuations are within ±1 kPa. When the equivalence ratio is 1.2, large pressure fluctuations up to ±13 kPa have been recorded. As shown in the pressure fluctuation spectrum, many peaks can be found in high frequency ranges such as 1680, 3500, 5085, 6760, and 8470 Hz.

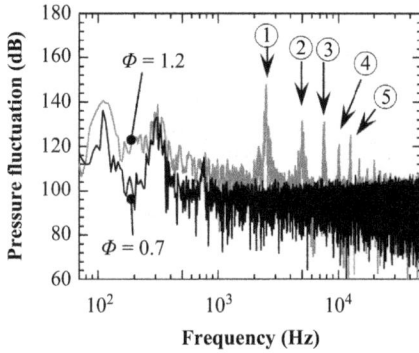

Figure 8.24. A pressure fluctuation spectrum of the 12-inch tubular flame burner (air flow rate, 1400 m³/h). (*Source*: Shimokuri, D., Shiraga, Y., Ishizuka, S., Ishii, K., Toh, H., High-Frequency Oscillatory Combustion in Tubular Flame Burners, Transaction of JSME 75(2009)1149–1156.[22] (Used by permission of JSME.)

According to a simple analysis on the vibratory combustion in a cylindrical vessel, the acoustic mode frequencies are given as follows:

$$f = \frac{c}{2\pi}\sqrt{\left(\frac{\alpha_{nm}}{r}\right)^2 + \left(\frac{(2n_z-1)\pi}{2(l+\Delta l)}\right)^2} \qquad (8.17)$$

in which f is the acoustic mode frequency [Hz], c is the sound speed, α_{nm} is the mth root of the second kind of the Bessel equation $dJ_n/dx = 0$, where J_n is the nth Bessel function, n is the tangential mode number such as 0, 1, 2,..., m is the radial mode number such as 1, 2, ..., r is the radius of the burner, ℓ and $\Delta\ell$ are the axial burner length and correction length whether the tube is open or closed end, respectively, and n_z is the axial mode number such as 1, 2, ... Assuming the sound speed at the burned gas temperature is 865 m/s at 2050K, the acoustic frequencies are 1660 ($n = 1$, $m = 1$), 3450 ($n = 0$, $m = 1$), 4800 ($n = 1$, $m = 2$), 6310 ($n = 0$, $m = 2$), and 7680 ($n = 1$, $m = 3$) Hz, which are well coincident with those observed in experiments, that is, 1680, 3500, 5085, 6760, and 8470 Hz, respectively. Note that n=1 means the instability mode is asymmetric. Thus, it is interesting to note that the most intense frequency of 1680 Hz observed in the tubular flame burner corresponds to the asymmetric circumferential mode ($n = 1$) instability.

8.3.2 LIQUID FUELS

A tubular flame burner for liquid fuels has been also developed. For small burners, prevaporized combustion or liquid film combustion may be useful. For large burners with a diameter 4 to 12 inches, tangentially spraying-type combustion has been attempted to obtain tubular flame combustion. Figure 8.25 is a picture of the liquid fuel combustion test with a 2-MW tubular flame burner.[18] The fuel was kerosene. This burner is the same as that for propane mentioned in the previous section except that eight nozzles are set up on a line and kerosene is tangentially spayed into the burner. In most of conventional liquid fueled burners, only one nozzle is installed at the end wall and a liquid fuel is injected axially into the furnace. Usually, the spray is designed not to reach the furnace wall; otherwise the furnace wall will be wet and damaged. To increase the heat output, a nozzle of large flow rate is selected and the furnace is designed to be long enough so that the spray does not reach the end wall. In the tubular flame burner, however, the fuel needs to be sprayed tangentially. The distance to the wall is less than the

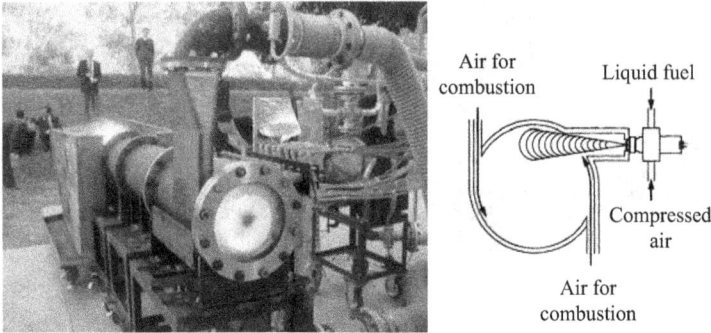

Figure 8.25. Combustion test (fuel, kerosene; burner diameter, 12 inches, 8 nozzles, 2MW). (*Source*: Ishizuka, S., Shimokuri, D., Ishii, K., Okada, K., Takashi, K., Suzukawa, Y., Development of Practical Combustors Using Tubular Flames, Journal of the Combustion Society of Japan 51-156(2009)104–113.[18] Used by permission of the Combustion Society of Japan.)

burner diameter. Therefore, instead of a large nozzle, a number of small nozzles are needed. In addition, to get a fine droplet to shorten the vaporizing time, an internal mixing twin-fluid atomizer has been chosen.

Up to now, banker-A and banker-C fuels have been successfully burned with 4- to 12-inch diameter burners. Interesting features of a liquid-fuel tubular flame burner are (1) blue flame combustion is possible at lean condition, and (2) emission of soot and dust is very little, for example, less than 0.005 g/m^3(stp) for a 2.3-MW tubular flame burner with kerosene or banker-A. It seems that a continuous flame zone is established due to the laminarization effect in a strong rotating flow field, and hence, it is very rare for fuel droplets to pass without complete vaporization and combustion, which leads to low emission of particulate matters.

8.3.3 SOLID FUELS

Although in the form of pulverized powder, a solid fuel can be used in a tubular flame burner. Figure 8.26 shows pictures of pulverized wood biomass powder and its appearance when fed with air, and the diameter distribution of the pulverized powder (mean diameter is 26 µm in this case).

Table 8.6 shows the compositions (wt%) and some properties of the biomass fuel. It contains water 9.9% by weight, and also N, S, and ash, each of which causes pollutants such as NOx, SOx, and particulate matter PM.

The burner adopted was 5 inches in diameter and 455 mm in length. The burner had two rectangular slits of 3 mm wide and 200 mm long through which air and biomass powder were injected. The burner had another tangential slit of 50 mm length and 3 mm width close to the closed end, through which a propane–air mixture was injected to assist the biomass powder combustion. The combustion load was varied up to 0.2 MW, while the pilot ratio, which is defined as the heat input of the pilot fuel (propane) to that of the sum of the main fuel (wood powder) and the pilot fuel (propane), was from 0 to 20%. The wood powders were classified

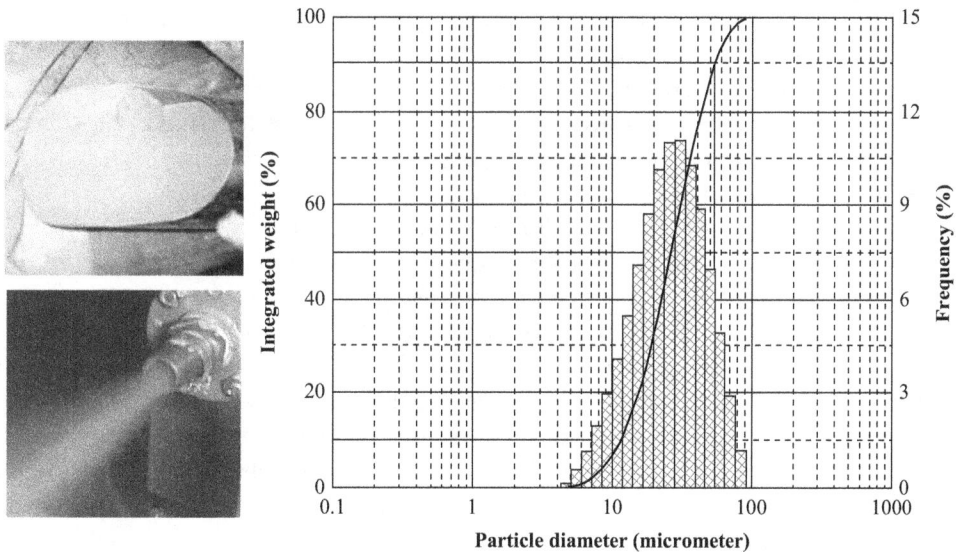

Figure 8.26. Pictures of pulverized biomass powder (left) and diameter distribution of wood biomass powder (right). (*Source*: Ishizuka, S., Shimokuri, D., Ishii, K., Okada, K., Takashi, K., Suzukawa, Y., Development of Practical Combustors Using Tubular Flames, Journal of the Combustion Society of Japan 51-156(2009)104–113.[18] with permission of the Combustion Society of Japan, and NEDO Progress Report No. 05001052-0, May 2006[26] with permission of NEDO.)

Table 8.6. Composition of the wood biomass. (*Source*: NEDO Progress Report, "A Burner System of Next Generation", No.05001052-0, May 2006.[26] Used by permission of NEDO.)

		Wood Biomass Mean Diameter 26, 45 μm	Wood Biomass Mean Diameter 75 μm
Composition wt (%)	C	48.2	46.1
	H	5.0	5.2
	N	0.7	0.4
	O	38.1	42.2
	Burnable S	0.04	0.01
	Cl	0.1	0.03
	Ash	7.8	6.1
	Total S	0.19	0.02
	total	100	100
Water (wt%) @ as received		9.9	5.6
Theoretical air (m³stp/kg)® wet base		4.38	4.09
Theoretical exhaust gas (m³stp/kg)@ wet base		4.91	4.67
Lower calorific value (MJ/kg) @wet base		15.3	14.2

Source: "NEDO Progress Report No. 05001052-0, Table 5-2-3-1, March 2006," published by the New Energy and Industrial Technology Development Organization (NEDO).

into three groups based on mean diameter, 26, 45, and 75 μm, and the combustion tests were individually made.

Figure 8.27 shows an example of the burning of the middle 45 μm class fuel under conditions of 0.15 MW and the equivalence ratio of 0.8 in the 5-inch diameter burner.[18] It is seen that combustion was successfully made. What is most important is that even if the pilot ratio is zero, that is, without an auxiliary gaseous fuel, self-sustained combustion is possible.

The interaction between solid particles and flame is one of the very fundamental problems in combustion research, and this problem has been experimentally and numerically studied using strained premixed and non-premixed flames and inert and combustible particles.[23–25] It has been shown that large inert particles can cause more effective flame cooling compared with smaller particles, and that at low strain rates combustible particles can effectively burn within the gaseous flame zone and thus enhance the overall reactivity, whereas at high strain rates, the particles are rapidly transported through the flame to have no or minor effect on the gaseous flame, and so on. Thus, the combustion of biomass particles in the tubular flame burner should be studied in more detail fundamentally. However, at present, the following knowledge has been obtained for practical use.[18,26]

(1) Stable combustion has been obtained for wide ranges up to 0.2 MW under stoichiometric and fuel-lean conditions.
(2) The wood biomass pulverized to a mean diameter of 45 μm can be burned stably without the assistance of the pilot flame at an equivalence ratio of 0.8.
(3) CO emission is less than 10 ppm at an equivalence ratio of 0.9. This value is less than the allowed level for a burner used in industrial furnaces.

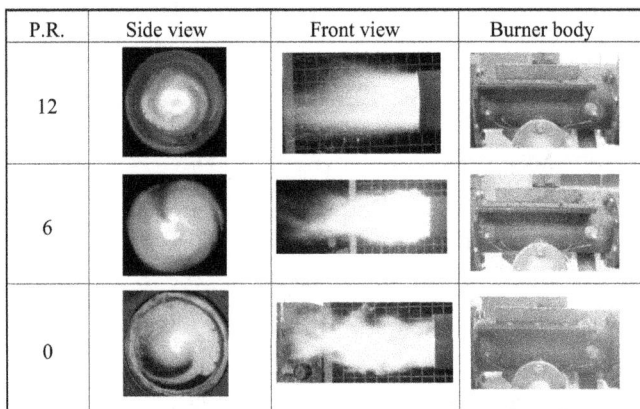

Figure 8.27. Combustion test of wood biomass particles in a 5-inch diameter tubular flame burner under various pilot ratios (0.15 MW; 45 μm; equivalence ratio, 0.8). (*Source*: Ishizuka, S., Shimokuri, D., Ishii, K., Okada, K., Takashi, K., Suzukawa, Y., Development of Practical Combustors Using Tubular Flames, Journal of the Combustion Society of Japan 51-156(2009)104–113.[18] Used by permission of the Combustion Society of Japan.)

(4) NOx emissions are 350–450 ppm under 150–200 kW load independently of equivalence ratio or flame temperature. Most of them are fuel NO. The NOx conversion ratio of the nitrogen in the wood biomass is approximately 30%.

(5) The combustion efficiency is approximately 97.5 % for the coarsely pulverized (mean diameter 75 μm) wood biomass with the pilot ratio of 12%. The value seems to be sufficient for an industrial burner of solid fuel combustion.

8.4   COMPACTNESS

Tubular flame burner is simple in structure. Therefore, the burner can be made compact. In this section, three examples are briefly introduced: the burners for fuel-processing system for PEFC, for heating a hollow fastening bolt, and for a superheated steam generator.

8.4.1   FUEL-PROCESSING SYSTEM FOR POLYMER ELECTROLYTE FUEL CELL

Co-generation systems with the use of PEFC have come into widespread use in Japan. PEFC needs a fuel-processing system to make hydrogen from a city gas that consists mainly of methane. Although there are several types of fuel-processing systems, a tubular flame burner installed in one of the fuel-processing systems is briefly introduced here.[27,28]

Figure 8.28 shows a schematic of the fuel-processing system.[18,26] The fuel-processing system consists of three catalytic processes: (1) steam reforming reaction of hydrocarbon ($CH_4 +$

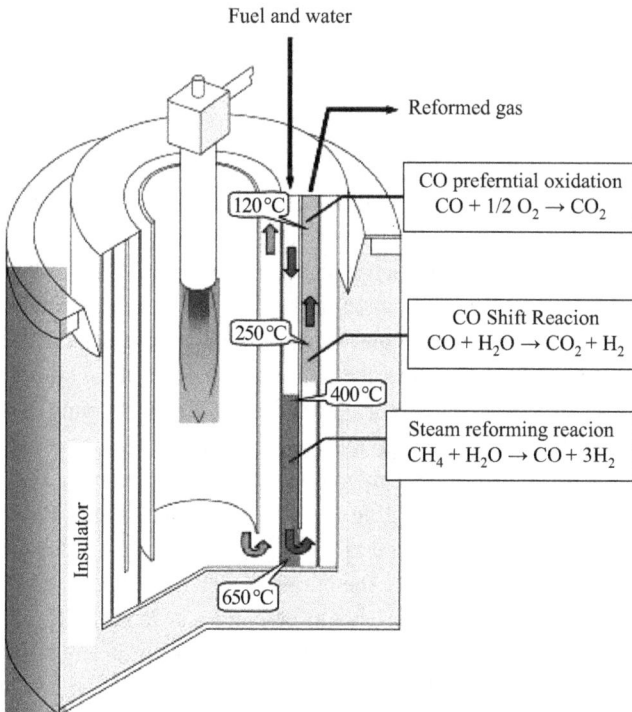

Fuel and water

Reformed gas

120 °C

250 °C

400 °C

650 °C

Insulator

CO preferntial oxidation
$CO + 1/2\ O_2 \rightarrow CO_2$

CO Shift Reacion
$CO + H_2O \rightarrow CO_2 + H_2$

Steam reforming reacion
$CH_4 + H_2O \rightarrow CO + 3H_2$

Figure 8.28. Schematic of the fuel-processing system. Three kinds of catalysts, installed in the multilayered reactors, are thermally controlled by a burner. (*Source*: Ishizuka, S., Shimokuri, D., Ishii, K., Okada, K., Takashi, K., Suzukawa, Y., Development of Practical Combustors Using Tubular Flames, Journal of the Combustion Society of Japan 51–156(2009)104–113.[18] Used by permission of the Combustion Society of Japan.)

$H_2O \rightarrow CO + 3H_2$), (2) CO shift reaction ($CO + H_2O \rightarrow CO_2 + H_2$), and (3) CO preferential oxidation ($CO + 0.5O_2 \rightarrow CO_2$). The catalyzer works most effectively in the steam reforming process between 650 and 400°C, while the catalyzer in the second CO shift reaction works around 250°C, and the catalyzer in the third CO preferential oxidation works around 120°C. The concentration of CO should be less than 10 ppm; otherwise the fuel cell is damaged severely. This fuel-processing system has a multilayer structure, at the middle of which a tubular flame burner is installed to keep the temperatures at appropriate values for the catalytic processes.

For 1 kW output, several combinations of burner component sizes have been attempted. They were the burner diameters of 14.9, 20, and 25 mm; the burner lengths of 7.5, 15, and 22.5 mm; the slit numbers of 2, 3, and 4; the slit widths of 0.8 and 1.2 mm for a constant slit length of 10 mm; and the back space between the upstream edge of the slit and the closed end plate of 3, 7.5, and 15 mm. The tubular flame burner which gave the best performance was 14.9 mm in diameter, 10.5 mm in length with a 3 mm back space, with two slits, one for fuel and one for air, that is, rapidly mixed-type combustion, 0.8 mm wide and 10 mm long. The swirl number roughly calculated was about 22.[26]

The technical merit of adopting a tubular flame burner is that it offers a very uniform temperature distribution. The performance of the catalyzers can be optimized and, in addition, the emissions of CO and NOx can be well controlled and reduced.

Figure 8.29 shows the radial temperature distributions of the conventional non-premixed co-annular swirl burner and a non-premixed tubular flame burner, which are found at 15 mm below the edges of the burner.[27] It is seen that in the conventional swirl burner the temperature becomes less than 1000 K in the Y-direction indicated while the temperature reaches 2000 K around the center, whereas in the tubular flame burner, the temperatures are almost within the 1200–1400 K range. A very uniform temperature distribution could be obtained in the tubular flame burner, which enables to optimize the emission of NOx below 50 ppm at 5% O_2 in a wide range of equivalent ratios (see Fig.8.30).[27]

8.4.2 HOLLOW FASTENING BOLT

To fasten a high-pressure vessel firmly, a shrinkage fitting method is used. Figure 8.31 shows, for example, pictures of a steam turbine usually operated under high pressure conditions and hollow fastening bolts to close the vessel. Usually, an electric heater is inserted into the hollow space to heat the bolt to have an adequate relative elongation to the casing. It takes, however, a few hours of heating; it also needs a large amount of electricity and a large facility for the electricity supply. Then, an attempt has been made to heat the bolt with a tubular flame burner.

According to a numerical analysis on heat transfer in a M80 fastening bolt, it is found that a 20 kW heating with a burned gas of 1328°C at the equivalence ratio 0.67 of a propane–air mixture can heat the bolt 300°C higher than the casing temperature, within one-fourth the time which a conventional electric heater needs.[29] In this heating method with a burned gas, however, the surface temperature of the bolt should not exceed 500°C; otherwise the quality of the bolt will never be guaranteed. The space for installing the burner is also limited, as seen in the enlarged photograph of Fig.8.31. Namely, the burner should be compact, detachable, and workable independently of the direction whether it is set horizontally, vertically, or in any other direction. The allowance is 150 mm in height and 280 mm in diameter in the present case.

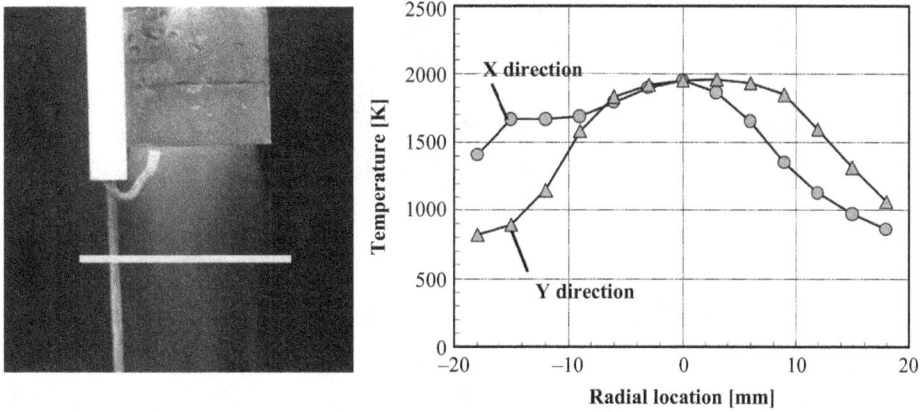

(a) The conventional burner using a non-premixed co-annular swirl flame.

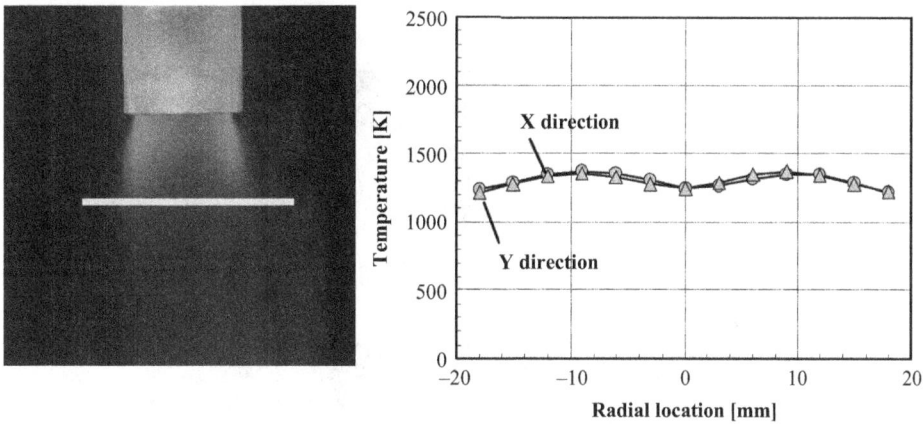

(b) The improved burner using a non-premixed tubular flame.

Figure 8.29. Temperature distribution of the gas in the radial direction at 15 mm below from the edge of the burners. Fuel is city gas and equivalent ratio is 0.83. (*Source*: Yagi, K., Atarashiya, K., Nojima, S., Development of a Fuel Processing System for Fuel Cell Systems Using a Tubular Flame Burner, Abstracts of Work-In-Progress Posters, Thirty-first International Symposium on Combustion, p.211, 2006.[27])

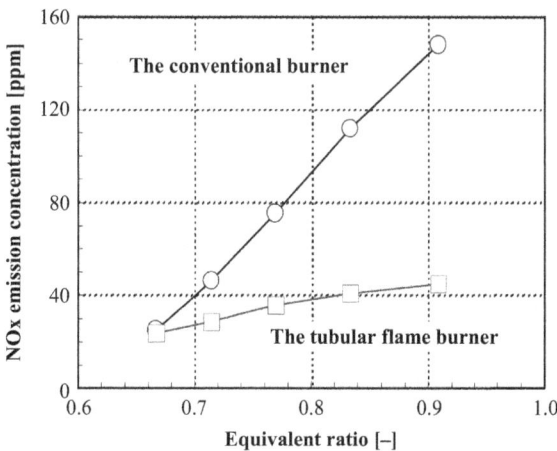

Figure 8.30. NOx emission characteristics. (*Source*: Yagi, K., Atarashiya, K., Nojima, S., Development of a Fuel Processing System for Fuel Cell Systems Using a Tubular Flame Burner, Abstracts of Work-In-Progress Posters, Thirty-first International Symposium on Combustion, p.211, 2006.[27])

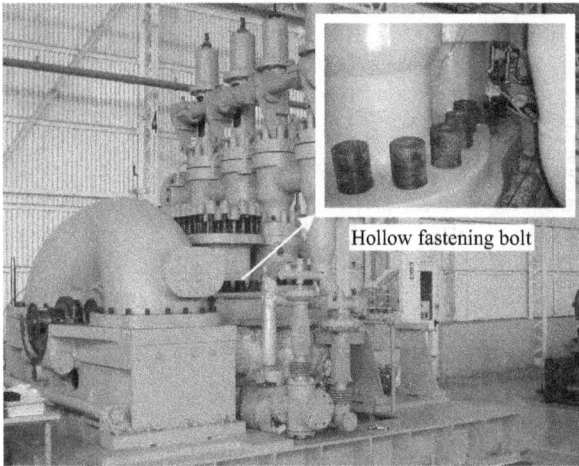

Figure 8.31. Outlook of a high-pressure steam turbine and hollow fastening bolts.

Figure 8.32. A tubular flame burner for heating a hollow fastening bolt.

To meet these requirements, a tubular flame burner has been developed. The burning velocity of the lean propane–air mixture is low, resulting in a large burner diameter and a long burner length. In order to make the burner compact, the burner is designed to have two sections; in the first section, a stoichiometric mixture of higher burning velocity is burned and in the second section, the burned gas is diluted with an excess air to reduce the temperature to the appropriate value. The final design for the burner was 50 mm in inner diameter, with two 3-mm wide and 50-mm long slits for the stoichiometric mixture and two 3-mm wide and 30-mm long slits for the excess air. The total length of the burner used is about 200 mm. Figure 8.32 shows the pictures of the burner and the flame, which were taken through a quartz window placed at the closed end. It is seen that a blue tubular flame is established and an inserted thermocouple is heated by the burned gas and radiated. The burner wall temperature is kept low due to an injection of a combustible gas of room temperature. The mixture was ignited with an electric spark placed near the cylindrical wall.

Figure 8.33 shows a schematic of apparatus for the combustion test.[29] The burner is mounted on a hollow M80 bolt, whose outer diameter and length are 76 and 634 mm, respectively. The

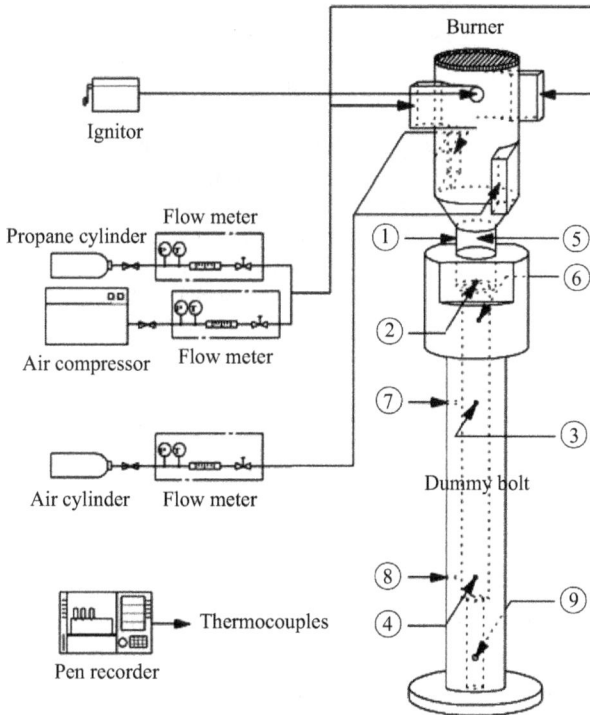

Figure 8.33. Schematic of temperature measurements. (*Source*: Oba, D., Ishizuka, S., Okamoto, K., Shimokuri, D., Atarashiya, K., Katagake, M., Development of a Tubular Flame Burner for Heating a Fastening Bolt, Proceedings of the forty-fourth Symposium (Japanese) on Combustion, pp.526–527, 2006.[29] Used by permission of the Combustion Society of Japan.)

hollow space in the bolt is 21 mm in diameter and 429 mm in length, which is further connected to a narrower hollow space of 15 mm in diameter and 205 mm in length.

Ten thermocouples were inserted into this bolt and temperatures were determined. The burned gas temperature diluted with the excess air was determined at position "1" just downstream the excess air slits on the center axis. Burned gas temperatures were also determined at positions "6" and "9", which were positioned at distances, 50 and 47 mm from the top and bottom ends of the bolt, respectively, for which the heads of the thermocouples were protruded to a distance of 3 mm from the inner wall surface. The inner temperatures of the bolt were determined at positions "2", "3", and "4", which were positioned at distances 15, 184, and 414 mm from the top end of the bolt, respectively, for which the heads of the thermocouples were immersed just behind 1 mm from the inner surface of the bolt. The inner temperatures of the bolt were also determined at position "7" and "8", which were positioned at distances,184 and 414 mm from the top end of the bolt, respectively, for which the heads of the thermocouples were immersed 10 mm from the outer surface of the bolt.

Figure 8.34 shows the time histories of the temperatures.[29] After ignition, the burned gas temperature at position "5" goes up to about 1200°C, while the inner bolt temperatures gradually increase. The mean inner bolt temperature at positions "3", "4", "7", and "8" exceeds about 300°C in about 10 minutes. Thus, it was elucidated that an adequate elongation of the bolt could be successfully obtained within a short operation time with a tubular flame burner.

It should be, however, noted that in its early stage of development, vibratory combustion often occurred after ignition, and on occasion, flame was extinguished. In order to understand the vibratory combustion, the burner was connected with a straight pipe of 0.75 inch in inner diameter and 100 to 1100 mm in length and the pressure fluctuations were determined.

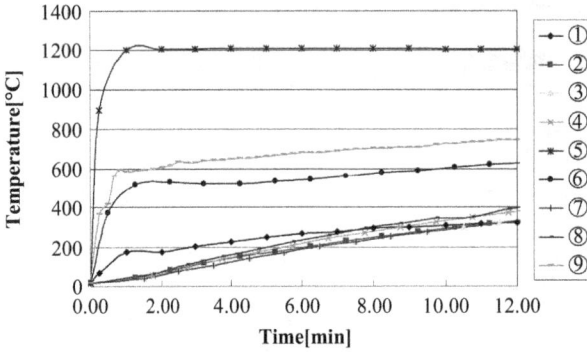

Figure 8.34. Time histories of temperatures. (*Source*: Oba, D., Ishizuka, S., Okamoto, K., Shimokuri, D., Atarashiya, K., Katagake, M., Development of a Tubular Flame Burner for Heating a Fastening Bolt, Proceedings of the forty-fourth Symposium (Japanese) on Combustion, pp.526–527, 2006.[29] Used by permission of the Combustion Society of Japan.)

Frequency analyses have shown that the characteristic frequencies f correspond to those of the Helmholtz resonator, which are given as

$$f = \frac{1}{2\pi}\sqrt{\frac{a^2 S}{VL}} \tag{8.18}$$

in which V denotes the volume of the combustion chamber, L and S respectively denote the length and cross-sectional area of the throat in the Helmholtz resonator, which are assumed to be equal to those of the pipe, respectively, and a is the sound velocity of the burned gas.[30]

After careful investigation, it was found that the fuel supply was easily plugged by the pressure increase in the chamber since the flow rate of propane was much smaller than the air flow rate. Then, the fuel supply line was improved to reduce the volume in the line as little as possible, which ensured a steady fuel supply, resulting in a successful formation of a stable tubular flame in the burner.

8.4.3 SUPERHEATED STEAM GENERATOR

Nowadays, superheated steam is widely used for cooking in family kitchens, restaurants, and delicatessen shops as well. Electric heater is convenient for generating superheated steam; however, superheated steam needs a large amount of heat to vaporize cold water and further heat up to a desired temperature such as 250°C. Thus, the electricity cost becomes very high and sometimes electric power failure happens since the maximum electricity power supply available is quite limited in houses or restaurants.

City gas has an advantage over electricity in that the cost is lower and it does not need so large a facility. Figure 8.35 shows an example of a grill for household use, in which a super-heated steam generator is installed.[18] To make the steam generator compact for the household use, a tubular flame burner is used. The superheated steam generator is 130 mm high, 250 mm wide, and 60 mm deep, while the tubular flame formed is 20 mm in diameter and 150 mm in length. Superheated steam of 250°C can be obtained at a rate of 1 kg/h with a 3-kW tubular flame in this kitchen grill.

Concerning a tubular-flame steam generator, more fundamental studies have been conducted using prototype generators.[31–33] Figure 8.36 schematically shows a recent prototype

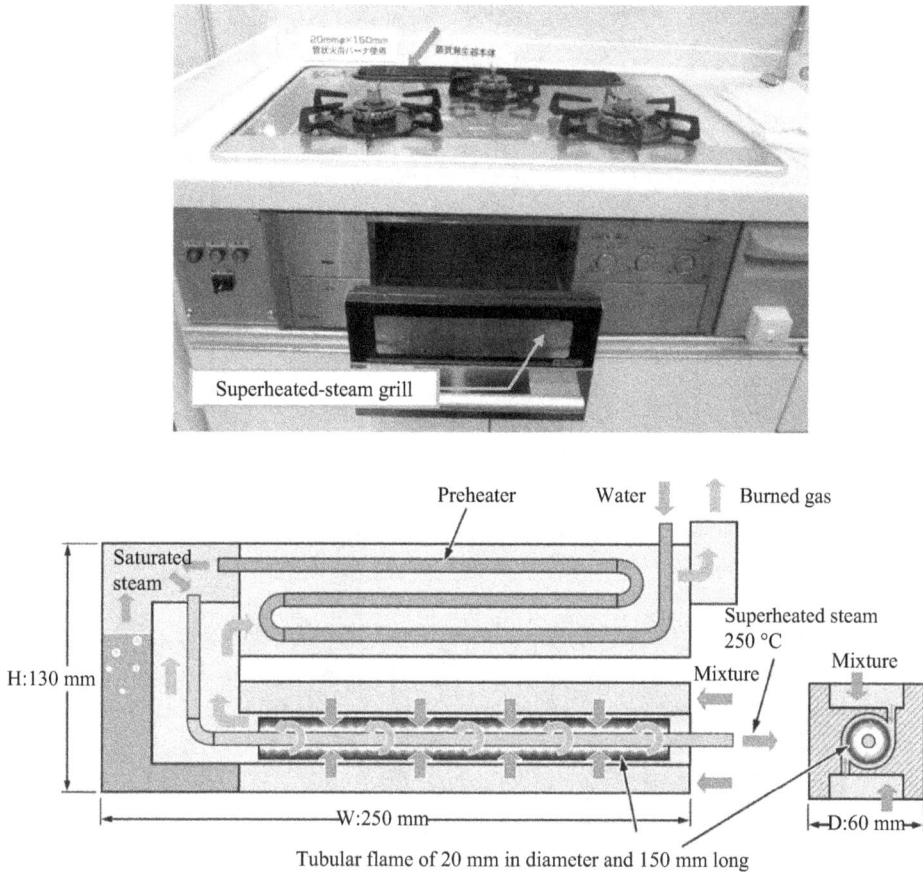

Figure 8.35. Tubular flame burner for superheated steam generator.[18] (*Source*: Ishizuka, S., Shimokuri, D., Ishii, K., Okada, K., Takashi, K., Suzukawa, Y., Development of Practical Combustors Using Tubular Flames, Journal of the Combustion Society of Japan 51–156(2009)104–113.[18] Used by permission of the Combustion Society of Japan.)

micro superheated steam generator and Fig. 8.37 shows an example of the combustion test.[33] In this generator, a tubular flame burner is set at the left end and a combustible mixture is tangentially injected through two slits of 2 mm wide and 25 mm long into a 40-mm inner diameter tube of quartz glass, which permits viewing. The fuel used is 13A, which consists of 89% methane, 7% ethane, 3% propane, and 1% butane. The lower heating value is 40.4 MJ/m^3(stp).

For an appropriate condition of the heat input Q and the excess air ratio λ, a tubular flame of finite length is formed inside the burner. Figure 8.37 shows the photograph under conditions of $Q = 5$ kW, $\lambda = 1.5$, flow rate water 90 cc/min. At the right open end of the quartz tube, an internally finned tube of aluminum-based alloy is attached, and with high efficiency, this tube is heated by the passing hot burned gas.

Both the quartz tube and the finned tube are surrounded by a polycarbonate tube of 105 mm in inner diameter, which also permits viewing and also works as an evaporator. The tube is eccentrically placed with respect to the burner tube, and set at an angle of 5 degrees to the horizontal line in order to collect vaporized water, that is, steam, efficiently.

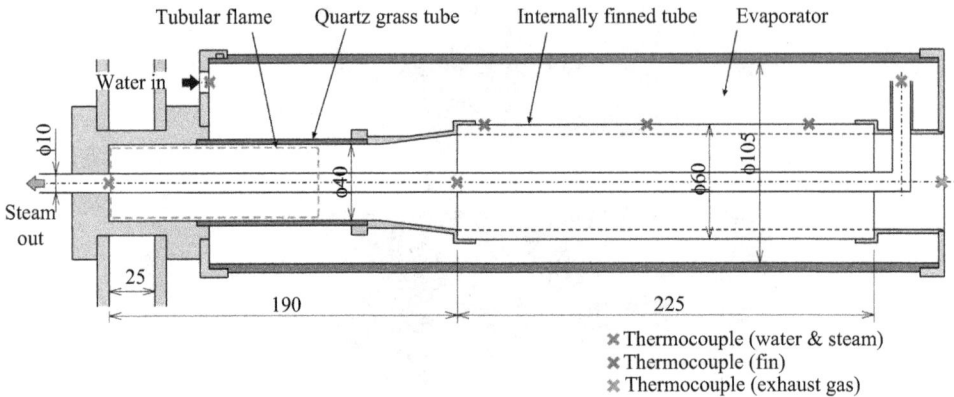

Figure 8.36. Prototype of a micro superheated steam generator. (*Source*: Matsumoto, R., Kobayashi, Y., Ozawa, M., Kegasa, A., Shiraga, Y., Takemori, T., Hisazumi, Y., Katsuki, M., Development of Superheated Steam Generator Using Tubular Flame, Proceedings of the 22nd International Symposium on Transport Phenomena ISTP-22 CD-ROM No.135, Delft, Netherlands.[33] Courtesy of Professor Ryosuke Matsumoto, Kansai University.)

Figure 8.37. Combustion test of a micro superheated steam generator. (Courtesy of Professor Ryosuke Matsumoto, Kansai University.)

Water is introduced from the left end of the evaporator, heated and vaporized by heat through the wall of the finned tube, and stored accordingly at the upper part of the evaporator. Then, the steam is introduced into a stainless steel tube of 10 mm inner diameter, which passes through the finned tube and also through the hot burned gas region of the tubular flame. Passing through the stainless tube, the steam is heated first by the finned tube and then further heated by the tubular flame. Finally, the superheated steam gets out from the left end of the micro superheated steam generator.

Figure 8.38 shows temperatures measured for the inlet water, the wall of the finned tube, steam in the evaporator, and the exhaust gas.[33] The inlet water temperature is 25.3°C and the temperature of the steam in the evaporator is 100°C since the pressure inside is atmospheric pressure. The wall temperature of the finned tube is highest at 131°C, although the exhaust gas temperature is much higher. The temperature of the steam increases in the hot burned gas region of the tubular flame and superheated water of 173°C is successfully obtained at the outlet of the steam generator.

The emissions of the exhaust gas are low enough, 31.7 ppm of CO and 7.4 ppm of NOx, for the household use. According to a heat balance analysis based on the lower heating value,

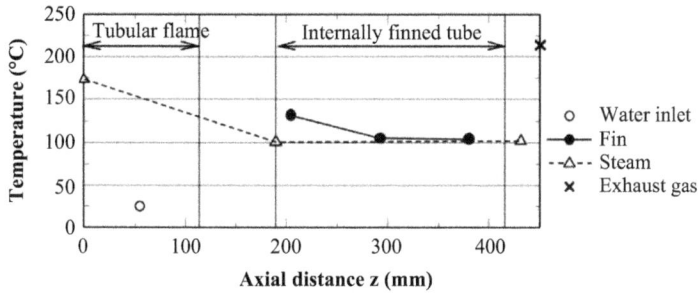

Figure 8.38. Temperature profile of the prototype micro superheated steam generator. (*Source*: Matsumoto, R., Kobayashi, Y., Ozawa, M., Kegasa, A., Shiraga, Y., Takemori, T., Hisazumi, Y., Katsuki, M., Development of Superheated Steam Generator Using Tubular Flame, Proceedings of the 22nd International Symposium on Transport Phenomena ISTP-22 CD-ROM No.135, Delft, Netherlands.[33] Courtesy of Professor Ryosuke Matsumoto, Kansai University.)

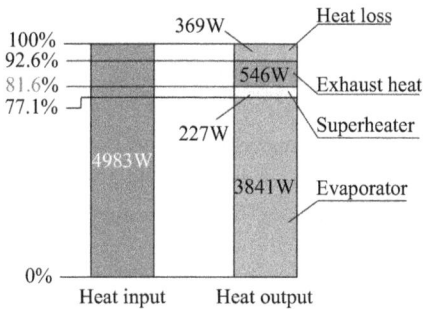

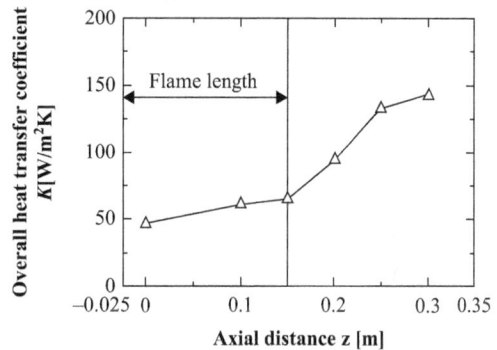

Figure 8.39. Heat balance of the prototype micro superheated steam generator. (*Source*: Matsumoto, R., Kobayashi, Y., Ozawa, M., Kegasa, A., Shiraga, Y., Takemori, T., Hisazumi, Y., Katsuki, M., Development of Superheated Steam Generator Using Tubular Flame, Proceedings of the 22nd International Symposium on Transport Phenomena ISTP-22 CD-ROM No.135, Delft, Netherlands.[33] Courtesy of Professor Ryosuke Matsumoto, Kansai University.)

Figure 8.40. Overall heat transfer coefficient ($Q = 5.0\,kW$, $\lambda = 1.4$). (*Source*: Kobayashi, Y., Matsumoto, R., Ozawa, M., Kegasa, A., Takemori, Y., Hisazumi, Y., Katsiki, M., Funagoshi, H., Development of Superheated-Steam Generator Using Tubular Flame, Transaction of JSME 77(2011)997–1001.[31] Used by permission of JSME.)

of the heat input of 4983 W, 3841 W was used in the evaporator, 546 W was carried out as an energy loss by the exhaust gas, and only 227 W was used in the superheat of the steam. The rest 369 W seems to be heat loss (see Fig.8.39).[33] To increase the temperature of the superheated steam, more heat is transferred in the burned gas region of the tubular flame or in the finned tube section.

Figure 8.40 shows variations of the overall heat transfer coefficient around a stainless steel pipe of 6 mm in outer diameter, which is placed in the middle of a tubular flame burner of 28 mm in inner diameter.[31] A finned tube is not attached in this case. A tubular flame is formed between the axial distance $Z = 0$ and 0.15 m. It is seen that the heat transfer coefficient is slightly increased as the end of the tubular flame is approached and the values are around 50 W/(m^{2}K) in the flame, whereas it is further increased and becomes about three times as high as the values in the flame zone. The reason why the heat transfer coefficient is smaller in the tubular flame region is considered due to less radial velocity by a centrifugal force of rotation. A device to increase the overall heat transfer coefficient in the flame zone or an improvement in the finned tube in the micro superheated steam generator is indispensable to get a higher temperature of the superheated steam.

8.5 GEOMETRY

Tubular flame has a peculiar geometry; its inside is occupied by a hot burned gas, which can be used for various purposes. In this section, three examples are introduced, that is, stabilizing a flame in a high-velocity stream; heating a gas stream which may carry pollutants, solid particles, liquid droplets, and so on; and heating a cylindrical wall such as the head of the Stirling engine.

8.5.1 FLAME STABILIZATION

As a new technique to stabilize a flame in a high-speed stream, a tubular flame can be used. Figure 8.41 shows conventional ways for flame stabilization. The first method is to insert a bluff body in a high-speed stream and stabilize a flame in the rear stagnation region where a hot burned gas is recirculating.[34] In this method, a large pressure loss of the main stream occurs. The second method is to use an opposing jet to stabilize a flame around the stagnation region of

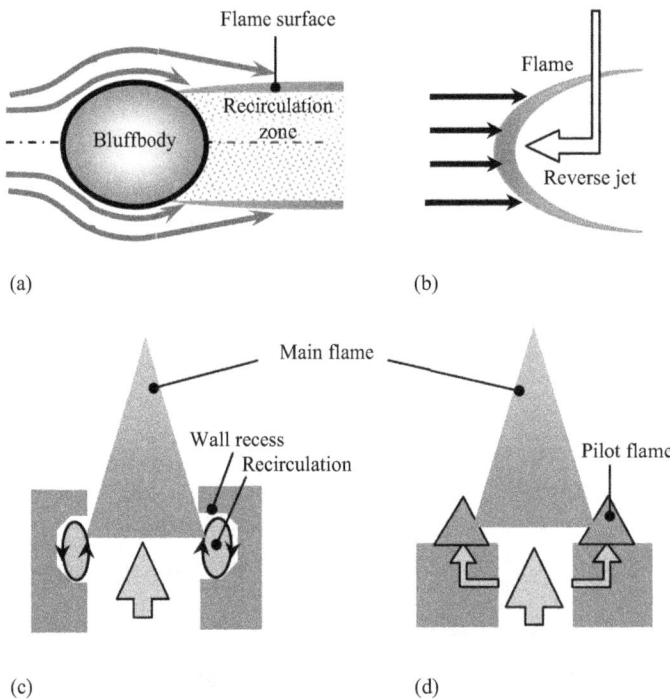

Figure 8.41. Methods of flame stabilization, (a) bluff body, (b) opposing jet, (c) recess wall, and (d) pilot flame. (*Source*: Shimokuri, D. Ishizuka, S., Flame Stabilization with a Tubular Flame, Proceedings of the Combustion Institute 30(2005)399–406.[34] Used by permission of Elsevier.)

low velocities, but there exists a large pressure loss as well. Other methods are to use a recess wall and a pilot flame. In these methods, the pressure loss may be small; however, the flames are prone to blow off because both the recirculating flow in the recess wall and the pilot flame are disturbed directly by the main high-speed stream.

A swirling tubular flame, however, is robust aerodynamically; hence, it is expected that the flame can anchor a flame of the main combustible stream. To elucidate the validity of this technique, an experiment has been conducted.

Figure 8.42 shows the experimental setup. The main burner consists of a diffuser, a settling chamber, and a contraction nozzle that ensures a flat velocity profile at the exit of the nozzle.[34] The exit diameter is 18 mm. To stabilize the main flame, a tubular flame burner is mounted on the nozzle. The inner diameter of the tubular flame burner is 50 mm. A combustible mixture is tangentially injected from four tangential slits, 2 mm wide and 30 mm long, and a tubular flame is established along the inner wall of the burner. It should be noted that the main nozzle diameter is 18 mm while the inner diameter of the tubular flame burner is 50 mm. There exists a 16-mm recess at the exit of the main nozzle. For burned gas analyses and photographing, a stainless or quartz tube of 100 mm inner diameter and 500 mm length is mounted on the burner with an exchangeable plate of 15 mm thickness. The total length of the inner wall is 60 mm.

There are four parameters in this experiment; the mean nozzle velocity and the equivalence ratio of the main stream, V_n and Φ_n, respectively, and the tangential velocity and the equivalence ratio of the tubular flame, v_t and ϕ_t, respectively. Figure 8.43 is a photograph when a stoichiometric methane–air mixture is axially injected at a velocity of 130 m/s, while a

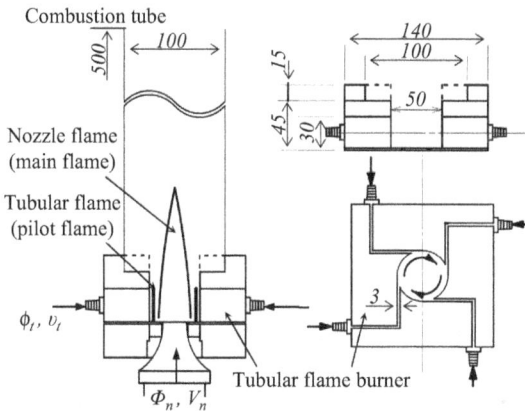

Figure 8.42. Schematic of the experimental apparatus. (*Source*: Shimokuri, D. Ishizuka, S., Flame Stabilization with a Tubular Flame, Proceedings of the Combustion Institute 30(2005)399–406.[34] Used by permission of Elsevier.)

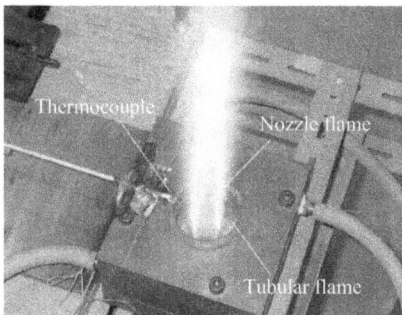

Figure 8.43. A photograph showing the nozzle flame anchored by a tubular flame ($V_n = 130$ m/s, $\Phi_n = 1.0$; $v_t = 8.5$ m/s; $\phi_t = 1.0$). (*Source*: Shimokuri, D. Ishizuka, S., Flame Stabilization with a Tubular Flame, Proceedings of the Combustion Institute 30(2005)399–406.[34] Used by permission of Elsevier.)

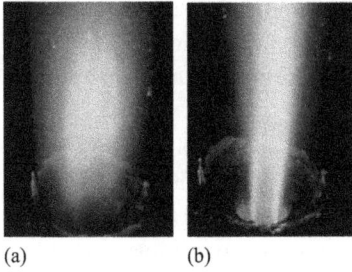

(a)   (b)

Figure 8.44. Appearance of the ducted combustion (a) without and (b) with a tubular flame (V_n = 90 m/s, Φ_n = 1.0, and ϕ_t = 1.0). (*Source*: Shimokuri, D. Ishizuka, S., Flame Stabilization with a Tubular Flame, Proceedings of the Combustion Institute 30(2005)399–406.[34] Used by permission of Elsevier.)

stoichiometric methane–air mixture is tangentially injected at 8.5 m/s. It is seen that the nozzle flame can be anchored with the tubular flame.[34]

Figure 8.44 shows the appearance of the ducted combustion with and without a tubular flame;[34] in this experiment a quartz tube of 100 mm diameter and 500 mm length is mounted on the burner. The velocity and equivalence ratio of the main stream are 90 m/s and 1.0, respectively, while the tangential velocity and the equivalence ratios are 8.5 m/s and 1.0, respectively. It is seen that the combustion is unstable without the tubular flame; however, a main flame is successfully stabilized with the tubular flame. Burned gas analyses with gas chromatography indicate that an almost chemical equilibrium condition can be achieved at 50 m/s, but not at 90 and 130 m/s. Since the energy input relative to the main flame is just 6.6% at 130 m/s, the present tubular burner is not enough to burn all the unburned gas completely at high velocities. However, it may be said that the tubular flame has a potential for stabilizing a flame in a high-speed stream.

8.5.2 HEATING PROCESS

In industries, there are various heating processes. For example, one is just to heat an air stream by a hot burned gas, although the air may be vitiated, and the other is to feed particles with air or inert gas into a hot atmosphere to melt or vaporize them. There are waste gas treatment equipments and chemical scrubber systems, in which direct combustion is used to remove the odor or pollutants in the waste gases. Tubular flame burners have an advantage in geometry since it can be installed in line with such a process. Now, several tubular flame burners are under development for these purposes. Here, basic features of fundamental interest in the tubular flame burner are briefly introduced.

Figure 8.45 shows counterflow- and swirl-type tubular flame burners for heating air stream.[35,36] The air to be heated enters from the left end, passes through the hot gas region of the tubular flame, and then exits from the right end. The inner diameter D and the heating length L (the length of a porous cylinder in the counterflow-type burner and tangential slits in the swirl-type burner) are the same for both the burners, 50 mm and 40 mm, respectively. By varying the slit width and the numbers used, the swirl intensity can be varied; the swirl numbers calculated are 3.1 for four 4-mm slits, 6.1 for four 2-mm slits, and 12.2 for two 2-mm slits. At the center, air is fed through a half-inch pipe (inner diameter 16 mm) from the left end of the burners, while a quartz tube of 50 mm in inner diameter and 300 mm length is attached at the right end of the burners.

Figure 8.46 shows the appearance of flames obtained by the two burners at lean, stoichiometric, and rich conditions.[35,36] In the case of the non-swirling, porous cylinder burner, the

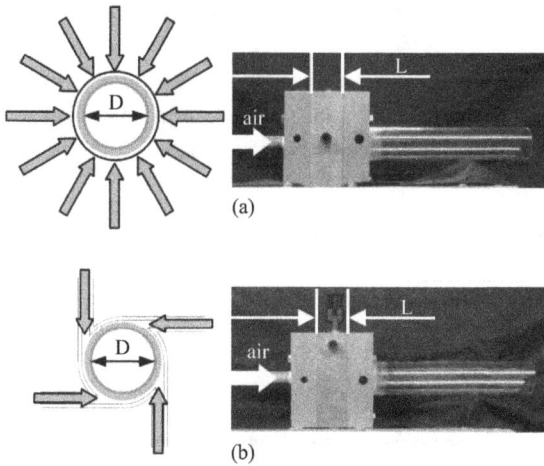

Figure 8.45. (a) Counterflow-type porous cylinder burner and (b) swirl-type burner (inner diameter D = 50 mm, length of the porous cylinder or the tangential slit L = 40 mm). (*Source*: Hu, J., Shimokuri, D., Ishizuka, S., A fundamental Study on the Heating with a Tubular Flame, Proceedings of the Fiftieth Symposium (Japanese) on Combustion, pp.424–425, 2012.[35] Used by permission of the Combustion Society of Japan.)

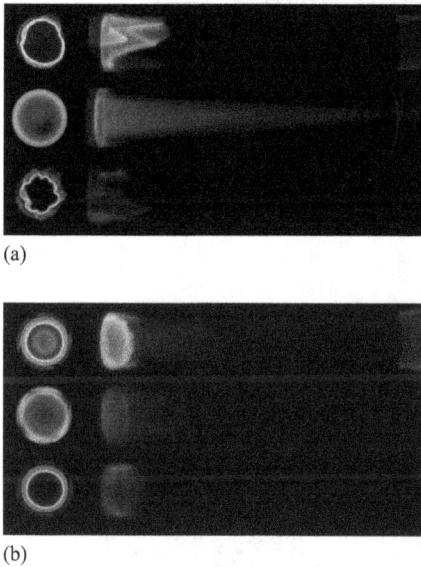

Figure 8.46. Flames appearances (Q_{air} = 5.0m^3/h (a) a counterflow burner (S = 0) and (b) a swirl-type burner (S = 6.1), Φ = 1.4 (upper), 1.0 (middle), 0.6(lower). (*Source*: Hu, J., Shimokuri, D., Ishizuka, S., A fundamental Study on the Heating with a Tubular Flame, Proceedings of the Fiftieth Symposium (Japanese) on Combustion, pp.424–425, 2012.[35] Used by permission of the Combustion Society of Japan.)

ranges in fuel concentration and in flow rate for which a tubular shaped flame is obtained are quite limited. As shown in Fig.8.46(a), at lean and rich conditions, the flames are deformed from the cylindrical shape, although the flame is cylindrical at the stoichiometric condition because the injection velocity is less than the burning velocity (the mean injection velocity normal to the wall is 24 cm/s, while the burning velocity of the stoichiometric mixture is about 40 cm/s). On the other hand, as shown in Fig.8.46(b), a cylindrical flame can be obtained for wide ranges in fuel concentration and in flow rate in the swirl-type burner. The flame diameters are almost equal to the burner inner diameter, and the flame length is increased as the mixture equivalence ratio becomes smaller or larger than unity.

Instead of a quartz tube, a stainless tube of the same inner diameter and length was attached to the burner, and temperatures were determined by inserting a thermocouple (Pt/Pt-13%, wire diameter 0.1 mm, SiO$_2$ coated) through three holes of 4 mm in diameter, which were positioned at the distance Z, 125, 200, and 275 mm from the beginning of the injection port, as shown in Fig.8.47.

Figure 8.48 shows the radial temperature distributions determined at positions Z being 125 mm and 275 mm.[35,36] The positions, r = 0 and 25 mm, correspond with the wall and the axis of symmetry, respectively. In these measurements, a stoichiometric methane–air mixture is

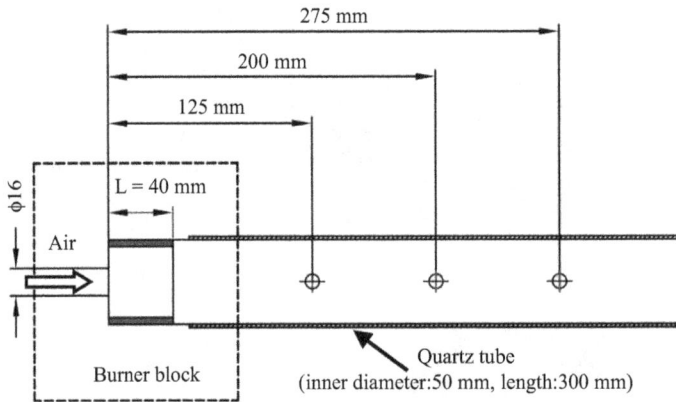

Figure 8.47. Schematic of temperature measurements. (*Source*: Hu, J., Shimokuri, D., Ishizuka, S., A fundamental Study on the Heating with a Tubular Flame, Proceedings of the Fiftieth Symposium (Japanese) on Combustion, pp.424–425, 2012.[35] Used by permission of the Combustion Society of Japan.)

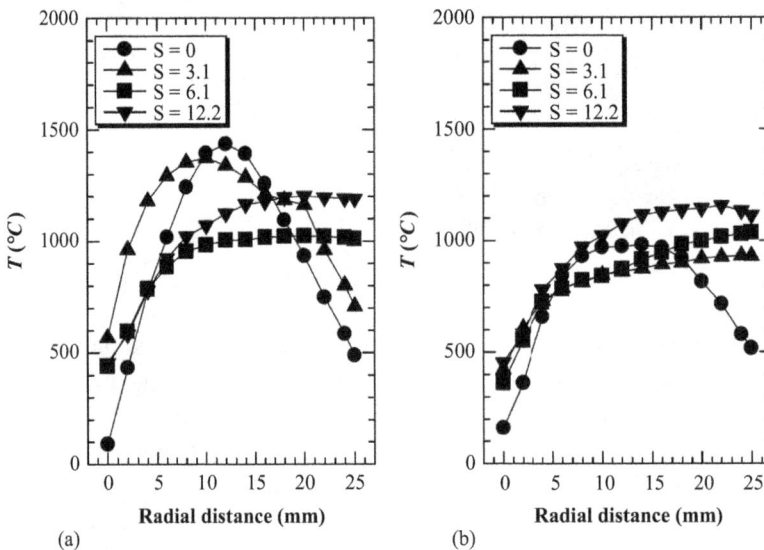

Figure 8.48. Radial temperature distributions (half-inch pipe: Q_{air} = 4.0m³/h, tubular flame burners: Q_{air} = 5.0 m³/h, Φ = 1.0, (a) Z = 125 mm, (b) Z = 275mm). (*Source*: Hu, J., Shimokuri, D., Ishizuka, S., A fundamental Study on the Heating with a Tubular Flame, Proceedings of the Fiftieth Symposium on Combustion, pp.424–425, 2012 (in Japanese).[35] Used by permission of the Combustion Society of Japan.)

injected from the tubular flame burners with an air flow rate of 5.0 m³/h (stp), while the air to be heated is injected from the half-inch pipe with a flow rate of 4.0 m³/h (stp). It is seen that in the case of the nonswirling burner, the temperature of the air stream at the center (r = 25 mm) is low at Z = 125 mm, and even downstream at Z = 275 mm, the temperature still remains low; whereas, in the swirling burner, the temperature of the air stream is increased at Z = 125 mm as the swirl number is increased; that is, around 700, 1000, and 1200°C for Sw = 3.1, 6.1, and 12.2, respectively. Downstream at Z = 275 mm, the temperature is further increased to 900°C for Sw = 3.1; however, those for Sw = 6.1 and 12.2 have remained almost unchanged compared with their values at Z = 125 mm, which suggests that the temperature has already reached the final value at Z = 125 mm for Sw = 6.1 and 12.2. The reason that the temperature for Sw = 12.2 is higher than that for Sw = 6.1 is because the temperature is more swiftly raised up by the tubular flame, and hence, heat loss with the stainless tube used in this measurement is minimized for Sw = 12.2. In the case of the nonswirling burner, the temperature at the center increases a little between Z = 125 and 275 mm, and the maximum temperature is rather decreased to 500°C at Z = 275 mm. In industrial processes, the tube will be lagged; however, the temperature will not reach the final temperature designed with the nonswirling, counterflow-type tubular flame burner in the presence of heat loss.

Concerning the rapid temperature rise in the swirling tubular flames, an interesting result has been obtained experimentally. Figure 8.49 is a picture of flow visualization at Z = 125 mm, taken with a PIV system, under a cold flow condition. The flow rate of air injected from the half-inch pipe is 0.03 m³/h (stp). The flow rate of air injected from the swirl-type tubular flame burner is 0.3 m³/h (stp) and the swirl number is 12.2. It should be noted that only the air from the half-inch pipe is seeded with MgO particles. It is seen that the outer, nonseeded swirling air is involved into the inner, seeded air stream, and the surface area between the two streams is extensively increased. Mixing and heat exchange between the two streams must be enhanced significantly, and this may result in the rapid temperature rise observed in the swirling tubular

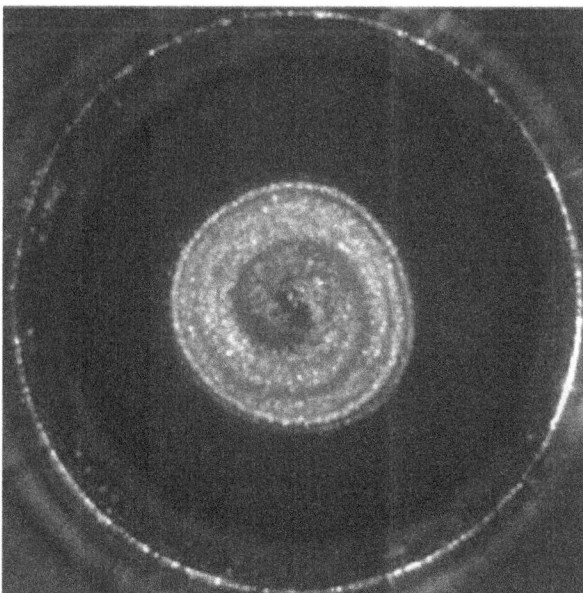

Figure 8.49. Flow visualization (Z = 125 mm, cold flow, half-inch pipe: Q_{air} = 0.03 m³/h (stp) seeded with MgO, swirl-type burner: Q_{air} = 0.3 m³/h (stp), Sw = 12.2).

flames. As in the helical annular mixer,[37] stretching and chaos may play an important role in the heating and mixing process in a vortex flow with a swirling tubular flame. Further investigation is under way.

In conclusion, the swirling tubular flame has an advantage on heating process in that (1) a tubular-shaped flame can be obtained for wide ranges in fuel concentration and in flow rate, independently of the burner-oriented direction; (2) the heat transfer rate is enhanced with swirl, and hence, the length of the heating process can be shortened.

8.5.3 STIRLING ENGINE

As a power source for emergency power supply, the Stirling engine has received considerable attention due to its high thermal efficiency as well as its fuel diversity as an external combustion engine. In order to heat the head of Stirling engine uniformly, a tubular flame has an advantage in geometry. Very recently, a tubular flame burner has been developed for a 1-kW electric power output Stirling engine using methane, propane, and a biogas of 60% methane and 40% CO_2 as fuel.[38,39]

Figure 8.50 shows the schematic of the tubular flame burner developed.[38,39] Since the outer diameter of the Stirling engine is 165 mm including fins, the inner diameter of the tubular flame is decided to be 200 mm; eight rectangular slits of 2 mm width and 8 mm length are made at the periphery. This burner is installed at the outside of the Stirling engine, as schematically shown in Fig.8.51.

Figure 8.52 shows the appearance of flame at representative equivalence ratios Φ while keeping the heat input at a constant of 10 kW, namely, keeping the flow rate of methane (Fig.8.52(a)) and propane (Fig.8.52(b)) at 1.0 and 0.4 m^3/h, respectively.[38,39] Although individual flames are anchored at the exits of the slits and separated under near-stoichiometric conditions, they are merged into a cylindrical flame under very lean conditions near extinction.

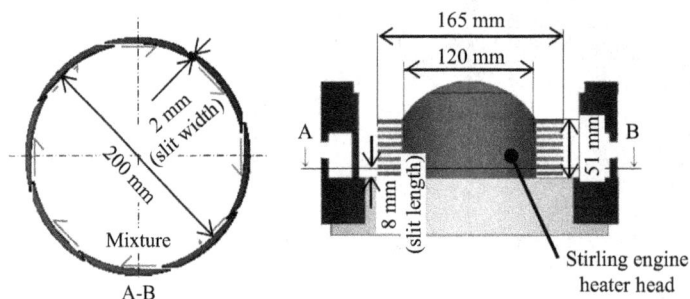

Figure 8.50. A tubular flame burner for a 1-kW Stirling Engine.
(*Source*: Kakehashi, Y., Shimokuri, D., Ishizuka, S., Toki, F., Murata, Y., Saito, M., Development of Tubular Flame Burner for a Stirling Engine, Proceedings of the 17th National Symposium on Power and Engine Systems (SPES2012), pp.175–176, 2012.[38] Used by permission of JSME.)

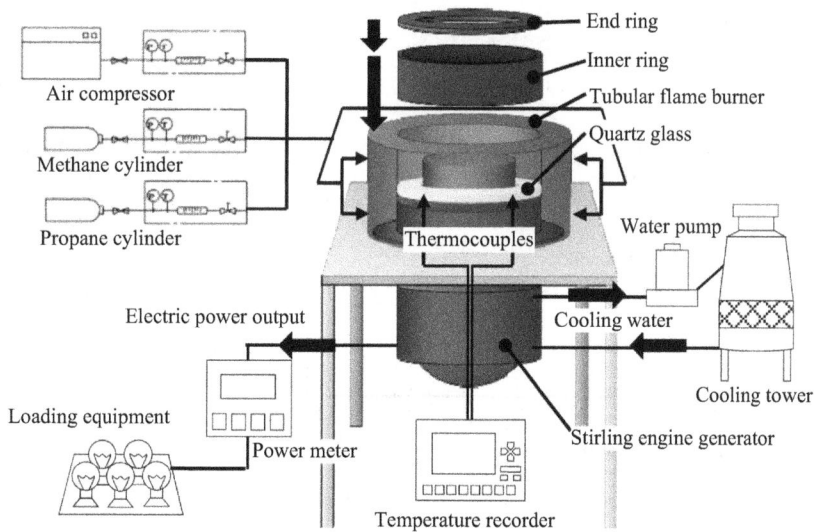

Figure 8.51. Schematic of the apparatus. (*Source*: Kakehashi, Y., Shimokuri, D., Ishizuka, S., Toki, F., Murata, Y., Saito, M., Development of Tubular Flame Burner for a Stirling Engine, Proceedings of the 17th National Symposium on Power and Engine Systems (SPES2012), pp.175–176, 2012.[38] Used by permission of JSME.)

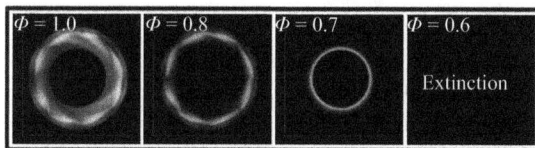

(a)

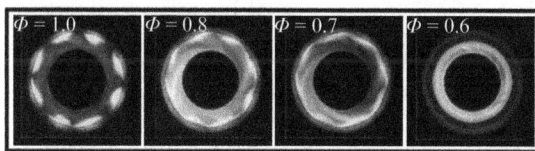

(b)

Figure 8.52. Appearance of flames ((a) methane, (b) propane, heat input, 10 kW). (*Source*: Kakehashi, Y., Shimokuri, D., Ishizuka, S., Toki, F., Murata, Y., Saito, M., Development of Tubular Flame Burner for a Stirling Engine, Proceedings of the 17th National Symposium on Power and Engine Systems (SPES2012), pp.175–176, 2012.[38] Used by permission of JSME.)

Figure 8.53 shows a mapping of stable flame region, in which the equivalence ratio at extinction is plotted against the heat input.[38,39] It is seen that stable combustion can be obtained even near the lean flammability limits for both methane and propane as in the tubular flame burner (see Table 1.1). This is amazing since the flame is short in length as compared with its diameter, and hence, it is appropriate to call the flame a ring flame, instead of a tubular flame.

Figure 8.54 shows the variations of the head temperature, electric power generation, and thermal efficiency with the equivalence ratio using the heat input as a parameter.[38,39] An important finding is that although the flame temperature takes its maximum around the stoichiometric condition, the heater head temperature, the electric power generation, and the thermal efficiency are increased with decreasing the equivalence ratio under a constant heat-input condition. These increases are attributed to an increase in the heat transfer coefficient to the head of the Stirling

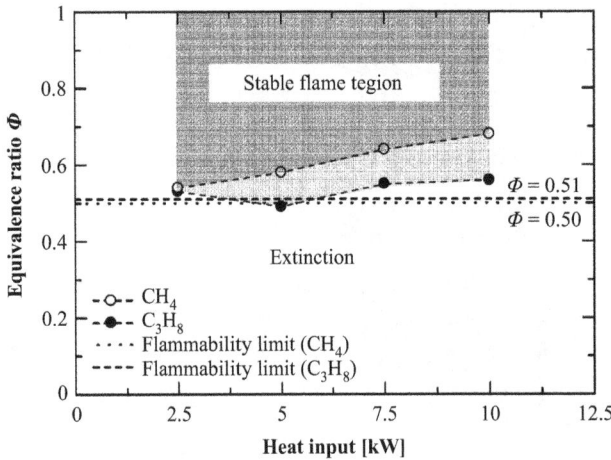

Figure 8.53. Stable flame region (*Source*: Kakehashi, Y., Shimokuri, D., Ishizuka, S., Toki, F., Murata, Y., Saito, M., Development of Tubular Flame Burner for a Stirling Engine, Proceedings of the 17th National Symposium on Power and Engine Systems (SPES2012), pp.175–176, 2012.[38] Used by permission of JSME.)

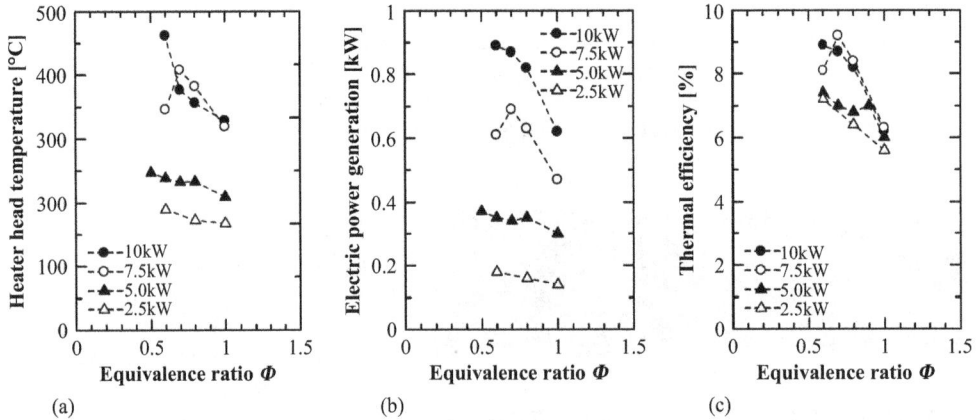

Figure 8.54. Variations of (a) heater head temperature, (b) electric power generation, and (c) thermal efficiency with equivalence ratio. (*Source*: Kakehashi, Y., Shimokuri, D., Ishizuka, S., Toki, F., Murata, Y., Saito, M., Development of Tubular Flame Burner for a Stirling Engine, Proceedings of the 17th National Symposium on Power and Engine Systems (SPES2012), pp.175–176, 2012.[38] Used by permission of JSME.)

engine because the air flow rate is increased with decreasing the equivalence ratio under a constant heat-input condition. Thus, in heating the finned Stirling engine, heat transfer is more important than the flame temperature in increasing the performance of the Stirling engine. The maximum thermal efficiency is 9.2% in the present system; however, it has been raised up to 12% by adopting a self-recirculating-type tubular flame burner.[40]

ACKNOWLEDGMENTS

The research project "A Burner System of Next Generation (development of an advanced tubular burner system)" was commissioned by the New Energy and Industrial Technology

Development Organization (NEDO) as a Fundamental Technology Research Facilitation Program (Private Sector Fundamental Technology Research Support Scheme) from September 2002 to March 2007. The author greatly appreciates NEDO for the financial support. The author sincerely acknowledge great contributions of many participants in this program; Messrs. Yoshitsugu Iino (General Manager in this program), Yutaka Suzukawa, Kuniaki Okada, Kouichi Takashi, and Dr. Munehiro Ishioka of JFE steel Co. Ltd.; Messrs Kenji Atarashiya and Katsuki Yagi of Mitsubishi Heavy Industries Ltd.; Mr. Nobuo Higuchi of Hirakawa Guidom Co. Ltd.; Profs. Takao Tsuboi and Kazuhiro Ishii of Yokohama National University and Dr. Daisuke Shimokuri of Hiroshima University. Before this project, prototype tubular flame burners were preliminarily developed, for which Mr. Shinichiro Fukushima and Dr. Minoru Suzuki (now Professor of Toho University) of NKK Co. Ltd. contributed extensively. Their efforts are highly appreciated.

The author also acknowledges helpful and fruitful discussions on a variety of tubular flame burners with Prof. Hidemi Toh of Kanazawa Institute of Technology, Messrs Tomoyoshi Sasaki of MHI Turbo-Techno Co. and Makoto Katagake of Mitsubishi Heavy Industries Ltd., Dr. Akeshi Kegasa of Osaka Gas company (now Professor of Osaka University), Prof. Ryosuke Matsumoto of Kansai University, Professor Emeritus Masashi Katsuki of Osaka University, and President Masamichi Saito of Promaterials, Inc.

REFERENCES

1. Schlichting, H. 1968. *Boundary-layer theory*, 6th ed., 79. New York: McGraw-Hill.
2. Jost, W., J. Krug, and L. Sieg. *Observations of Disturbed Flames, Fourth Symposium (International) on Combustion*, 535–7. Baltimore: Williams and Wilkins. DOI: 10.1016/S0082-0784(53)80074-3.
3. Hagiwara, R., M. Okamoto, S. Ishizuka, H. Kobayashi, A. Nakamura, and M. Suzuki. 2000. Combustion characteristics of a tubular flame burner for methane. *Transaction of JSME* 66:3226–32 (in Japanese).
4. Yamaoka, I. and H. Tsuji. 1985. *Determination of Burning Velocity Using Counterflow Flames, Twentieth Symposium (International) on Combustion*, 1883–92. DOI: 10.1016/S0082-0784(85)80687-1.
5. Ishizuka, S., T. Motodamari, and D. Shimokuri. 2007. Rapidly-mixed combustion in a tubular flame burner. *Proceedings of the Combustion Institute* 31:1085–92. DOI: 10.1016/j.proci.2006.07.128.
6. Shimokuri, D., Y.-Y. Zhang, and S. Ishizuka. 2007. PIV measurements on a 2-inch tubular flame burner. In *Proceedings of the Sixth Asia-Pacific Conference on Combustion*, 154–7.
7. Ishizuka, S. 1993. Characteristics of tubular flames. *Progress in Energy and Combustion Science* 19:187–226. DOI: 10.1016/0360-1285(93)90015-7.
8. Beer, J. M. and C. A. Chigier. 1972. *Combustion aerodynamics*. London: Applied Science Publication.
9. Beer, J. M., N. A. Chigier, T. W. Davies, and K. Bassindale. 1971. Laminarization of turbulent flames in rotating environments. *Combustion and Flame* 16:39–45. DOI: 10.1016/S0010-2180(71)80009-3.
10. Zawadzki, A. and J. Jarosinski. 1983. Laminarization of flames in rotating flow. *Combustion Science and Technology* 35:1–13. DOI: 10.1080/00102208308923700.
11. Shi, B., D. Shimokuri, and S. Ishizuka. 2013. Methane/Oxygen combustion in a rapidly mixed type tubular flame burner. *Proceedings of the Combustion Institute* 34:3369–77. DOI: 10.1016/j.proci.2012.06.133.
12. Poinsot, T. and D. Veyhante. 2001. *Theoretical and numerical combustion*, 57. Philadelphia: Edwards.

13. Kee, R. J., J. F. Grcar, M. D. Smooke, and J. A. Miller. 1985. A FORTRAN program for modeling steady laminar one-dimensional premixed flames, SAND 85-8240, Sandia National Laboratories Report.

14. Kee, R. J., G. Dixon-Lewis, J. Warnatz, M. E. Coltrin, and J. A. Miller. 1986. A FORTRAN computer code package for the evaluation of gas-phase multicomponent transport properties, Technical Report SAND86-8246, Sandia National Laboratories.

15. Saito, N., Y. Ogawa, S. Saso, C. Liao, and R. Sakei. 1996. Flame-extinguishing concentrations and peak concentrations of N_2, Ar, CO_2 and their mixtures for hydrocarbon fuels. *Fire Safety Journal* 27 (3):185–200. DOI: 10.1016/S0379-7112(96)00060-4.

16. Ogawa, Y., N. Saito, and C. Liao. 1998. Burner diameter and flammability limit measured by tubular flame burner. *Proceedings of the Combustion Institute* 27:3221–7. DOI: 10.1016/S0082-0784(98)80186-0.

17. Liao, C., N. Saito, S. Saso, and Y. Ogawa. 1996. Flammability limits of combustible gases and vapors measured by a tubular flame method. *Fire Safety Journal* 27 (1):49–68. DOI: 10.1016/S0379-7112(96)00021-5.

18. Ishizuka, S., D. Shimokuri, K. Ishii, K. Okada, K. Takashi, and Y. Suzukawa. 2009. Development of practical combustors using tubular flames. *Journal of the Combustion Society of Japan* 51–156:104–13 (in Japanese).

19. Ishioka, M., K. Okada, and S. Ishizuka. 2006. Development of tubular flame burner for by-product fuel gases in steel works. *Journal of the Combustion Society of Japan* 48–145:250–6 (in Japanese).

20. NEDO Progress Report, 2004. A burner system of next generation, No.03002451-0, (in Japanese).

21. Ishizuka, S. 2002. Flame propagation along a vortex axis. *Progress in Energy and Combustion Science* 28:477–542. DOI: 10.1016/S0360-1285(02)00019-9.

22. Shimokuri, D., Y. Shiraga, S. Ishizuka, K. Ishii, and H. Toh. 2009. High-frequency oscillatory combustion in tubular flame burners. *Transaction of JSME* 75:1149–56 (in Japanese).

23. Andac, M. G., F. N. Egolfopoulous, C. S. Campbell, and R. Lauvergne. 2000. Effects of inert dust clouds on the extinction of strained, laminar flames at normal- and micro-gravity. *Proceedings of the Combustion Institute* 28:2921–9. DOI: 10.1016/S0082-0784(00)80717-1.

24. Andac, M. G., F. N. Egolfopoulous, and C. S. Campbell. 2002. Effects of combustible dust clouds on the extinction behavior of strained, laminar premixed flames in normal gravity. *Proceedings of the Combustion Institute* 29:1487–93. DOI: 10.1016/S1540-7489(02)80182-1.

25. Andac, M. G., F. N. Egolfopoulous, C. S. Campbell, and J. C. Lee. 2005. Effects of combustible dust clouds on premixed flame extinction in normal- and micro-gravity. *Proceedings of the Combustion Institute* 30:2369–77. DOI: 10.1016/j.proci.2004.08.256.

26. NEDO Progress Report, 2006. A burner system of next generation, No.05001052-0 (in Japanese).

27. Yagi, K., K. Atarashiya, and S. Nojima. 2006. Development of a fuel processing system for fuel cell systems using a tubular flame burner. In Abstracts of work-in-progress posters in *Thirty-first International Symposium on Combustion*, 211.

28. Yagi, K., K. Nakagawa, and S. Nojima. 2007. Development of a tubular-flame burner for a fuel processing system. In *Book of Abstracts, European Congress of Chemical Engineering (ECCE-6*, Copenhagen, 16–20.

29. Oba, D., S. Ishizuka, K. Okamoto, D. Shimokuri, K. Atarashiya, and M. Katagake. 2006. Development of a tubular flame burner for heating a fastening bolt. In *Proceedings of the Forty-Fourth Symposium (Japanese) on Combustion*, 526–7 (in Japanese).

30. Oba, D., S. Ishizuka, D. Shimokuri, K. Atarashiya, and M. Katagake. 2007. Development of a tubular flame burner for heating a fastening bolt (Second Report). In *Proceedings of the Forty-Fifth Symposium on Combustion*, 202–3 (in Japanese).

31. Kobayashi, Y., R. Matsumoto, M. Ozawa, A. Kegasa, Y. Takemori, Y. Hisazumi, M. Katsiki, and H. Funagoshi. 2011. Development of superheated-steam generator using tubular flame. *Transaction of JSME* 77:997–1001 (in Japanese).

32. Funagoshi, H., R. Matsumoto, M. Ozawa, and M. Katsuki. Micro steam generator using tubular flame. In *Proceedings of 7th World Conference on Experimental Heat Transfer, Fluid Mechanics and Thermodynamics*, Krakow, Poland: CD-ROM HT-8.

33. Matsumoto, R., Y. Kobayashi, M. Ozawa, A. Kegasa, Y. Shiraga, T. Takemori, Y. Hisazumi, and M. Katsuki. Development of superheated steam generator using tubular flame. In *Proceedings of the 22nd International Symposium on Transport Phenomena*, Delft, Netherlands: ISTP-22 CD-ROM No.135.

34. Shimokuri, D. and S. Ishizuka. 2005. Flame stabilization with a tubular flame. *Proceedings of the Combustion Institute* 30:399–406. DOI: 10.1016/j.proci.2004.08.007.

35. Hu, J., D. Shimokuri, and S. Ishizuka. 2012. A fundamental study on the heating with a tubular flame. In *Proceedings of the Fiftieth Symposium (Japanese) on Combustion*, 424–5 (in Japanese).

36. Hu, J., B. Shi, D. Shimokuri, and S. Ishizuka. 2013. An experimental study on the heating process with a tubular flame. In *Proceedings of the Ninth Asia-Pacific Conference on Combustion*, 119.

37. Ottino, J. M. 1989. *The kinematics of mixing: Stretching, chaos, and transport* Cambridge: University Press.

38. Kakehashi, Y., D. Shimokuri, S. Ishizuka, F. Toki, Y. Murata, and M. Saito. 2012. Development of tubular flame burner for a Stirling engine. In *Proceedings of the 17th National Symposium on Power and Engine Systems (SPES2012)*, 175–6 (in Japanese).

39. Kakehashi, Y., D. Shimokuri, S. Ishizuka, F. Toki, Y. Murata, and M. Saito. 2012. *A Tubular Flame Burner for a 1 kW Stirling Engine. Work-In-Progress Poster at the Thirty-fourth International Symposium on Combustion*, W1P137.

40. Kakehashi, Y., D. Shimokuri, S. Ishizuka, F. Toki, Y. Murata, and M. Saito. 2012. Development of self-recirculation type tubular flame burner for a Stirling engine. In *Proceedings of the 15th Symposium on Stirling Cycle*, 35–6 (in Japanese).

INDEX

THIS TITLE IS FROM OUR MECHANICAL ENGINEERING GROUP COLLECTION.
OTHER TITLES OF INTEREST MIGHT BE...

Automotive Sensors
By John Turner, Joe Watson

Centrifugal and Axial Compressor Control
By Gregory K. McMillan

Virtual Engineering
By Joe Cecil

Reduce Your Engineering Drawing Errors:
Preventing the Most Common Mistakes
By Ronald Hanifan

Chemical Sensors: Fundamentals of Sensing
Materials Volume 2: Nanostructured Materials
By Ghenadii Korotcenkov

Biomedical Sensors
By Deric P. Jones, Joe Watson

Chemical Sensors: Comprehensive Sensor
Technologies Volume 4: Solid State Devices
By Ghenadii Korotcenkov

Acoustic High-Frequency Diffraction Theory
By Frederic Molinet

Chemical Sensors: Comprehensive Sensor
Technologies Volume 5: Electrochemical
and Optical Sensors
By Ghenadii Korotcenkov

Chemical Sensors: Comprehensive Sensor
Technologies Volume 6: Chemical Sensors
Applications
By Ghenadii Korotcenkov

Bio-Inspired Engineering
By Chris Jenkins

Chemical Sensors: Simulation and Modeling
Volume 1: Microstructural Characterization
and Modeling of Metal Oxides
By Ghenadii Korotcenkov

The Essentials of Finite Element Modeling and
Adaptive Refinement: For Beginning Analysts to
Advanced Researchers in Solid Mechanics
By John O. Dow

Chemical Sensors: Simulation and Modeling
Volume 2: Conductometric-Type Sensors
By Ghendaii Korotcenkov

Aerospace Sensors
By Alexander Nebylov

Chemical Sensors: Simulation and Modeling
Volume 3: Solid-State Devices
By Ghenadii Korotcenkov

Chemical Sensors: Simulation and Modeling
Volume 4: Optical Sensors
By Ghenadii Korotcenkov

Classical and Modern Engineering Methods in Fluid
Flow and Heat Transfer: An Introduction
for Engineers and Students
By Abram Dorfman

PEM Fuel Cells: Thermal and Water
Management Fundamentals
By Yun Wang, Ken S. Chen, Sun Chan Cho

Chemical Sensors: Simulation and Modeling
Volume 5: Electrochemical Sensors
By Ghenadii Korotcenkov

Announcing Digital Content Crafted by Librarians

Momentum Press offers digital content as authoritative treatments of advanced engineering topics, by leaders in their fields. Hosted on ebrary, MP provides practitioners, researchers, faculty and students in engineering, science and industry with innovative electronic content in sensors and controls engineering, advanced energy engineering, manufacturing, and materials science. **Momentum Press offers library-friendly terms:**

- perpetual access for a one-time fee
- no subscriptions or access fees required
- unlimited concurrent usage permitted
- downloadable PDFs provided
- free MARC records included
- free trials

The **Momentum Press** digital library is very affordable, with no obligation to buy in future years.

For more information, please visit **www.momentumpress.net/library** or to set up a trial in the US, please contact **mpsales@globalepress.com**.